Global Sustainability

Md. Faruque Hossain

Global Sustainability

Springer

Md. Faruque Hossain
Kennesaw State University
Marietta, GA, USA

ISBN 978-3-031-34574-6 ISBN 978-3-031-34575-3 (eBook)
https://doi.org/10.1007/978-3-031-34575-3

This Springer imprint is published by the registered company Springer Nature Switzerland AG
The registered company address is: Gewerbestrasse 11, 6330 Cham, Switzerland

To

Famia
Faria
Shafin
Nowshin

I simply love them more than their imagination. Hope at least one of them will be the captain planet to fulfill my dream – Build a Better World.

Contents

About the Author

Md. Faruque Hossain has more than 20 years of industry experience in the field of sustainability research, development, and project management under global top public agencies and fortune-listed companies. He worked and/or consulted in diverse global top tier companies to conduct research and development for million-dollar to over billion-dollar projects for ensuring global sustainability. Hossain also worked for the NYC Department of Citywide Administrative Services' Senior Management team and interacted with the heads of all public agencies and highest level government officials of local, state, federal, and international organization leaders for developing global sustainability policy. During his tenure in NYC Department of Environmental Protection as Acting Director, Hossain managed a world-class team of scientists, consultants, architects, engineers, and contractors from AECOM, Fluor, and Skanska and maintained highest level professional relationship to conduct global sustainability research and practice. Hossain received his Ph.D. from Hokkaido University, did post-graduate research in Chemical Engineering at the University of Sydney, and Executive Education in Architecture at Harvard University. He is the editor of several International Journals of Global Sustainability-related field. Dr. Hossain is renowned as the industry leader and notable scientist to conduct innovative research and project development for energy, environment, building, and infrastructure sustainability field for building a better Earth. He has hundreds of world-class publications in very high impact journals, and he wrote 5 books (published by Springer and Elsevier) and 82 book chapters (published by Springer, Elsevier, and Taylor & Francis) in the field of global sustainability. Currently, he is working at the School of Architecture and Construction Management at Kennesaw State University as an Assistant Professor and simultaneously running his own company "Green Globe Technology" with his motto to practice sustainability for building a better planet.

Part I
Introduction

Chapter 1
Survival Period of Mankind on Earth

Introduction

Since the 1960s, massive development of industrialization and the misuse of the natural resources throughout the world accumulates atmospheric CO_2 concentration heavily, which certainly will be dangerous for mankind to take fresh breath in the near future [3, 7]. Numerous studies revealed that current accumulation of CO_2 into the atmosphere is 400 ppm, and it is increasing such a rapid rate that it will reach soon at the toxic level of 1200 CO_2 that may result in severe respiratory problems in human beings or worst death throughout the world [4, 8, 13]. So, it is the time without a doubt to make the global environment green by reducing CO_2 emissions that will confirm the versatility, adaptability, and manageability of our mother Earth, which won't result in maladjustment simultaneously, but will be presentable as a sustainable world for our future generation to take a fresh breath. Thus, in the research, a detailed calculation of global CO_2 emission from all sources on Earth and sequestration of CO_2 by all sinks on this planet have been estimated to evaluate the net increasing rate of CO_2 into the atmosphere to give an advance warning to the mankind for the forthcoming environmental vulnerability due to the heavy CO_2 accumulation into the atmosphere. Simply, this study will help the global scientific community, policy makers, and leaders to take this forthcoming danger seriously to mitigate global CO_2 immediately in order to console the forthcoming deadly respiratory problem that will eliminate the mankind on Earth.

M. F. Hossain, *Global Sustainability*, https://doi.org/10.1007/978-3-031-34575-3_1

Methods and Simulation

CO_2 Emissions

The decadal increasing rate in CO_2 emissions due to all industrial development globally was estimated from the difference between two consecutive decades from the period 1960s, 1970s, 1980s, 1990s, 2000s, 2010s, and 2020s, and then it was converted into yearly growth rate divided by past year emission to the current year emissions by using the following equation:

$$\mathrm{FF} = \left[\frac{E_{\mathrm{FF}\,(t_{0+1})} - E_{\mathrm{FF}\,(t_0)}}{E_{\mathrm{FF}\,(t_0)}}\right] * 100\%\mathrm{yr}^{-1} \tag{1.1}$$

Here, this simple calculation is being analysed to determine the increasing rate of CO_2 emissions per year. To precisely estimate the CO_2 increasing rate considering each decadal period, a leap-year factor is also being applied to determine net yearly increasing rate of CO_2 (E_{Ff}) by using its logarithm equal to the below equation:

$$\mathrm{Ff} = \frac{1}{E_{\mathrm{FF}}} \frac{\mathrm{d}(\ln E_{\mathrm{FF}})}{\mathrm{d}t} \tag{1.2}$$

Here, the net CO_2 emission increasing rates have been calculated accounting multi-decadal time scales by integrating a nonlinear function into $\ln(E_{\mathrm{FF}})$ in Eq. (1.2) to calculate eventually yearly increasing rate of CO_2 into the atmosphere [1, 5, 11]. Thus, the algorithm of E_{FF} of this equation is being fitted into MATLAB algorithm E_{FF} to confirm the precise increasing rate of CO_2 yearly.

Similarly, the CO_2 emission calculated here (E_{LUC}) due to the misuse of all natural resources throughout the world was calculated by implementing dynamic global environmental modelling (DGVM) simulations in MATLAB considering the difference between two consecutive decadal periods of 1960s, 1970s, 1980s, 1990s, 2000s, 2010s, and 2020s [2, 18, 19]. Then, a time series is being implemented in this simulation by allocating the dynamic emission of CO_2 due to the misuse of all natural resources throughout the world within 2 consecutive years, and then it was converted into yearly growth rate divided by past year emission to the current year emissions by using the following equation:

$$\mathrm{LUC} = \left[\frac{E_{\mathrm{LUC}\,(t_{0+1})} - E_{\mathrm{LUC}\,(t_0)}}{E_{\mathrm{LUC}\,(t_0)}}\right] * 100\%\mathrm{yr}^{-1} \tag{1.3}$$

Here, the equation is being calculated in yearly CO_2 emissions growth rate [12, 15, 19]. However, to precisely determine the increasing rate of CO_2 in multiple decades, a leap-year factor is also being applied to ensure the net yearly increasing rate of carbon dioxide (E_{LUC}), which is expressed by the following equation:

$$\mathrm{LUC} = \frac{1}{E_{\mathrm{LUC}}} \frac{\mathrm{d}(\ln E_{\mathrm{LUC}})}{\mathrm{d}t} \tag{1.4}$$

Here, the CO_2 emission increasing rates have been estimated corresponding to all decadal time scales by applying a nonlinear function in ln(E_{LUC}) in Eq. (1.4) to determine annual CO_2 emission into the atmosphere [16, 17]. Thus, the algorithm of E_{FF} is being integrated into MATLAB to confirm the precise emission rate of CO_2 from all natural resources misuses.

Finally, the global total CO_2 emission from all industrial development and misuse of natural resources per year have been calculated by combing all four equations (Eqs. 1.1, 1.2, 1.3, and 1.4) as follows:

$$\begin{aligned} \mathrm{FL} = & \left[\frac{E_{\mathrm{FF}}\,(t_{0+1}) - E_{\mathrm{FF}}\,(t_0)}{E_{\mathrm{FF}}\,(t_0)}\right] * 100\%\mathrm{yr}^{-1} + \frac{1}{E_{\mathrm{FF}}} \frac{\mathrm{d}(\ln E_{\mathrm{FF}})}{\mathrm{d}t} \\ & + \left[\frac{E_{\mathrm{LUC}}\,(t_{0+1}) - E_{\mathrm{LUC}}\,(t_0)}{E_{\mathrm{LUC}}\,(t_0)}\right] * 100\%\mathrm{yr}^{-1} + \frac{1}{E_{\mathrm{LUC}}} \frac{\mathrm{d}(\ln E_{\mathrm{LUC}})}{\mathrm{d}t} \end{aligned} \tag{1.5}$$

Thereafter, the global CO_2 sequestration considering all (1) ocean sink and (2) terrestrial sink available throughout the world has been calculated from 1960 to 2029 by conducting 10 years period each experimental data set and then converting it into the time period for an average annual rate.

CO_2 Sink

Consequently, the CO_2 sequestered by the ocean is being calculated for the past years and the next years from the decadal set of 1960s, 1970s, 1980s, 1990s, 2000s, 2010s, and 2020s by implementing all oceans' carbon sink cycle models [9, 14]. This approach is being implemented to accurately analyse the physio-biological processes of global oceans' direct involvement in CO_2 sequestering by the ocean surfaces and its fauna [6, 10]. Thus, the oceans' CO_2 sink is being determined accurately by dividing the individual yearly values with the previous year's value; therefore, the oceanic CO_2 sequestration per year (t) in GtC yr^{-1} is being calculated as follows:

$$S_{\mathrm{OCEAN}}(t) = \frac{1}{n} \sum\nolimits_{m=1}^{m=n} \frac{S^{m}_{\mathrm{OCEAN}}\,(t)}{S^{m}_{\mathrm{OCEAN}(t10-t1)}} \tag{1.6}$$

Here n is the number of oceans, m is the factors involving CO_2 sequestration, and t represents the period.

Then the absorption of CO_2 per year by terrestrial vegetation and the Earth are also being determined to determine the total CO_2 sequestration by the land (S_{LAND}) similarly from the decadal set of 1960s, 1970s, 1980s, 1990s, 2000s, 2010s, and 2020s and convert it into annual rate. Here, the net CO_2 sink by land is being calculated as follows:

$$S_{LAND} = E_{FF} + E_{LUC} - (G_{ATM} + S_{OCEAN}) \tag{1.7}$$

Here, S_{LAND} is calculated from the remainder of the estimates where G_{ATM} is the present of CO_2 into the atmosphere, (E_{FF}) is the carbon from industrial development, and the E_{LUC} is the CO_2 from the misuse of all natural resources throughout the world [2, 19].

Then, the computation of S_{LAND} in Eq. (1.7) is being utilized to determine E_{LUC} by subtracting the (G_{ATM} + S_{OCEAN}) CO_2.

Subsequently, the total CO_2 sequestration in a year period has been calculated by combing these two equations (Eqs. 1.6 and 1.7) as follows:

$$\begin{aligned} S_{OCEAN}(t) + S_{LAND} = \frac{1}{n} \sum\nolimits_{m=1}^{m=n} \frac{S^{m}_{OCEAN}(t)}{S^{m}_{OCEAN^{(t10-t1)}}}) \\ +(E_{FF} + E_{LUC} - (G_{ATM} + S_{OCEAN}) \end{aligned} \tag{1.8}$$

Atmospheric CO_2 Concentration (G_{ATM}) Increasing Rate

Finally, the net yearly increasing rate of the atmospheric CO_2 concentration is being determined yearly from the variation of the total CO_2 emission and total CO_2 sequestration each year.

Results and Discussion

CO_2 Emission

The average global CO_2 emissions from 1960 to 2029 during this time scale showed that total CO_2 emissions from combined industrial development and the misuse of all natural resources throughout the world are at an annual average of 1.7 GtC yr^{-1} of 1.7 ± 0.7 GtC yr^{-1} per decade in the 1960s (1960–1969), annual average of 2.2 GtC yr^{-1} of 1.7 ± 0.8 GtC yr^{-1} per decade in the 1970s (1970–1979), annual average of 1.5 GtC yr^{-1} of 1.6 ± 0.8 GtC yr^{-1} per decade in the 1980s (1980–1989), annual average of 2.45 GtC yr^{-1} of 2.6 ± 0.8 GtC yr^{-1} per decade in the 1990s (1990–1999), annual average of 2.45 GtC yr^{-1} of 2.6 ± 0.8 GtC yr^{-1} per decade

Mean (GtC Yr $^{-1}$)	1960–1969	1970–1979	1980–1989	1990–1999	2000–2009	2010–2019	2020-2029
Total CO_2 Emission (Inustrial Development and Misuse of Natural Resources)							
DGVM simulations and the mean Average data of DEP, DOE, IPCC, CFC, CDIAC, IEA, UNEP, NOAA, and NASA for each decadal period of CO_2 emissions	1.7 ± 0.7	1.7 ± 0.8	1.6 ± 0.8	2.6 ± 0.8	2.6 ± 0.8	3.26 ± 0.5	3.26 ± 0.5
Net CO_2 emissions rate (%) per year	1.7	2.2	1.5	2.45	2.45	3.26	3.26
Total CO_2 Sink (Ocean and Terrestrial Vegetation and Land)							
DGVM simulations and the Mean Average data of DEP, DOE, IPCC, CFC, CDIAC, IEA, UNEP, NOAA, and NASA for each decadal period of CO_2 sequestrtion	1.5 ± 0.5	1.3 ± 0.5	1.4 ± 0.6	1.6 ± 0.4	1.15 ± 0.5	1.15 ± 0.5	1.0 ± 0.5
Net CO_2 sequestration rate (%) per year	1.5	1.3	1.4	1.4	1.15	1.15	1.15
Annual Increaing Rate of Atmospheric CO_2 (%)							
G_{ATM}	0.2	0.9	0.1	1.05	1.3	2.11	2.11

Fig. 1.1 The results from DGVM simulation in MATLAB, implemented from the data of DEP, DOE, IPCC, CFC, CDIAC, IEA, UNEP, NOAA, and NASA to confirm the yearly increasing rate of atmospheric CO_2 (%). The results described the variation of the total CO_2 emissions from industrial development and misuse of all natural resources throughout the world and the total CO_2 sink (ocean and land) from the years of 1960–1969, 1970–1979, 1980–1989, 1990–1999, 2000–2009, 2010–2019, and 2020–2029 shown in GtCyr^{-1}

in the 2000s (2000–2009), and annual average of 3.26 GtC yr^{-1} of 3.26 ± 0.5 GtC yr^{-1} per decade in the 2010s (2010–2019) and expected to be increased to annual average of 3.26 GtC yr^{-1} of 3.26 ± 0.5 GtC yr^{-1} per decade in the 2020s (2020–2029) (Fig. 1.1).

CO_2 Sink

Subsequently, the results of CO_2 sequestration by ocean and the terrestrial vegetation and land suggested that the average global CO_2 sink from 1960 to 2029 during this time scale showed that total CO_2 emissions from combined industrial development and the misuse of all natural resources throughout the world are at an annual average of 1.5 GtC yr^{-1} of 1.5 ± 0.2 GtC yr^{-1} per decade in the 1960s (1960–1969), annual average of 1.3 GtC yr^{-1} of 1.3 ± 0.5 GtC yr^{-1} per decade in the 1970s (1970–1979), annual average of 1.4 GtC yr^{-1} of 1.4 ± 0.6 GtC yr^{-1} per decade in the 1980s (1980–1989), annual average of 1.4 GtC yr^{-1} of 1.6 ± 0.4 GtC yr^{-1} per decade in the 1990s (1990–1999), annual average of 1.15 GtC yr^{-1} of 1.15 ± 0.5 GtC yr^{-1} per decade in the 2000s (2000–2009), and annual average of 1.15 GtC yr^{-1} of 1.15 ± 0.5 GtC yr^{-1} per decade in the 2010s (2010–2019) and expected to be increased to annual average of 1.15 GtC yr^{-1} of 1.15 ± 0.5 GtC yr^{-1} per decade in the 2020s (2020–2029) (Fig. 1.1).

Atmospheric CO_2 Concentration (G_{ATM}) Increasing Rate

Then, the rate of growth of the atmospheric CO_2 concentration is being calculated by comparing the decadal and individual annual values for 10 years periodical set, which suggested that the average global CO_2 annual growth from 1960 to 2029 is 0.2% at the decade 1960s, 0.9% at the decade 1970s, 0.1% at the decade 1980s, 1.15% at the decade 1990s, 1.3% at the decade 2000s, and 2.11% at the decade 2010s and expected to be 2.11% at the decade 2020s. The projected growth rate of atmospheric CO_2 concentration is presumably suggested that the increase at rate of CO_2 will remain the same 2.11% per year for the next several decades if we do not curb this acceleration of CO_2 emissions (Fig. 1.1).

Since the current CO_2 concentration into the atmosphere is 400 ppm and is growing at a rate of 2.11% per year, the following equations confirmed that it will attain at a toxic level of 1200 ppm in 53 years.

$$1200 = 400(1 + .0211)^{\text{Year}} \tag{1.9}$$

$$3 = (1 + .0211)^{\text{Year}} \tag{1.10}$$

$$\text{Log } 3 = \text{Year Log}(1.0211) \tag{1.11}$$

$$\text{Year} = 52.61 = 53 \text{ (round figure)} \tag{1.12}$$

Consequently, all human beings on Earth will be in serious breathing problem due to the toxic level of CO_2 into the atmosphere. Simply, it is an urgent demand to reduce the CO_2 emission globally to mitigate the forthcoming deadly breathing problem for mankind as well as secure a better planet for our next generation.

Conclusion

The total global CO_2 emissions due to the industrial development and the misuse of all natural resources throughout the world estimated for the past several decades as well as total CO_2 sink by ocean and land were also calculated to determine the increasing rate in CO_2 into the atmospheric each year. The yearly increasing rate of the atmospheric CO_2 accumulation over the last several years was confirmed by simulated estimate, which revealed that it is increasing at a rate of 2.11% yearly. If the current annual CO_2 growth rate is not copped now, the atmospheric CO_2 accumulation shall indeed reach at a toxic level of 1200 ppm in 53 years, which may result in the extinction of the entire human race due to the severe respiratory problem from the planet Earth.

Acknowledgement The author, Md. Faruque Hossain, declares that any findings, predictions, and conclusions described in this article are solely performed by the author and it is confirmed that there is no conflict of interest for publishing this research paper in a suitable journal and/or publisher.

References

1. Achard, F., Beuchle, R., Mayaux, P., Stibig, H., Bodart, C., Brink, A., & Eva, H. D. (2014). Determination of tropical deforestation rates and related carbon losses from 1990 to 2010. *Global Change Biology, 20*(8), 2540–2554.
2. Ballantyne, A., Alden, C., Miller, J., Tans, P., & White, J. (2012). Increase in observed net carbon dioxide uptake by land and oceans during the past 50 years. *Nature, 488*(7409), 70–72.
3. Bauer, J. E., Cai, W., Raymond, P. A., Bianchi, T. S., Hopkinson, C. S., & Regnier, P. A. (2013). The changing carbon cycle of the coastal ocean. *Nature, 504*(7478), 61–70.
4. Betts, R. A., Jones, C. D., Knight, J. R., Keeling, R. F., & Kennedy, J. J. (2016). El niño and a record CO2 rise. *Nature Climate Change, 6*(9), 806–810.
5. Canadell, J. G., Le Quéré, C., Raupach, M. R., Field, C. B., Buitenhuis, E. T., Ciais, P., et al. (2007). Contributions to accelerating atmospheric CO2 growth from economic activity, carbon intensity, and efficiency of natural sinks. *Proceedings of the National Academy of Sciences, 104*(47), 18866–18870.
6. Chevallier, F. (2015). On the statistical optimality of CO2 atmospheric inversions assimilating CO2 column retrievals. *Atmospheric Chemistry and Physics, 15*(19), 11133–11145.
7. Davis, S. J., & Caldeira, K. (2010). Consumption-based accounting of CO2 emissions. *Proceedings of the National Academy of Sciences, 107*(12), 5687–5692.
8. Erb, K., Kastner, T., Luyssaert, S., Houghton, R. A., Kuemmerle, T., Olofsson, P., & Haberl, H. (2013). Bias in the attribution of forest carbon sinks. *Nature Climate Change, 3*(10), 854–856.
9. Hossain, M. F. (2016). Theory of global cooling. *Energy, Sustainability and Society, 6*, 24.
10. Hossain, M. F. (2017). Green science: Independent building technology to mitigate energy, environment, and climate change. *Renewable and Sustainable Energy Reviews, 73*, 695–705.
11. Houghton, R. A. (2007). Balancing the global carbon budget. *Annual Review of Earth and Planetary Sciences, 35*(1), 313–347.
12. Jain, A. K., Meiyappan, P., Song, Y., & House, J. I. (2013). CO2 emissions from land-use change affected more by nitrogen cycle, than by the choice of land-cover data. *Global Change Biology, 19*(9), 2893–2906.
13. Li, W., Ciais, P., Wang, Y., Peng, S., Broquet, G., Ballantyne, A. P., et al. (2016). Reducing uncertainties in decadal variability of the global carbon budget with multiple datasets. *Proceedings of the National Academy of Sciences, 113*(46), 13104–13108.
14. Liu, Z., Guan, D., Wei, W., Davis, S. J., Ciais, P., Bai, J., et al. (2015). Reduced carbon emission estimates from fossil fuel combustion and cement production in China. *Nature, 524*(7565), 335–338.
15. Mason Earles, J., Yeh, S., & Skog, K. E. (2012). Timing of carbon emissions from global forest clearance. *Nature Climate Change, 2*(9), 682–685.
16. Morimoto, S., Goto, D., Murayama, S., Fujita, R., Tohjima, Y., Ishidoya, S., & Maksyutov, S. (2021). Spatio-temporal variations of the atmospheric greenhouse gases and their sources and sinks in the arctic region. *Polar Science, 27*, 100553.
17. Prietzel, J., Zimmermann, L., Schubert, A., & Christophel, D. (2016). Organic matter losses in German alps forest soils since the 1970s most likely caused by warming. *Nature Geoscience, 9*(7), 543–548.
18. Schwietzke, S., Sherwood, O. A., Bruhwiler, L. M., Miller, J. B., Etiope, G., Dlugokencky, E. J., & White, J. W. (2016). Upward revision of global fossil fuel methane emissions based on isotope database. *Nature, 538*(7623), 88–91.
19. Stephens, B. B., Gurney, K. R., Tans, P. P., Sweeney, C., Peters, W., Bruhwiler, L., et al. (2007). Weak northern and strong tropical land carbon uptake from vertical profiles of atmospheric CO2. *Science, 316*(5832), 1732–1735.

Part II
Sustainable Energy

Chapter 2
Harvesting Global Renewable Energy

Introduction

The conventional energy consumption for powering modern civilization throughout the world is indeed accelerating the finite level of the current fossil fuel reserve of 36,630 EJ [10]. The global fossil fuel energy consumption was 283 EJ/Yr in the year 1980, 348 EJ/Yr in the 1995, 405 EJ/Yr in 2005, and 515 EJ/Yr in 2015 and will reach 610 EJ/Yr in 2025, 705 EJ/Yr in 2035, 860 EJ/Yr in 2045, and 990 EJ/Yr in 2050 [10, 27]. The utilization of fossil fuel globally in the year 2018 was 2.236×10^{20} EJ, which is responsible for creating 8.01×10^{11} tons of CO_2 into the atmosphere and accounted for the acceleration of deadly climate change rapidly [6, 17, 24]. Consequently, adverse environmental impacts such as acidic rain, flood, and climate change are occurring unpredictably throughout the world [27, 29]. A recent study shows that the concentration of CO_2 in the atmosphere is current 400 ppm, which needs to be lowered to a standard level grade of 300 ppm CO_2 for the wellness of clean breathing and respiratory system for all mammals [32, 34]. Another research revealed that greenhouse gas emission accelerates the global diurnal mean temperature fluctuating rapidly posing a serious threat to the natural ecosystem and mankind's well-being due to the utilization of burning fossil fuel since it creates radioactive CO_2 into the atmosphere in a certain period of time [21, 27].

Unfortunately, the consumption of conventional energy currently is still accelerating rapidly throughout the world; the situation shall remain unchanged until a renewable source of energy is developed to utilize sustainable energy [4, 7, 13]. Simply it is an urgent demand to develop sustainable energy technology to mitigate fossil fuel consumption where the "new source that fulfills the requirement of the present without disrupting the ability of future demand of energy to fulfill the complete needs for the future generations." Hence, renewable energy such as solar energy utilization globally can be an interesting source to fulfill the net energy requirement throughout the world. It is a natural renewable energy source generated

M. F. Hossain, *Global Sustainability*, https://doi.org/10.1007/978-3-031-34575-3_2

by the sun, which is created by nuclear fusion that takes place in the sun that is constantly flowing away from the sun to the solar system, and part of it reaches Earth, which is a tremendous source of clean and renewable energy [14, 17, 19]. If only 0.001% of the annual solar energy coming on Earth is used, it will meet the net energy demand for the whole planet, which is clean and abundant everywhere. In this study, research has been performed to harvest the total global solar energy reaching Earth to mitigate global net energy need, which is clean and environmentally friendly.

Material, Methods, and Simulation

Calculation of Net Solar Energy on Earth

The total Earth's surface is clarified by characterizing various directional angles considering the Cartesian coordinate system, where x denotes to skyline convention, y denotes east-west, and z denotes for zenith in order to measure the total solar irradiance during the day entire year (Fig. 2.1). The position of the celestial body in this framework is thus chosen by h, which is denoted for height, and A, which is denoted for the azimuth angle, while the central framework utilized it as the convenience factor, which is z hub. It focuses on the North Pole, and the y hub indistinguishably focuses on the horizon of the skylight, and the x pivot is opposite to both of North Pole and horizon. Therefore, the angles and coordinate frequencies are being encountered mathematically by calculating the latitude and longitude to implement correct angles to trap the solar irradiance most efficiently. Here, the zero point of latitude is considered the primary meridian that controls the function of the meridian in the Eastern Hemisphere and Western Hemisphere angle of the Earth's surface and so does the north of the equator in the Northern Hemisphere, and south of the equator is the Southern Hemisphere that is also being controlled by this Earth surface modeling to trap to solar energy more efficiently. Finally, the δ and ω point hours are being clarified accurately considering this analytical Cartesian coordinates to decide the position of solar irradiance vector to clarify the solar energy reaching the Earth's surface to determine net solar energy calculation into the Earth's surface [14, 15, 33].

Once the angle of the Earth's surface is modeled, then the Earth's surface is considered as the net areal dimension of solar energy emission by considering the peak hours of solar radiation generated from the sun [16, 19]. Thus, the radiation of the solar irradiance released by the sun and accosted by the Earth is computed by the solar constant, which is defined by the measurement of the solar energy flux density perpendicular to the ray direction per unit area per unit time [2, 3, 5]. Thus, the calculation of this amount of net solar energy includes all types of radiation scattered and reflected ones both being modeled by using MATLAB software to calculate total global solar radiation emission on Earth.

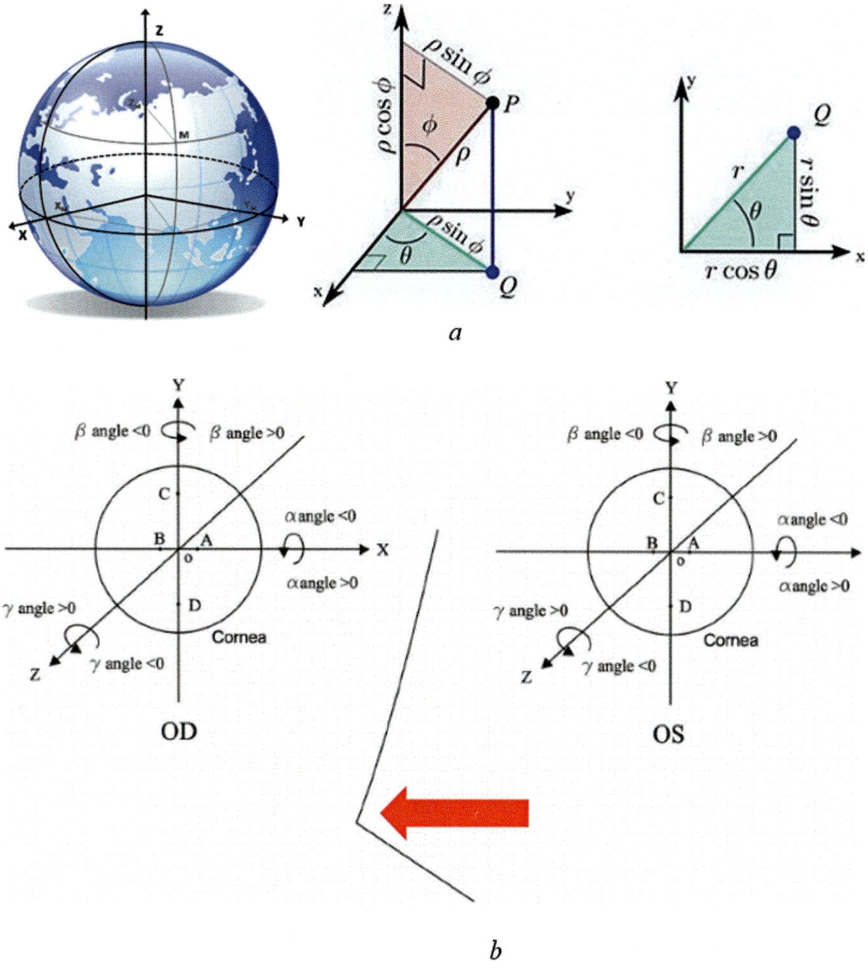

Fig. 2.1 (**a**) Cartesian coordinates clarification of the southern axis of x, the western axis of y, and the zenith axis of z clarified to calculate the total solar energy reaching Earth considering the average energy density of sunlight on the surface of Earth is 1366 W/m^2 by implementing the diameter of the Earth is 10,000,000 of the meridians at the North Pole to the equator and the radius of Earth is $2/\pi \times 107$ m. The location of this celestial body is analyzed by determining two angles of $\sin \theta$ and $\cos \theta$. (**b**) The longitudinal and latitudinal equatorial angles have been clarified where the convention z-axis point denotes the North Pole and the east-west axis y-axis denotes the identical angle of the horizon

Then, the sunlight is being clarified as the motion of the photon flux by considering the first function of the fundamental solar thermal energy and anti-reflective coatings of solar cells, and then it has modified into the second order of function of the solar energy [20, 23, 25]. The integration of these two functions is computed by implementing the solar quantum dynamics, which is clarified as the most acceptable quantum technology to calculate the net solar energy emission on Earth [34, 35]. Just

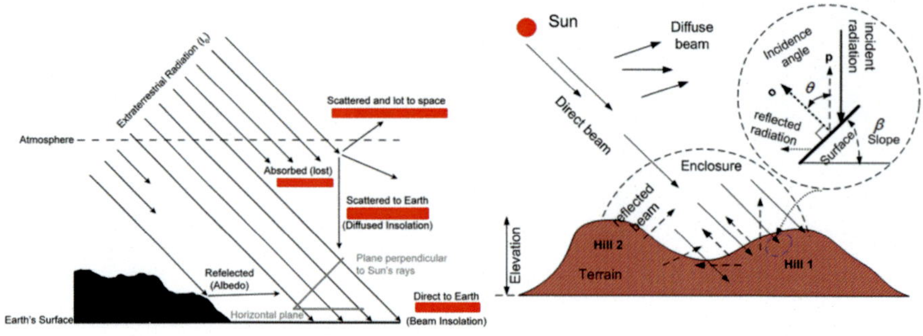

(a) (b)

Fig. 2.2 (**a**) shows the emission of solar energy on the Earth's surface, (**b**) various kinds of irradiance on the Earth surface; direct beam, reflected beam, and diffuse beam at various angles

because the Earth's surface can emit solar energy accurately at a given temperature of approximately 700 °C where the energy density of the solar radiation is derived from the peak solar irradiance generation from a single solar photon flux [21, 27] (Fig. 2.2).

The estimation of global solar irradiance calculation on the Earth's surface is being further clarified considering the three background solar data calculation by using *Pyrheliometer* to measure direct beam irradiance approaching from the sun and radius of the Earth's surface [3]. Then, the *pyranometer* has also been utilized to determine net hemispherical irradiance beam along with the diffused beam on horizon, and thus, the net global solar radiation (W/m^2) is being determined considering the horizon by a pyranometer, which is denoted as:

$$I_{\text{tot}} = I_{\text{beam}} \cos\theta + \mathrm{I}_{\text{diffuse}}$$

where θ represents the zenith angle that has been implemented to calculate the net solar energy reaching on Earth by the clarification of electron energy level of hydrogen (Fig. 2.3).

This measurement is being then confirmed against modest pyrheliometers using the thermo-coupler detector and the PV detectors recorder considering the determination of wavelength of the solar spectrum [2, 6]. Eventually *photoelectric sunshine recorder* has been used, which is intermittent and varied by the solar irradiance intensity [9]. Since the solar radiation is related to the photon charge, therefore, the attributes of photon energy on Earth surface are being computed considering the quantum flow of photon radiation in global scale by using MATLAB 9.0 Classical Multidimensional Scaling [1]. Consequently, a computational model of photon radiation is being quantified to demonstrate the solar energy generation from sunlight considering radiation emission. Thereafter, the mode of the solar quanta absorbance by Earth surface is being determined by the peak solar radiation output tracking into the Earth surface [3, 28]. Naturally, the induced solar irradiance is, thereafter, computed by the Earth surface area by implementing the parameters of

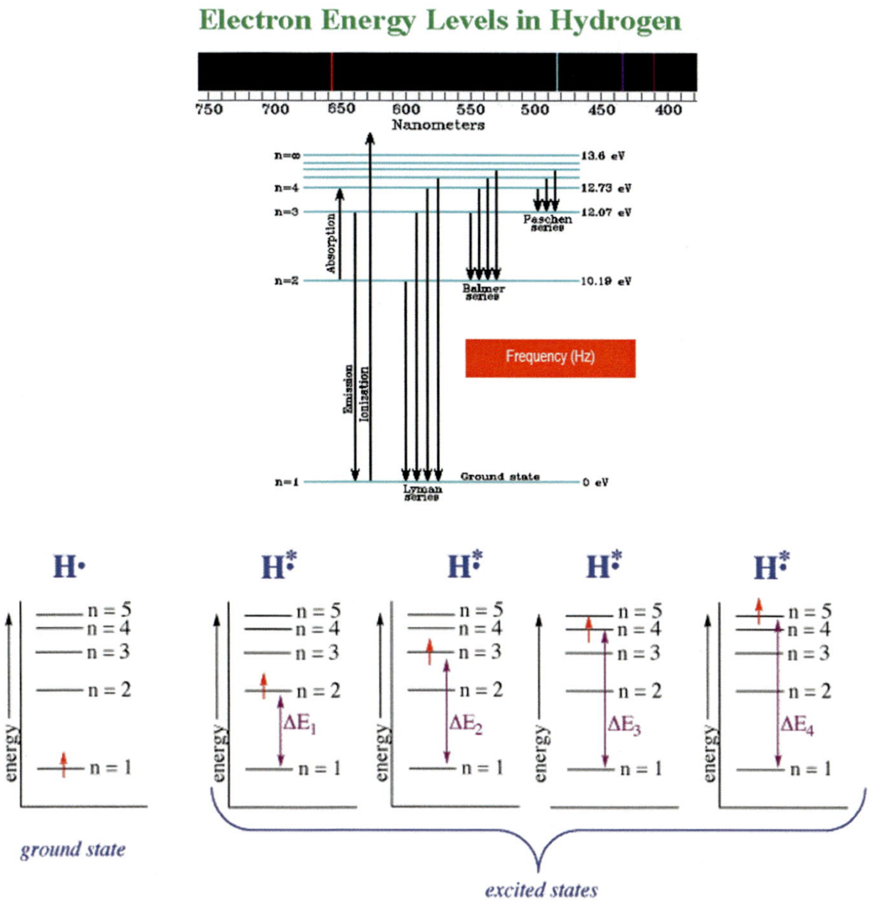

Fig. 2.3 The solar energy state hydrogen depicting the absorption and emission modes and energy deliberation rate revealing the energy density by clarifying the optimum solar irradiance deliberation from a photon particle at various wavelength and frequencies of 10^{17} Hz to 10^{14} Hz. The electron energy level H2 is being determined by conducting a down path transition involves in accost of a solar energy with respect ground state and excited state of solar energy

solar energy proliferation on it and transformation rate of solar energy into electricity energy generation. Thus, the accurate calculation of the current–voltage (*I–V*) characteristic is being subsequently conducted by the conceptual model of net solar radiation intake into the Earth surface by computing the net active solar volt (I_{v+}) generation into the Earth surface [11, 13, 22].

Then, the mathematical determination of the net current formation via I_{pv} on Earth surface has been modeled out, by calculating *I–V–R* correlation ship within the Earth surface in order to use this energy commercially throughout the world (Fig. 2.4).

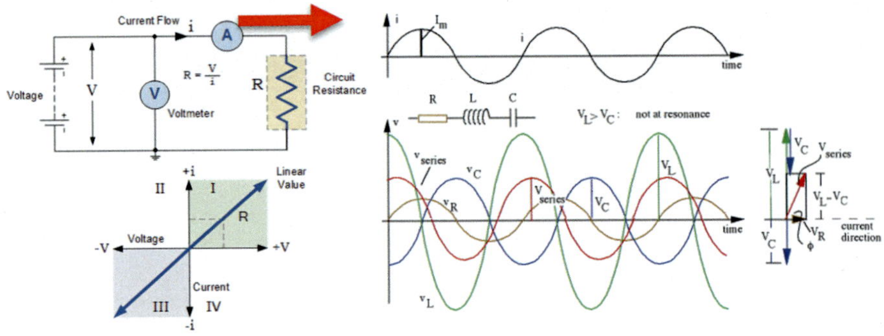

Fig. 2.4 The conceptual circuit diagram of the whole Earth surface depicted the (**a**) net photo-physical current generation into the Earth surface by detailing model of *I–V–R* relationship in order to use electricity throughout the world, respectively

Hence, the following equation is being computed as the energy deliberation from the Earth surface, whose origin is the photon irradiance and ambient temperature of the solar energy:

$$P_{pv} = \eta_{pvg} A_{pvg} G_t \tag{2.1}$$

Here, η_{pvg} denotes the Earth surface performance rate, A_{pvg} denotes the Earth surface array (m^2), and G_t denotes the photon irradiance intaking rate on the plane (W/m^2) of Earth surface, and thus, the η_{pvg} could be rewritten as follows:

$$\eta_{pvg} = \eta_r \eta_{pc} [1 - \beta(T_c - T_{c\ ref})] \tag{2.2}$$

where η_{pc} denotes the energy formation efficiency, when maximum power point tracking (MPPT) is being implemented, which is close to 1, β denotes the temperature cofactor (0.004–0.006 per °C), η_r denotes the mode of energy efficiency, and $T_{c\ ref}$ denotes the condition of temperature at °C. The reference Earth temperature ($T_{c\ ref}$) can be rewritten by calculating from the equation below:

$$T_c = T_a + \left(\frac{\text{NOCT} - 20}{800}\right) G_t \tag{2.3}$$

where T_a denotes the encircling temperature in °C, G_t denotes the solar radiation in Earth surface (W/m^2), and thus, it denotes the modest optimum Earth temperature in °C. Considering this temperature condition, the net solar radiation Earth surface can be calculated by the equation below:

$$I_t = I_b R_b + I_d R_d + (I_b + I_d) R_r \tag{2.4}$$

The solar energy here is necessarily working as a conceptual *P–N* junction superconductor in order to form electricity through the Earth surface, which are is interlinked in a parallel series connection [5, 12, 39]. Thus, a unique conceptual circuit model, as shown in Fig. 2.4, with respect to the N_s series of Earth surface and N_p parallel arrays has been computed by the following Earth surface solar energy equation based on current and volt relationship

$$I = N_p\left[I_{ph} - I_{rs}\left[\exp\left(\frac{q(V + IR_s)}{\mathrm{AKTN_s}} - 1\right)\right]\right] \tag{2.5}$$

where

$$I_{rs} = I_{rr}\left(\frac{T}{T_r}\right)^3 \exp\left[\frac{E_G}{AK}\left(\frac{1}{T_r} - \frac{1}{T}\right)\right] \tag{2.6}$$

Hence, in Eqs. (2.5) and (2.6), q denotes the electron charge (1.6×10^{-19} C), K denotes Boltzmann's constant, A denotes the diode standardized efficiency, and T denotes the Earth temperature (K). Accordingly, I_{rs} denotes the Earth surface reversed current motion at T, where T_r denotes the Earth condition temperature, I_{rr} denotes the reversed current at T_r, and E_G denotes the photonic band-gap energy of the superconductor utilized for the Earth surface. Thus, the photonic current I_{ph} will be generated in accordance with the Earth surface temperature and radiation condition that can be expressed by:

$$I_{ph} = \left[I_{SCR} + k_i(T - T_r)\frac{S}{100}\right] \tag{2.7}$$

Here, I_{SCR} denotes the current motion considering the optimum temperature of the Earth and solar radiation dynamic on the Earth surface, k_i denotes the short-circuited current motion, and S denotes the solar radiation calculation in a unit area (mW/cm^2). Subsequently, the I–V features of the Earth surface shall be deformed from the conceptual model of the circuit that can be expressed as:

$$I = I_{ph} - I_D \tag{2.8}$$

$$I = I_{ph} - I_0\left[\exp\left(\frac{q(V + R_sI)}{AKT} - 1\right)\right] \tag{2.9}$$

I_{ph} denotes the photonic current dynamic (A), I_D denotes the diode-originated current dynamic (A), I_0 denotes the inversed current dynamic (A), A denotes the diode-induced constant, q denotes the charge of the electron (1.6×10^{-19} C), K denotes Boltzmann's constant, T denotes the Earth temperature (°C), R_s denotes the series resistance (Ω), R_{sh} denotes the shunt resistance (Ω), I denotes the cell current motion (A), and V denotes the Earth voltage motion (V). Therefore, the net current flow into the Earth surface can be determined by conducting the following equation:

$$I = I_{PV} - I_{D1} - \left(\frac{V + IR_S}{R_{SH}}\right) \tag{2.10}$$

where

$$I_{D1} = I_{01}\left[\exp\left(\frac{V + IR_s}{a_1 V_{T1}}\right) - 1\right] \tag{2.11}$$

Here, I and I_{01} are being denoted as the reversed current flow into the conceptual circuit and V_{T1} and V_{T2} are being denoted as the optimum thermal volts into the circuit. Thus, the circuit standard factors are being presented by a_1 and a_2, and then it has normalized the mode of Earth surface by expressing following equation:

$$v_{oc} = \frac{V_{oc}}{cK\,T/q} \tag{2.12}$$

$$P_{max} = \frac{\frac{V_{oc}}{cK\,T/q} - \ln\left(\frac{V_{oc}}{cK\,T/q} + 0.72\right)}{\left(1 + \frac{V_{oc}}{K\,T/q}\right)}\left(1 - \frac{V_{oc}}{\frac{V_{oc}}{I_{SC}}}\right) \times \left(\frac{V_{oc0}}{1 + \beta \ln \frac{G_0}{G}}\right)\left(\frac{T_o}{T}\right)^y I_{sc0}\left(\frac{G}{G_o}\right)^a \tag{2.13}$$

where ν_{oc} denotes the standard point of the open-circuit voltage, V_{oc} denotes the thermal voltage $V_t = nkT/q$, c denotes the constant current motion, K denotes Boltzmann's constant, T denotes the temperature into the Earth surface PV cell in Kelvin, α denotes the function that represents the nonlinear motion of photocurrents, q denotes the electron charge, γ denotes the function acting for all nonlinear temperature-voltage currents, and β denotes the Earth surface mode for specific dimensionless function for enhancing current flowing rate. Subsequently, Eq. (2.13) represents the peak energy generation from the Earth surface module, which is interlined in both series and parallel connection. Thus, the equation for the net energy formation into the array of N_s has been interlinked in series and N_p has been interlinked in parallel considering the power P_M of each mode of connection and that is finally expressed by using the following equation:

$$P_{array} = N_s N_p P_M \tag{2.14}$$

Modeling of Net Electricity Energy Generation from Total Solar Irradiance on Earth

To convert global solar energy into electricity energy, a model is also being prepared by integrating global Albanian symmetries of gauge field scalar [30, 37]. Naturally, the net solar energy particle will functionally be acted as the dynamic photons of particle T^{α} at the global symmetrical array of Earth surface by initiating the gauge field of $A_{\mu}^{\alpha}\,(x)$, and then, the local Albanian will subsequently be started to activate at the global U (1) phase symmetry to deliver net electricity energy [40]. Thus, the model is being considered as a complex vector field of $\Phi(x)$ of Earth surface where electric charge q will couple with the EM field of $A^{\mu}(x)$, and thus, the equation can be expressed by $\mathfrak{h}$:

$$\mathfrak{h} = -\frac{1}{4}\,F_{\mu\nu}F^{\mu\nu} + D_{\mu}\Phi^{*}\,D^{\mu}\Phi - V(\Phi^{*}\Phi) \tag{2.15}$$

where:

$$\begin{aligned} D_{\mu}\Phi(x) &= \partial_{\mu}\Phi(x) + iqA_{\mu}(x)\Phi(x) \\ D_{\mu}\Phi^{*}(x) &= \partial_{\mu}\Phi^{*}(x) - iqA_{\mu}(x)\Phi^{*}(x) \end{aligned} \tag{2.16}$$

and

$$V(\Phi^{*}\Phi) = \frac{\lambda}{2}(\Phi^{*}\Phi)^{2} + \mathrm{m}^{2}(\Phi^{*}\Phi) \tag{2.17}$$

Here $\lambda > 0$ $\mathrm{m}^2 < 0$; therefore $\Phi = 0$ is a local optimum vector quantity, while the minimum form of degenerated scalar circle is clarified as $\Phi = \frac{v}{\sqrt{2}} * \mathrm{e}^{i\theta}$,

$$v = \sqrt{\frac{-2\,\mathrm{m}^2}{\lambda}},\ \text{any real } \theta \tag{2.18}$$

Subsequently, the vector field Φ of the global Earth surface will form a nonzero functional value $\langle\Phi\rangle \neq 0$, which will simultaneously determine the U (1) symmetrical net solar energy generation. Therefore, the global U (1) net symmetrical electrical energy of $\Phi(x)$ will be delivered as expected value of $\langle\Phi\rangle$ by confirming the x-dependent state of the symmetrical $\Phi(x)$ array of Earth surface and can be expressed by the following equation:

$$\Phi(x) = \frac{1}{\sqrt{2}}\,\Phi_{\mathrm{r}}(x) * \mathrm{e}^{i\Theta(x)},\ \text{real } \Phi_{\mathrm{r}}(x) > 0, \text{real } \Phi(x) \tag{2.19}$$

Thus, the net calculation of the electricity energy generation from the net Earth surface solar energy is being determined considering the vector $\Phi(x) = 0$, and it is first-order function of $\langle \Phi \rangle \neq 0$, considering the peak level of solar energy emission on the Earth surface of $\Phi\langle x \rangle \neq 0$ [10, 21, 28]. Thus, the net electricity energy generation from the global solar energy calculation $\phi_r(x)$ and $\Theta(x)$ and its vector on the Earth surface field ϕ_r have been confirmed by conducting the following equation:

$$V(\phi) = \frac{\lambda}{8}\left(\phi_r^2 - v^2\right)^2 + \text{const,} \tag{2.20}$$

or the resultant electricity energy generation is shifted by its VEV, $\Phi_r(x) = v + \sigma(x)$,

$$\phi_r^2 - v^2 = (v+\sigma)^2 - v^2 = 2v\sigma + \sigma^2 \tag{2.21}$$

$$V = \frac{\lambda}{8}\left(2v\sigma - \sigma^2\right)^2 = \frac{\lambda v^2}{2} * \sigma^2 + \frac{\lambda v}{2} * \sigma^3 + \frac{\lambda}{8} * \sigma^4 \tag{2.22}$$

Simultaneously, the functional derivative $D_\mu \phi$ will become

$$D_\mu \phi = \frac{1}{\sqrt{2}}\left(\partial_\mu\left(\phi_r e^{i\Theta}\right) + iqA_\mu * \phi_r e^{i\Theta}\right) = \frac{e^{i\Theta}}{\sqrt{2}}\left(\partial_\mu \phi_r + \phi_r * i\partial_\mu \Theta + \phi_r * iqA_\mu\right) \tag{2.23}$$

$$\begin{aligned} \left|D_\mu \phi\right|^2 &= \frac{1}{2}\left|\partial_\mu \phi_r + \phi_r * i\partial_\mu \Theta + \phi_r * iqA_\mu\right|^2 \\ &= \frac{1}{2}\left(\partial_\mu \phi_r\right) + \frac{\phi_r^2}{2} * \left(\partial_\mu \Theta q A_\mu\right)^2 \\ &= \frac{1}{2}\left(\partial_\mu \sigma\right)^2 + \frac{(v+\sigma)^2}{2} * \left(\partial_\mu \Theta + qA_\mu\right)^2 \end{aligned} \tag{2.24}$$

Altogether,

$$\mathfrak{H} = \frac{1}{2}\left(\partial_\mu \sigma\right)^2 - v(\sigma) - \frac{1}{4} F_{\mu\nu} F^{\mu\nu} + \frac{(v+\sigma)^2}{2} * \left(\partial_\mu \Theta + qA_\mu\right)^2 \tag{2.25}$$

To determine the formation of this net electricity generation referred as ($\mathfrak{H}_{\text{sef}}$) into the Earth surface, the function of the electrostatic fields has been quantified by conducting the quadratic calculation and described by the following equation:

$$\mathfrak{H}_{\text{sef}} = \frac{1}{2}\left(\partial_\mu \sigma\right)^2 - \frac{\lambda v^2}{2} * \sigma^2 - \frac{1}{4} F_{\mu\nu} F^{\mu\nu} + \frac{v^2}{2} * \left(qA_\mu + \partial_\mu \Theta\right)^2 \tag{2.26}$$

Here this net electricity generation ($\mathfrak{h}_{free}$) function certainly will admit a realistic vector particle of positive mass$^2 = \lambda v^2$ integrating the areal $A_\mu\ (x)$ function and the electricity energy generation fields $\Theta(x)$ to confirm to determine the net electricity energy from the global solar energy calculation into the Earth surface (Fig. 2.1).

Results and Discussion

Calculation of Net Solar Energy on Earth

To calculate the net solar energy on the Earth surface, the net irradiance of photon emission has been calculated by integrating Eqs. (2.22) and (2.23). Necessarily, the functional Earth surface area $J(\omega)$, the photonic quantum field, and the unit area $J(\omega)$ are being calculated considering the constant irradiance coupling point and the Weisskopf-Wigner approximation mechanism to confirm the accurate solar energy reaching on the Earth surface [31] (Fig. 2.5).

The computed results show that the distribution of solar irradiance on the Earth's sphericity and orbital parameters is the application of the unidirectional beam incident to a rotating sphere of Milankovitch cycles from spherical Earth law of cosines:

$$\cos(c) = \cos(a)\cos(b) + \sin(a)\sin(b)\cos(C) \tag{2.27}$$

where a, b, and c are being considered as the arc lengths, in radians, of the sides of a spherical triangle and C represents the angle of the vertex that has arc length of c. Determining this calculation of solar zenith angle Θ, the following equations are being clarified considering the application of spherical law of cosines:

$$C = h$$

$$c = \Theta$$

$$a = \frac{1}{2}\pi - \phi$$

$$b = \frac{1}{2}\pi - \delta$$

$$\cos(\Theta) = \sin(\phi)\sin(\delta) + \cos(\phi)\cos(\delta)\cos(h) \tag{2.28}$$

In order to simplify this equation, it has been further clarified as a general derivative as follows:

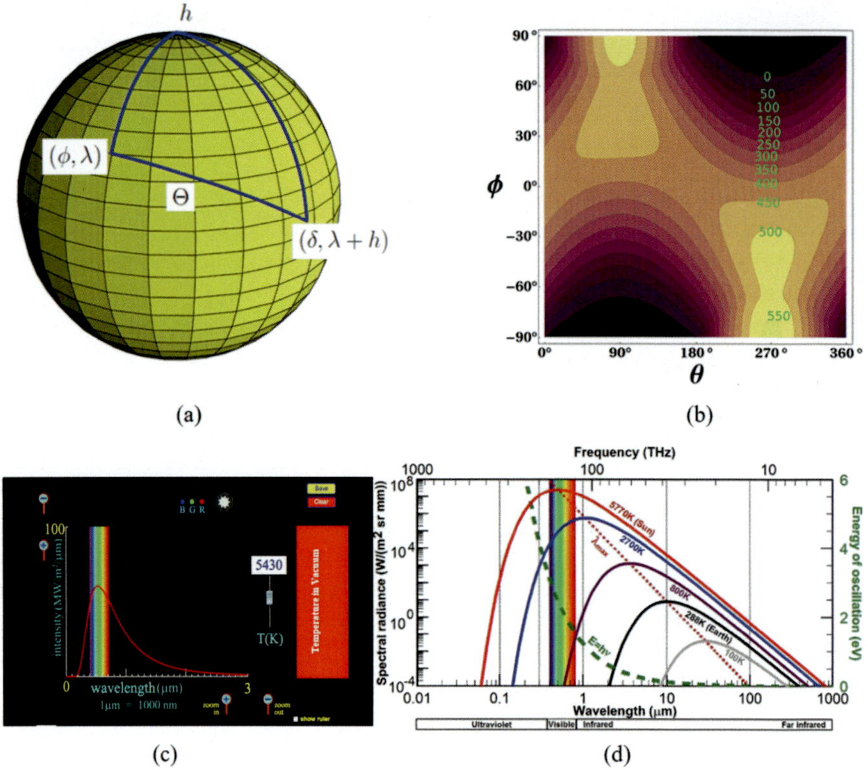

Fig. 2.5 (**a**) The sphere triangular for cosines is being clarified as the solar zenith angle Θ considering latitude φ and longitude λ, (**b**) the average daily irradiation at the top of the atmosphere, where θ is the polar angle of the Earth's orbit, and $\theta = 0$ at the vernal equinox, and $\theta = 90°$ at the summer solstice; φ is the latitude of the Earth. (**c**) shows the solar irradiance at various frequencies and (**d**) the peak temperature that suggests the calculative power to determine the net solar energy

$$\begin{aligned}\cos(\theta) = &\sin(\phi)\sin(\delta)\cos(\beta) + \sin(\delta)\cos(\phi)\sin(\beta)\cos(\gamma)\\ &+ \cos(\phi)\cos(\delta)\cos(\beta)\cos(h) - \cos(\delta)\sin(\phi)\sin(\beta)\cos(\gamma)\cos(h)\\ &- \cos(\delta)\sin(\beta)\sin(\gamma)\sin(h)\end{aligned}$$

where β denotes the angle from the horizon and γ denotes the azimuth angle.

The sphere of Earth from the sun here is represented denoted R_E where the average distance is represented as R_0, with approximation of one astronomical unit (AU). Here, the solar constant is being represented as S_0 where the solar irradiance density onto a Earth plane tangent is calculated as:

$$Q = \begin{cases} S_o \dfrac{R_0^2}{R_E^2} & \cos(\theta)\cos(\theta) > 0 \\ 0 & \cos(\theta) \leq 0 \end{cases}$$

The mean Q over a day is the average of Q over one rotation or the hour angle progressing from $h = \pi$ to $h = -\pi$: Thus, the equation has been rewritten as:

$$Q^{-\text{day}} = -\frac{1}{2\pi}\int_{\pi}^{-\pi} Q\mathrm{d}h$$

Since h_0 is the hour angle when Q becomes positive, thus, it could occur at sunrise when $\Theta = 1/2\ \pi$ or for h_0 as a solution of

$$\sin(\phi)\sin(\delta) + \cos(\phi)\cos(\delta)\cos\left(h_{0)}\right. = 0$$

or

$$\cos(h_0) = -\tan(\phi)\tan(\delta)$$

Once $\tan(\varphi)\tan(\delta) > 1$, then the sun does not set and the sun is already risen at $h = \pi$, so $h_0 = \pi$. Then the $\tan(\varphi)\tan(\delta) < -1$, the sun does not rise and

$$\mathrm{Q}^{-\text{day}} = 0$$

$\frac{R_0^2}{R_E^2}$ is nearly constant over the course of a day and can be taken outside the integral

$$\begin{aligned} \int_{\pi}^{-\pi} Q\mathrm{d}h &= \int_{h_0}^{-h_0} Q\mathrm{d}h = S_0\frac{R_0^2}{R_E^2}\int_{h_0}^{-h_0}\cos(\theta)\mathrm{d}h \\ &= S_0\frac{R_0^2}{R_E^2}\left[h\sin(\phi)\sin(\delta) + \cos(\phi)\cos(\delta)\sin(h)\right]_{h=h_0}^{h=-h_0} \\ &= -2S_0\frac{R_0^2}{R_E^2}\left[h_0\sin(\phi)\sin(\delta) + \cos(\phi)\cos(\delta)\sin\left(h_{0)}\right.\right] \end{aligned}$$

Therefore,

$$Q^{-\text{day}} = \frac{S_0}{\pi}\frac{R_0^2}{R_E^2}\left[h_0\sin(\phi)\sin(\delta) + \cos(\phi)\cos(\delta)\sin\left(h_{0)}\right.\right]$$

Since the θ is being considered as the conventional polar angle describing a planetary orbit, thus, $\theta = 0$ at the vernal equinox and the declination δ as a function of orbital position would be

$$\delta = \varepsilon \sin(\theta)$$

where ε is the obliquity and the conventional longitude of perihelion ϖ shall be related to the vernal equinox, so the elliptical orbit can be rewritten as:

$$R_{\mathrm{E}} = \frac{R_0}{1 + e\cos(\theta - \omega)}$$

or

$$\frac{R_0}{R_{\mathrm{E}}} = 1 + e\cos(\theta - \omega)$$

Here the ϖ, ε, and e are being calculated from astrodynamical laws, so that a consensus of observations of Q¯day can be determined from any latitude φ and θ. However, $\theta = 0°$ is considered as duration of the vernal equinox, $\theta = 90°$ is exactly the time of the summer solstice, $\theta = 180°$ is exactly the time of the autumnal equinox, and $\theta = 270°$ is exactly the time of the winter solstice. Therefore, the equation can be simplified for irradiance on a given day as follows:

$$Q = S_0\left(1 + 0.034\cos\left(2\pi\frac{n}{365.25}\right)\right)$$

where n is the number of a day of the year, and thus, the solar characteristics for both theoretical function of optimum and modular to generate electricity can be shown per unit area (Fig. 2.6).

Eventually, a peak high frequency cutoff Ω_{C} of solar irradiance is being calculated to keep away the bifurcation of DOS from the Earth surface. Necessarily, a tipped high-frequency cutoff of Earth surface Ω_{d} is being determined by controlling the positive DOS in 2D and 1D of the photon irradiance. Hence pi_2 (x) acted as an algorithm function and e_{rfc} (x) acts as an additional function [25, 36]. Thus, the DOS of Earth surface here represented as $\varrho_{\mathrm{PC}}(\omega)$, which is determined by calculating photonic energy frequencies of Maxwell's rules into the Earth surface [21]. For a 1D on Earth surface, the represented DOS is thus being expressed as $\varrho_{\mathrm{PC}}(\omega) \propto \frac{1}{\sqrt{\omega - \omega_{\mathrm{e}}}}\Theta(\omega - \omega_{\mathrm{e}})$, where $\Theta(\omega - \omega_{\mathrm{e}})$ represents the Heaviside functional step and ω_{e} e expressed the frequency of the net solar energy generation (Fig. 2.7).

This DOS is thus determined to confirm a 3D isentropic function in the Earth surface to acquire an accurate net qualitative state of solar energy by inducing the non-Weisskopf-Wigner mode of photons in the Earth surface [7, 18, 19]. Naturally, this 3D state will be the functional DOS into the PBE area of DOS: $\varrho_{\mathrm{PC}}(\omega) \propto \frac{1}{\sqrt{\omega - \omega_{\mathrm{e}}}}\Theta(\omega - \omega_{\mathrm{e}})$, and thus, it has been integrated to the net electricity (EF) vector of Earth surface to determine the net electricity energy generation accurately on the Earth surface [13, 37]. Considering the 2D and 1D, the photonic energy DOS are being clarified by the pure algorithm of divergence, which is close to the PBE, and thus expressed as $\varrho_{\mathrm{PC}}(\omega) \propto -[\mathrm{ln}|(\omega - \omega_0)/\omega_0| - 1]\Theta(\omega - \omega_{\mathrm{e}})$,

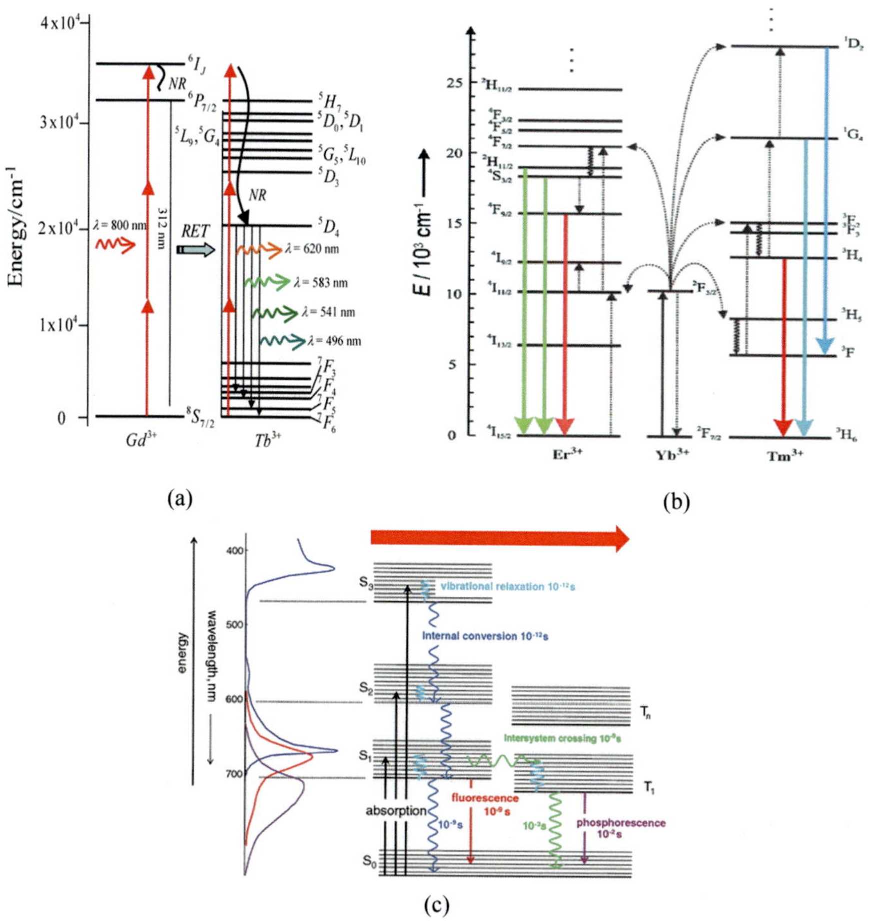

Fig. 2.6 (**a**) shows the above stands for the solar energy function of optimum working point for energy generation per unit area, (**b**) depicted the theoretical function of modular point of energy generation per unit area, and (**c**) energy generation at various wavelength of photon spectrum

where ω_e denoted the midpoint of tip algorithm. The functional area $J(\omega)$ is thus clarified as the photon energy generation on the Earth surface where the solar energy generation of $V(\omega)$ depends on the total solar irradiance on the Earth surface [22, 23],

$$J(\omega) = \varrho(\omega)|V(\omega)|^2 \tag{2.29}$$

Hence, the PB frequency ω_c and proliferative solar energy are being considered as the function $u(t, t_0)$ for photon energy generation in the relation $\langle a(t)\rangle = u(t, t_0)\langle a(t_0)\rangle$. It is therefore determined using the functional integral equation and expressed as

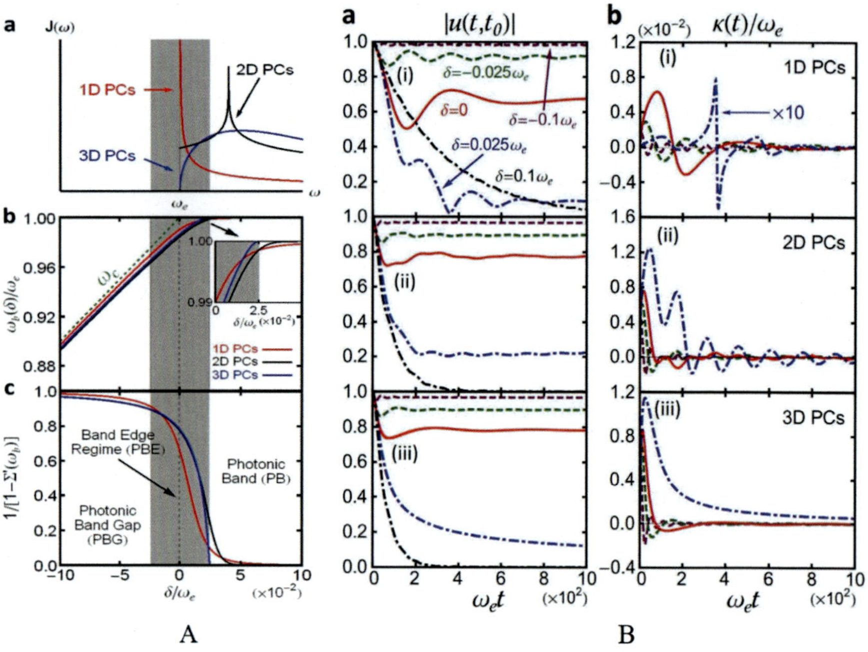

Fig. 2.7 (*A*) The structural composition of photon and rate of energy deliberation in the acting PV panel. (**a**) Functional area at different DOS magnitude of 1D, 2D, and 3D PV cells. (**b**) Photon frequency rate at the functional photonic band edge regime (PBE) and photonic band gap (PBG). (**c**) Photon's magnitude to deliver high energy into the functional photonic band edge regime (PBE) and photonic band gap (PBG). (*B*) Proliferation of photon dynamics into photovoltaic cells. (**a**) Considering the PB area, $<a(t)> = 5\, u(t, t_0) < a(t_0)>$, and (**b**) photon dynamic rate $k(t)$, functional vatable for (i) 1D, (ii) 2D and (iii) 3D quantum field into the PV cells. (Courtesy Figs. A and B: Ping-Yuan Lo, Heng-Na Xiong & Wei-Min Zhang (2015); Scientific Reports, volume 5, Article number: 9423)

$$u(t, t_0) = \frac{1}{1 - \Sigma'(\omega_b)} e^{-i\omega(t - t_0)} + \int_{\omega_e}^{\infty} d\omega \frac{J(\omega) e^{-i\omega(t - t_0)}}{[\omega - \omega_c - \Delta(\omega)]^2 + \pi^2 J^2(\omega)} \tag{2.30}$$

where $\Sigma'(\omega_b) = [\partial\Sigma(\omega)/\partial\omega]_{\omega = \omega_b}$ and $\Sigma(\omega)$ denoted the storage-induced PB photonic energy proliferations,

$$\Sigma(\omega) = \int_{\omega_e}^{\infty} d\omega' \frac{J(\omega')}{\omega - \omega'} \tag{2.31}$$

Here, the frequency ω_b in Eq. (2.17) denoted the photon energy frequency module in the PBG ($0 < \omega_b < \omega_e$), and thus it is calculated using the areal condition: $\omega_b - \omega_c - \Delta(\omega_b) = 0$, where $\lesssim \Delta(\omega) = \mathcal{P}\left[\int d\omega' \frac{J(\omega')}{\omega - \omega'}\right]$ is a primary-value integral.

Then, the net photon energy, considering the proliferation magnitude $|u(t, t_0)|$ have been calculated and are being shown in Fig. 2.6a for 1D, 2D, and 3D of Earth surface with respect to PBG function [20, 22, 38]. The solar energy dynamic rate $\kappa(t)$ is being depicted in Fig. 2.4b, neglecting the function $\delta = 0.1\ \omega_e$. The result revealed that emitted photons are being generated at a high rate once ω_c crosses from the PBG to PB area. Because the range in $u(t, t_0)$ is $1 \geq |u(t, t_0)| \geq 0$, the crossover area as related to the condition is being denoted as $0.9 \gtrsim |u(t \to \infty, t_0)| \geq 0$ where this corresponded to $-0.025\omega_e \lesssim \delta \lesssim 0.025\omega_e$, with a production rate $\kappa(t)$ within the PBG ($\delta < -0.025\omega_e$) and in the area of the PBE ($-0.025\omega_e \lesssim \delta \lesssim 0.025\omega_e$) of the Earth surface.

The generation of solar energy emission is almost exponential for $\delta \gg 0.025\omega_e$, which is a Markov factor. It is shown in Fig. 2.5a as the dash-doted black curves with $\delta = 0.1\omega_e$. In the crossover area ($-0.025\omega_e \lesssim \delta \lesssim 0.025\omega_e$), the PB frequency of the PBE of Earth surface sharply increases the mode of the emission of photon energy generation [8, 28, 41]. Thus, this proliferation of emitted solar photon confirms the net energy state photon in the Earth surface of the PBG where the photons are in a non-equilibrium photonic energy state [6, 17, 33].

Then, the solar irradiance on entire Earth surface is being clarified considering thermal variation with respect to the solar energy concentration function $v(t, t)$ by determining the non-equilibrium solar energy scattering and reflecting calculation globally as

$$(t, t) = \int_{t_0}^{t} dt_1 \int_{t_0}^{t} dt_2 u^*(t_1, t_0)\tilde{g}(t_1, t_2)u(t_2, t_0) \tag{2.32}$$

Here, the two-time correlation function of Earth surface $\tilde{g}(t_1, t_2) = \int d\omega J(\omega)\overline{n}(\omega, T)e^{-i\omega(t - t')}$ reveals the solar energy generation variations induced by the thermal relativistic condition of Earth surface, where $\overline{n}(\omega, T) = 1/\left[e^{h\omega/k_B T} - 1\right]$ is the proliferation of the photon energy emission in the Earth surface at the optimum temperature T and expressed as

$$v(t, t \to \infty) = \int_{\omega_e}^{\infty} d\omega \mathcal{V}(\omega)$$

with

$$\mathcal{V}(\omega) = \overline{n}(\omega, T)[\mathcal{D}_l(\omega) + \mathcal{D}_d(\omega)] \tag{2.33}$$

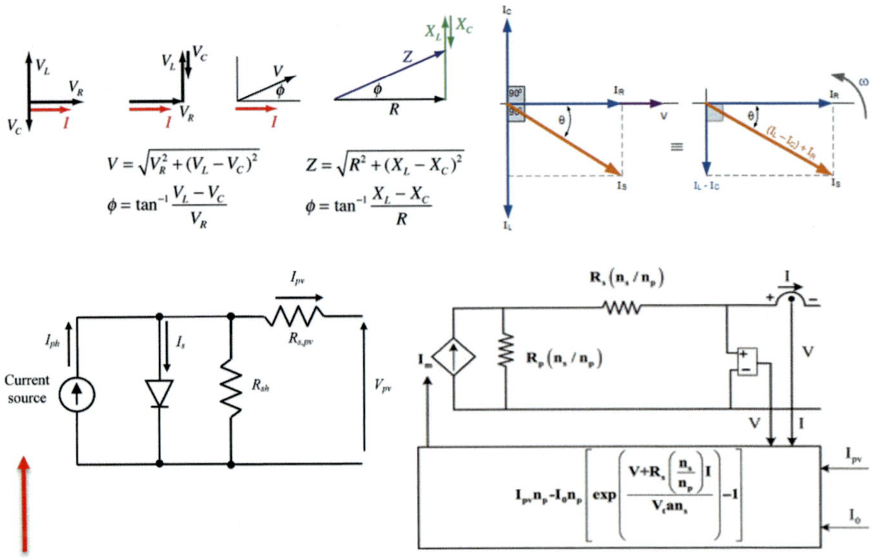

Fig. 2.8 (**a**) The scalar field of the Earth surface, (**b**) solar energy scalar field on Earth surface, (**c**) net electricity current energy generation on Earth from the total solar energy on Earth [10, 19]

Here, Eq. (2.18) is simplified to determine the non-equilibrium condition: $\mathcal{V}(\omega) = \overline{n}(\omega, T)\mathcal{D}_{\text{d}}(\omega)$. Under low-temperature conditions on Earth surface, Einstein's photon energy fluctuation dissipation is not dynamically viable at the PB on Earth surface but also connecting the photonic energy state that has been measured as the field intensity of solar energy induction $n(t) = \langle a^{\dagger}(t)a(t)\rangle = |u(t, t_0)|^2 n(t_0)v(t, t)$, where $n(t_0)$ represents the primary PB of Earth surface. Therefore, in Fig. 2.8, the plotted the net amount of photon energy s versus temperature on Earth surface has been clarified as the non-equilibrium proliferated photon energy generation, as shown by the solid-blue curve (Fig. 2.6). To be more specific, the first PB of Earth surface has been considered as the Fock state photon number n_0, i. e. $\rho(t_0) = |n_0\rangle\langle n_0|$, which is obtained mathematically through the quantum dynamics of the photon energy and then by solving the Eq. (2.33), respecting the state of net photon energy production at time t:

$$\rho(t) = \sum_{n=0}^{\infty} \mathcal{P}_n^{(n_0)}(t)|n_0\rangle\langle n_0| \tag{2.34}$$

$$\mathcal{P}_n^{(n_0)}(t) = \frac{[v(t,t)]^n}{[1+v(t,t)]^{n+1}}[1-\Omega(t)]^{n_0} \times \sum_{k=0}^{\min\{n_0,n\}} \binom{n_0}{k}\binom{n}{k}\left[\frac{1}{v(t,t)}\frac{\Omega(t)}{1-\Omega(t)}\right]^k \tag{2.35}$$

where $\Omega(t) = \frac{|u(t,t_0)|^2}{1+v(t,t)}$. Therefore, the result reveals that an electron state photon energy will evolve into different Fock states of $|n_0\rangle$ is $\mathcal{P}_n^{(n_0)}(t)$ on the Earth surface. The proliferation of net photon energy dissipation $\mathcal{P}_n^{(n_0)}(t)$ in the primary state $|n_0 = 5\rangle$ and steady-state limit $\mathcal{P}_n^{(n_0)}(t \to \infty)$ is thus shown in Fig. 2.8. Therefore, the generation of net photon energy generation on Earth surface emit will ultimately reach the thermal non-equilibrium state, which is being expressed as

$$\mathcal{P}_n^{(n_0)}(t \to \infty) = \frac{[\overline{n}(\omega_c, T)]^n}{[1+\overline{n}(\omega_c, T)]^{n+1}} \tag{2.36}$$

To probe this huge photon energy generation on Earth surface, a further calculation of the photon energy distribution within the quantum field of Earth surface has been conducted through the high-temperature coherent states and solving Eq. (2.17) considering the energy state of photons and expressed by

$$\rho(t) = \mathcal{D}[\alpha(t)]\rho_T[v(t,t)]\mathcal{D}^{-1}[\alpha(t)] \tag{2.37}$$

where $\mathcal{D}[\alpha(t)] = \exp\{\alpha(t)\alpha^\dagger - \alpha^*(t)\alpha\}$ denotes the displacement functions with $\alpha(t) = u(t, t_0)\alpha_0$ and

$$\rho_T[v(t,t)] = \sum_{n=0}^{\infty} \frac{[v(t,t)^n]}{[1+v(t,t)]^{n+1}} \mid n\rangle\langle n \mid \tag{2.38}$$

Here, ρ_T denotes a thermal state with an average particle quantum $v(t, t)$, where Eq. (2.11) suggests that the peak point photon energy generation state will be evolved into a thermal state [40], which is considered as the functional state of the photon $\mathcal{D}[\alpha(t)] \mid n\rangle$ in the Earth surface. Thus, the net photon energy generation calculation is being represented by the following equation:

$$\begin{aligned} \langle m|\rho(t)|n\rangle &= J(\omega) = e^{-\Omega(t)|\alpha_0|^2} \frac{[\alpha(t)]^m[\alpha^*(t)]^n}{[1+v(t,t)]^{m+n+1}} \\ &= \sum_{k=0}^{\min\{m,n\}} \frac{\sqrt{m!n!}}{(m-k)!(n-k)!k!}\left[\frac{v(t,t)}{\Omega(t)|\alpha_0|^2}\right]^k \end{aligned} \tag{2.39}$$

where the emission of the net photon energy ($\langle m| \rho(t)| n\rangle$) into the Earth surface and its conversion of photon energy into electricity $[1 + v(t, t)]^{m+n+1}$ and non-equilibrium condition $[\alpha(t)]^m[\alpha^*(t)]^n$ of the Earth surface have been calculated.

Modeling of Net Electricity Energy Generation from Total Solar Irradiance on Earth

To transform this tremendous amount of photon energy into electricity energy, the net solar energy is being computed on a conceptual model of series and parallel circuit of Earth surface. The conceptual Earth surface is being then hypothetically implemented into the I–V single-diode circuit of Earth surface in order to get the precise I–V relationship of the net solar energy reaches on Earth surface by calculating from the following equation:

$$I = I_L - I_O \left\{ \exp^{\left[\frac{q(V+I_{RS})}{AkTc}\right]} - 1 \right\} - \frac{(V + I_{RS})}{R_{Sh}} \tag{2.40}$$

Here, I_L denotes the photon formation current, I_O denotes the ideal current flow into the diode, RS denotes the resistance in a series, A denotes the diode function, k $(= 1.38 \times 10^{-23}\ \mathrm{W/m^2\ K})$ denotes Boltzmann's constant, q $(=1.6 \times 10^{-19}\mathrm{C})$ denotes the charge amplitude of the electron, and T_C denotes the Earth temperature. Consequently, the $I - q$ linked in the Earth surface is being varied into the diode cell, which is expressed as the dynamic current as follows [17, 26]:

$$I_O = I_{RS} \left(\frac{T_C}{T_{ref}} \right)^3 \exp \left[\frac{qEG\left(\frac{1}{T_{ref}} - \frac{1}{T_C} \right)}{KA} \right] \tag{2.41}$$

where I_{RS} denotes the dynamic current representing the functional transformation of solar radiation and qEG denotes the band-gap solar radiation into the conceptual Earth surface at different DOS dimensional modes of 1D, 2D, and 3D (Table 2.1).

Here, considering this conceptual Earth surface, the I–V relationship with the exception of I–V curve, a calculative result of linked I–V curves among all of the conceptual solar cells has been determined [36, 37]. Thus, the equation is being rewritten as follows in order to determine the V–R relationship much more accurately:

$$V = -IR_s + K \log \left[\frac{I_L - I + I_O}{I_O} \right] \tag{2.42}$$

where K denotes the constant $\left(= \frac{AkT}{q} \right)$ and I_{mo} and V_{mo} are being denoted as the net current and voltage in the conceptual Earth surface. Subsequently, the relationship among I_{mo} and V_{mo} shall remain motional in the I–V Earth surface and that can be written as:

Table 2.1 The solar energy in various DOS dimensional modes reaching on Earth surface (*ES*) with corresponding various unit area $J(\omega)$ and self-energy induced reservoir of $\Sigma(\omega)$

Solar energy (ES)	Unit area $J(\omega)$ for different DOS	Solar energy correction in Earth surface $\Sigma(\omega)$
1D	$\frac{ES}{2\pi r}\frac{1}{\sqrt{\omega-\omega_e}}\Theta(\omega-\omega_e)$	$-\frac{ES}{\sqrt{2\omega_e-\omega}}$
2D	$-ES\left[\ln\left\|\frac{\omega-\omega_0}{2\omega_0}\right\|-1\right]\Theta(\omega-\omega_e)\Theta(\Omega_d-\omega)$	$ES\left[Li_2\left(\frac{\Omega_d-\omega_0}{\omega-\omega_0}\right)-Li_2\left(\frac{\omega_0-\omega_e}{\omega_0-\omega}\right)-\ln\frac{\omega_0-\omega_e}{\Omega_d-\omega_0}\ln\frac{\omega_e-\omega}{\omega_0-\omega}\right]$
3D	$ES\sqrt{\frac{2\omega-\omega_e}{\Omega_C}}\exp\left(-\frac{\omega-\omega_e}{\Omega_C}\right)\Theta(\omega-\omega_e)$	$ES\left[\pi\sqrt{\frac{\omega_e-\omega}{\Omega_C}}\exp\left(-\frac{2\omega-\omega_e}{\Omega_C}\right)\mathrm{erfc}\sqrt{\frac{\omega_e-\omega}{\Omega_C}}-\sqrt{2\pi r}\right]$

The functional C, η and χ act like coupled forces between the solar energy at Earth surface of 1D, 2D, and 3D areal surface (Courtesy: Ping-Yuan Lo, Heng-Na Xiong, & Wei-Min Zhang (2015); Scientific Reports, volume 5, Article number: 9423)

$$V_{\mathrm{mo}} = -I_{\mathrm{mo}}R_{\mathrm{Smo}} + K_{\mathrm{mo}} \log\left(\frac{I_{\mathrm{Lmo}} - I_{\mathrm{mo}} + I_{\mathrm{Omo}}}{I_{\mathrm{Omo}}}\right) \tag{2.43}$$

where I_{Lmo} denotes the photon-induced current, I_{Omo} denotes the dynamic current into the diode, R_{smo} denotes the resistance in series, and K_{mo} denotes the factorial constant.

Once all non-series (NS) cells are being interlinked in the series, then the series resistance is being calculated as the sum of each solar cell series resistance $R_{\mathrm{smo}} = N_{\mathrm{S}} \times R_{\mathrm{S}}$ current considering the functional coefficient of the constant factor $K_{\mathrm{mo}} = N_{\mathrm{S}} \times K$. The flow of current dynamics into the circuit is lined to the cells in a series connection [37]. Thus, the current dynamics in Eq. (2.40) remains the same in each part of $I_{\mathrm{omo}} = I_{\mathrm{o}}$ and $I_{\mathrm{Lmo}} = I_{\mathrm{L}}$. Thus, the mode of $I_{\mathrm{mo}} - V_{\mathrm{mo}}$ relationship for the N_{S} series of connected cells can be expressed by:

$$V_{\mathrm{mo}} = -I_{\mathrm{mo}}N_{\mathrm{S}}R_{\mathrm{S}} + N_{\mathrm{S}}K \log\left(\frac{I_{\mathrm{L}} - I_{\mathrm{mo}} + I_{\mathrm{o}}}{I_{\mathrm{o}}}\right) \tag{2.44}$$

Naturally, the current–voltage relationship can be further modified considering all parallel linked into N_{P} cell connection in all parallel mode and can be described as follows [28, 34]:

$$V_{\mathrm{mo}} = -I_{\mathrm{mo}}\frac{R_{\mathrm{S}}}{N_{\mathrm{P}}} + K \log\left(\frac{N_{sh}I_{\mathrm{L}} - I_{\mathrm{mo}} + N_{\mathrm{p}}I_{\mathrm{o}}}{N_{\mathrm{p}}I_{\mathrm{o}}}\right) \tag{2.45}$$

Since the photon-induced current primarily depends on the solar radiation and optimum temperature configuration, the net current dynamic is being calculated as:

$$I_{\mathrm{L}} = G\left[I_{\mathrm{SC}} + K_{\mathrm{I}}\left(T_{\mathrm{C}} - T_{\mathrm{ref}}\right)\right] * V_{\mathrm{mo}} \tag{2.46}$$

where I_{sc} denotes the current at 25 °C and $\mathrm{KW/m^2}$, K_{I} denotes Earth surface coefficient factor, T_{ref} denotes the optimum temperature, and G denotes the solar energy in $\mathrm{mW/m^2}$ [19, 41].

Finally, the electricity energy generation around the Earth surface has been computed in order to confirm the net emitted photon utilization by integrating local Albanian electric fields; thus, the global U (1) gauge field will allow to add a mass term of the functional particle of $\varnothing' \rightarrow e^{i\alpha(x)}\varnothing$. It is then further clarified by explaining the variable derivative of transformation law of scalar field using the following equation [21, 37]:

$$\partial_\mu \rightarrow D_\mu = \partial_\mu = ieA_\mu \qquad [\text{covariant derivatives}]$$

$$A'_\mu = A_\mu + \frac{1}{e}\partial_\mu \alpha \qquad [A_\mu \text{ derivatives}] \tag{2.47}$$

Here, the global U (1) gauge denotes the invariant local Albanian for a complex scalar field, which can be expressed as:

$$\mathfrak{H} = (D^{\mu})^{\dagger}\left(D_{\mu}\varnothing\right) - \frac{1}{4}F_{\mu v}F^{\mu v} - V(\varnothing) \tag{2.48}$$

The term $\frac{1}{4}F_{\mu v}F^{\mu v}$ is the dynamic term for the gauge field of the Earth surface and $V(\varnothing)$ denotes the extra term in the local Albanian, which is $V(\varnothing^{*}\varnothing) = \mu^{2}(\varnothing^{*}\varnothing) + \lambda\,(\varnothing^{*}\varnothing)^{2}$.

Therefore, the generation of local Albanian ($\mathfrak{H}$) under the perturbational function of the quantum field of the Earth surface has been confirmed by the calculation of mass-scalar particles ϕ_1 and ϕ_2 along with a mass variable of μ. In this condition $\mu^2 < 0$ had an infinite number of quantum which is being clarified by $\phi_1^2 + \phi_2^2 = -\mu^2/\lambda = v^2$ and the $\mathfrak{H}$ through the variable derivatives using further of shifted fields η *and* ξ defined the quantum field as $\phi_0 = \frac{1}{\sqrt{2}}\,[(v+\eta)+i\xi]$.

$$\text{Kinetic term}: \mathfrak{H}(\eta,\xi) = (D^{\mu}\phi)^{\dagger}(D^{\mu}\phi) = (\partial^{\mu} + ieA^{\mu})\phi^{*}\left(\partial_{\mu} - ieA_{\mu}\right)\phi \tag{2.49}$$

Thus, this expanding term in the $\mathfrak{H}$ associated to the scalar field of the Earth surface is suggesting that the net Earth surface field is prepared to initiate the net electricity energy generation into its quantum field of induced photon energy, respectively, at the normal, normalized, and normal modes [10, 19].

To determine this electricity energy, hereby, a non-variable function of readily dynamics has been implemented for the calculation of $\overline{\varphi}[s_0]$ to confirm the expected value of s_0 considering the Earth surface, which is expressed as follows:

$$\overline{\varphi}[s_0] = 2s_0\,(\ln 4s_0 - 2) + \ln 4s_0(\ln 4s_0 - 2) - \frac{(\pi^2-9)}{3} + s_0^{-1}\left(\ln 4s_0 + \frac{9}{8}\right) + \cdots (s_0 \gg 1) \tag{2.50}$$

$$\begin{aligned}\overline{\varphi}[s_0] = &\left(\frac{2}{3}\right)(S_0-1)^{\frac{3}{2}} + \left(\frac{5}{3}\right)(S_0-1)^{\frac{5}{2}} \\ &- \left(\frac{1507}{420}\right)(S_0-1)^{\frac{7}{2}}(1/2 \text{ instead of } 1)\end{aligned} \tag{2.51}$$

Then the final equation can be rewritten as where s_0 is the areal value of electricity energy generation into the Earth surface (1 m^2).

$$\overline{\varphi}[s_0] = \left(\frac{2}{3}\right)(S_0-1)^{\frac{3}{2}} + \left(\frac{5}{3}\right)(S_0-1)^{\frac{5}{2}} - \left(\frac{1507}{420}\right)(S_0-1)^{\frac{7}{2}} \tag{2.52}$$

The function $\overline{\varphi}[s_0]$ thus determined the net electricity energy generation from the total solar energy into the atmosphere by calculation of Earth's cross-sectional area of 127,400,000 km^2, and the net solar energy intercepted by the surface of Earth is

1.740 × 1017 W. Considering the seasonal and climate variation, the net power reaching the ground generally averages 200 W per square meter per day [26]. Thus, the average power reaching at any time by the Earth's surface is calculated as 127.4 × 106 × 106 × 200 = 25.4 × 1015 W or 25,400 TW, which is TW × 24 × 365 = 222,504,000 TWh. Since the net annual electrical energy (not the total energy) consumed in the world from all sources in 2019 was 22,126 TWh, the available solar energy is over 10,056 times the world's consumption.

Conclusion

Since the fossil fuel utilization throughout the world is getting finite level and the major contributor for climate change, usages of renewable energy such as solar energy reaching of Earth have been clarified as an interesting source of renewable energy to mitigate the global energy and environmental crisis. Simply, the solar energy, here, the radiant energy from the Sun is thus estimated in order to use this energy as the primary source potentially in every sector of our daily live in order to confirm energy security for everybody in the near future, which is clean to mitigate the global climate change completely.

Acknowledgments The author, Md. Faruque Hossain, declares that any findings, predictions, and conclusions described in this article are solely performed by the author and it is confirmed that there is no conflict of interest for publishing this research paper in a suitable journal and/or publisher.

References

1. Ben Messaoud, R. (2019). Extraction of uncertain parameters of double-diode model of a photovoltaic panel using simulated annealing optimization. *The Journal of Physical Chemistry C, 123*(48), 29096–29103.
2. Besharat, F., Dehghan, A. A., & Faghih, A. R. (2013). Empirical models for estimating global solar radiation: A review and case study. *Renewable and Sustainable Energy Reviews, 21*, 798–821.
3. Birnbaum, K. M., Boca, A., Miller, R., Boozer, A. D., Northup, T. E., & Kimble, H. J. (2005). Photon blockade in an optical cavity with one trapped atom. *Nature, 436*(7047), 87–90.
4. Brešar, B., Klavžar, S., & Škrekovski, R. (2003). Quasi-median graphs, their generalizations, and tree-like equalities. *European Journal of Combinatorics, 24*(5), 557–572.
5. Chang, D. E., Sørensen, A. S., Demler, E. A., & Lukin, M. D. (2007). A single-photon transistor using nanoscale surface plasmons. *Nature Physics, 3*(11), 807–812.
6. Danish, M. S. S., & Senjyu, T. S. (2020). Green building efficiency and sustainability indicators. In *Green building management and smart automation* (pp. 128–145). IGI Global.
7. Dayan, B., Parkins, A. S., Aoki, T., Ostby, E. P., Vahala, K. J., & Kimble, H. J. (2008). A photon turnstile dynamically regulated by one atom. *Science, 319*(5866), 1062–1065.
8. Douglas, J. S., Habibian, H., Hung, C., Gorshkov, A. V., Kimble, H. J., & Chang, D. E. (2015). Quantum many-body models with cold atoms coupled to photonic crystals. *Nature Photonics, 9*(5), 326–331.

9. Evaristo, J., Jasechko, S., & McDonnell, J. J. (2015). Global separation of plant transpiration from groundwater and streamflow. *Nature, 525*(7567), 91–94.
10. Fratzl, P., & Barth, F. G. (2009). Biomaterial systems for mechanosensing and actuation. *Nature, 462*(7272), 442–448.
11. Gleyzes, S., Kuhr, S., Guerlin, C., Bernu, J., Deleglise, S., Busk Hoff, U., et al. (2007). Quantum jumps of light recording the birth and death of a photon in a cavity. *Nature, 446*(7133), 297–300.
12. Guerlin, C., Bernu, J., Deleglise, S., Sayrin, C., Gleyzes, S., Kuhr, S., et al. (2007). Progressive field-state collapse and quantum non-demolition photon counting. *Nature, 448*(7156), 889–893.
13. Guo, Y., Al-Jubainawi, A., & Ma, Z. (2019). Performance investigation and optimisation of electrodialysis regeneration for LiCl liquid desiccant cooling systems. *Applied Thermal Engineering, 149*, 1023–1034.
14. Hossain, M. (2018). Transforming dark photons into sustainable energy. *International Journal of Energy and Environmental Engineering, 9*(1), 99–110.
15. Hossain, M. (2019). Green technology: Transformation of transpiration vapor to mitigate global water crisis. *Polytechnica, 2*(1), 26–29.
16. Hossain, M. (2022). Photon application in the design of sustainable buildings to console global energy and environment. In *Sustainable design for global equilibrium* (pp. 125–142). Springer.
17. Hossain, M. F. (2016). Solar energy integration into advanced building design for meeting energy demand and environment problem. *International Journal of Energy Research, 40*(9), 1293–1300.
18. Hossain, M. F. (2017). Green science: Independent building technology to mitigate energy, environment, and climate change. *Renewable and Sustainable Energy Reviews, 73*, 695–705.
19. Hossain, M. F. (2018). Green science: Advanced building design technology to mitigate energy and environment. *Renewable and Sustainable Energy Reviews, 81*, 3051–3060.
20. Hossain, M. F. (2018). Photon energy amplification for the design of a micro PV panel. *International Journal of Energy Research, 42*(12), 3861–3876.
21. Hossain, M. F. (2019). Advanced building design. In *Sustainable design and build*. Butterworth-Heinemann.
22. Hossain, M. F. (2019). RETRACTED: Natural mechanism to console global water, energy, and climate change crisis. *Sustainable Energy Technologies and Assessments, 35*, 347–353.
23. Hossain, M. F. (2019). RETRACTED: Theoretical mechanism to breakdown of photonic structure to design a micro PV panel. *Energy Reports, 5*, 649–657.
24. Jasechko, S., Sharp, Z. D., Gibson, J. J., Birks, S. J., Yi, Y., & Fawcett, P. J. (2013). Terrestrial water fluxes dominated by transpiration. *Nature, 496*(7445), 347–350.
25. Joannopoulos, J. D., Villeneuve, P. R., & Fan, S. (1997). Photonic crystals: Putting a new twist on light. *Nature, 386*(6621), 143–149.
26. Langer, L., Poltavtsev, S. V., Yugova, I. A., Salewski, M., Yakovlev, D. R., Karczewski, G., et al. (2014). Access to long-term optical memories using photon echoes retrieved from semiconductor spins. *Nature Photonics, 8*(11), 851–857.
27. Maxwell, R. M., & Condon, L. E. (2016). Connections between groundwater flow and transpiration partitioning. *Science, 353*(6297), 377–380.
28. Newell, J. R. (2009). A story of things yet-to-be: The status of geology in the united states in 1807. *Geological Society, London, Special Publications, 317*(1), 213–217.
29. Park, K., Marek, P., & Filip, R. (2017). Qubit-mediated deterministic nonlinear gates for quantum oscillators. *Scientific Reports, 7*(1), 1–8.
30. Sayrin, C., Dotsenko, I., Zhou, X., Peaudecerf, B., Rybarczyk, T., Gleyzes, S., & Brune, M. (2011). Real-time quantum feedback prepares and stabilizes photon number states. *Nature, 477*(7362), 73–77.
31. Shen, H. Z., Xu, S., Cui, H. T., & Yi, X. X. (2019). Non-markovian dynamics of a system of two-level atoms coupled to a structured environment. *Physical Review A, 99*(3), 032101.
32. Tame, M. S., McEnery, K. R., Özdemir, Ş. K., Lee, J., Maier, S. A., & Kim, M. S. (2013). Quantum plasmonics. *Nature Physics, 9*(6), 329–340.

33. Tu, M. W., & Zhang, W. (2008). Non-markovian decoherence theory for a double-dot charge qubit. *Physical Review B, 78*(23), 235311.
34. Vargas, R., Carvajal, D., Madriz, L., & Scharifker, B. R. (2020). Chemical kinetics in solar to chemical energy conversion: The photoelectrochemical oxygen transfer reaction. *Energy Reports, 6*, 2–12.
35. Wheeler, T. D., & Stroock, A. D. (2008). The transpiration of water at negative pressures in a synthetic tree. *Nature, 455*(7210), 208–212.
36. Xiao, Y., Li, M., Liu, Y., Li, Y., Sun, X., & Gong, Q. (2010). Asymmetric Fano resonance analysis in indirectly coupled microresonators. *Physical Review A, 82*(6), 065804.
37. Yan, W., & Fan, H. (2014). Single-photon quantum router with multiple output ports. *Scientific Reports, 4*(1), 1–6.
38. Yan, W., Huang, J., & Fan, H. (2013). Tunable single-photon frequency conversion in a sagnac interferometer. *Scientific Reports, 3*(1), 1–6.
39. Yang, L., Wang, S., Zeng, Q., Zhang, Z., Pei, T., Li, Y., & Peng, L. (2011). Efficient photovoltage multiplication in carbon nanotubes. *Nature Photonics, 5*(11), 672–676.
40. Zhang, W., Lo, P., Xiong, H., Tu, M. W., & Nori, F. (2012). General non-markovian dynamics of open quantum systems. *Physical Review Letters, 109*(17), 170402.
41. Zhu, Y., Hu, X., Yang, H., & Gong, Q. (2014). On-chip plasmon-induced transparency based on plasmonic coupled nanocavities. *Scientific Reports, 4*(1), 1–7.

Chapter 3
Deconstruction of Photon Structure to Produce Clean Energy

Introduction

Clean energy is presently in demand to create a better planet for future generations. Conventional energy consumption by building sectors releases 8.01×10^{11} tons of CO_2 per year and causes nearly 40% of global warming. Therefore, in this study, I have proposed an advanced micro photovoltaic (PV) panel that utilizes Bose-Einstein photon distribution theory to convert a single photon into multiple photons to mitigate the total energy demand of a building and significantly reduce global warming. Photonic band-gap (PBG) structures have been studied in the last decade at the *nano*scale to obtain better knowledge of the characteristic dispersion properties of photons [1, 3]. Some interesting features of photonic structure have been identified; for example, when photons are induced by PBGs, spontaneous photonic inhibition occurs due to electron level energy state photons [4, 6]. For the last several decades, quantum optics has been extensively explored under zero-degree temperature and point break quantum electrodynamics (QED) considering photonic band edges (PBEs). Previous work has suggested [10, 14] that the emitted photons are at a slightly higher energy level in a dynamic equilibrium state [3, 8]. Although these findings are very interesting, the photonic non-equilibrium dynamics at the *nano* point break in the PBG state have not yet been studied under extreme relativistic conditions. To activate this dormant photon into an activated energy state, optical circuit networks consisting of *nano* point breaks and waveguides have been proposed for use under extreme relativistic conditions to confirm the creation of point defects. This would provide a mechanism for incorporation of PBG waveguide defect arrays into PV panels. Therefore, the quantum dynamics of photons will be activated by point defects and PBG waveguides under extreme relativistic conditions. Consequently, the point break photons in a PV semiconductor array will not obey Bose-Einstein photon distribution theory. The dormant state photon will be broken down in these regimes and will create what I term HnP^{-} in the PV panel. To calculate the energy conversion created by utilizing these HNEPs in the PV panel,

M. F. Hossain, *Global Sustainability*, https://doi.org/10.1007/978-3-031-34575-3_3

I have developed a model of a PV module for a PV panel design using the MATLAB/Simulink software package. The model is explained using a detailed mathematical calculation considering a PV as a circuit including the PV current origin and a single diode to confirm the PV module behaviour under extreme relativistic parameters for calculating the solar energy transformation rate. The calculations reveal that if only a mere 0.0008% of a building's curtain wall is designed as an extreme relativistic conditioned PV panel, it will capture sufficient solar irradiance to exponentially convert enough energy to satisfy the total energy demand of the building.

Methods and Simulations

Photon Dynamics Transformation

The calculation of the photon dynamics was conducted at the nanoscale via point break waveguides rooted within a photovoltaic semiconductor circuit, particularly at the extreme relativistic state. For this particular calculation, both point break and photovoltaic semiconductor have been treated as waveguides for photon reservoirs. Subsequently, within the photovoltaic panel, the *nano* point break flaws, purely, fulfil electronic dynamics for unceasing conditions of photon manufacture at atomic spectra, bearing in mind contour maps; therefore, the expression of Hamiltonian is as under:

$$H = \sum \omega_{ci} a_i^\dagger a_i + \sum_{\mathrm{K}} \omega_k b_k^\dagger b_k + \sum_{ik} \left(V_{ik} a^\dagger b_k + V_{ik}^* b_k^\dagger a_i \right) \tag{3.1}$$

where $a_i \left(a_i^\dagger \right)$ is the nano point break mode driver, $b_k \left(b_k^\dagger \right)$ is the driver of the photon nanostructure photodynamic modes, and V_{ik} is the photonic mode magnitude amid the photon *nano*structure and *nano* breakpoints [2, 11, 13] (Fig. 3.1).

Considering into account the first photonic structure as within an equilibrium condition, the entire photonic reservoir structure is incorporated in view of excited coherent condition photons within the photovoltaic semiconductor; it is expressed as the below equation [5, 7, 9]:

$$\begin{aligned} \rho(t) &= -i \left[H_{\mathrm{c}}'(t), \rho(t) \right] + \sum_{ij} \left\{ k_{ij}(t) \left[2 a_j \rho(t) a_i^\dagger - a_i^\dagger a_j \rho(t) - \rho(t) a_i^\dagger a_j \right] \right. \\ &\quad \left. + k_{ij}(t) \left[a_i^\dagger \rho\ (t) a_j + a_j \rho(t) a_i^\dagger - a_i^\dagger a_j \rho(t) - \rho(t) a_j a_i^\dagger \right] \right\} \end{aligned} \tag{3.2}$$

In this case, $\rho(t)$ is the photon attenuated density within breakpoint conditions; $H_{\mathrm{c}}'(t) = \sum_{ij} \omega_{cij}'(t) a_i^\dagger a_j$ is the point break re-standardized Hamiltonian with reference to the frequencies of point break $\omega_{cii}'(t) = \omega_{ci}'(t)$, as well as $\omega_{cij}'(t)$, which is the

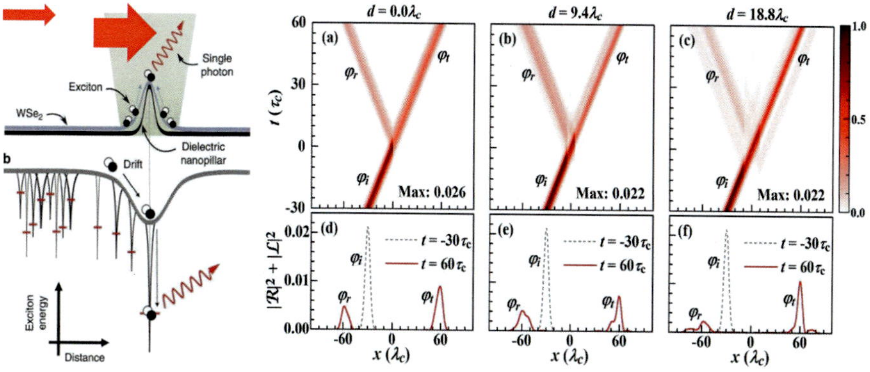

(A) (B)

Fig. 3.1 (**a**) The photonic mode for conversion of excited state energy in respect distance of the photonic frequencies. (**b**) (*a–c*) Contour maps of the photon probability densities, normalized to their maximum values in the maps as functions of x and t. (*d–f*) Probability distributions of the incident (φ_i), reflected (φ_r), and transmitted (φ_t) pulses

function instigated induced photon couplings amid the breakpoints. The factors $\tilde{\kappa}_{ij}(t)$ and $\kappa_{ij}(t)$ are considered a photonic dynamics within the photovoltaic semiconductor beneath the maximum relativistic states. The non-perturbative principle is purely the one that resolves time-reliant factors $\tilde{\kappa}_{ij}(t)$ and $\kappa_{ij}(t)$ and ω'_{cij}. In the case of photon reservoir, Hamiltonian is represented by $H_I = \sum_k \lambda_k x q_k$, with q_k and x being the secondary point break reservoir and the primary exclusive point break location, respectively. Taking into account a photon quantum dynamics, the entire Hamiltonian reservoir point break is revised as $H_I = \sum_k V_k \left(a^\dagger b_k + b_k^\dagger a + a^\dagger b_k^\dagger + a b_k\right)$ to approve the photonic dynamics magnitude in the point break. As a result, one characterizes the photonic dynamics by the dissipated factors of photon: $\tilde{\kappa}(t)$ and $\breve{\kappa}(t)$. These factors can therefore be determined as the below equations:

$$\omega'_c(t) = -\,\mathrm{Im}[u(t, t_0)/u(t, t_0)] \tag{3.3}$$

$$k(t) = -\,\mathrm{Re}[u(t, t_0)/u(t, t_0)] \tag{3.4}$$

$$\widehat{k}(t) = \dot{v}(t, t) + 2v(t, t)\,k(t) \tag{3.5}$$

With reference to the above equations, $u(t, t_0)$ is the photonic region of the point break and $v(t, t)$ is the photon dynamics as a result of the induced reservoir. The function $v(t, t)$ is explained further by the use of non-equilibrium dynamics theory and the following integral-differential equation [20, 21]:

$$\dot{u}(t,t_0) = -i\omega_c u\,(t,t_0) - \int_{t_0}^{t} dt' g(t-t')u(t',t_0) \tag{3.6}$$

$$v(t,t) = \int_{t_0}^{t} dt \int_{0}^{t} dt_2 u^*(t_1,t_0)\widehat{g}(t_1-t_2)u(t_2,t_0) \tag{3.7}$$

The primary frequency at the point break is represented by v_c. Consequently, in Eqs. (3.6) and (3.7) above, the integral functions can be utilized to calculate the bac-up function in the point breaks. The number of photons generated by the non-stable condition is exceptionally expressed per *unit* area of the photonic structure $J(\varepsilon)$ via the following connections:

$$g(t-t') = \int d\omega J(\omega) e^{-i\omega(t-t')} \text{ and } \tilde{g}(t-t') = \int d\omega J(\omega)\overline{n}(\omega,T) e^{-i\omega(t-t')},$$

where, in this case, $\overline{n}(\omega,T) = 1/\left[e^{\hbar\omega/k_B T} - 1\right]$ is the prime photon dynamics within the photovoltaic panel at a temperature (T). The clarification of the unit area $J(\varepsilon)$ is done in connection to the density of states $\varrho(\omega)$ production of photon within the photovoltaic at the V_k magnitude amid the photovoltaic and point break circuits:

$$J(\omega) = \sum_k |V_k|^2 \delta(\omega - \omega_k) = \varrho(\omega)|V(\omega)|^2 = [n * e(1+2n)]^4 \tag{3.8}$$

Lastly, the summary of the proliferation of the photon production is done with regard to the condition of photon dynamic within the photovoltaic panel, so that $V_k \rightarrow V(\omega)$ and i of V_{ik} may be calculated at the one diode mode breakpoint. Here the production of the non-equilibrium photon ($J(\omega)$, given that Eq. (3.8) can be determined exactly through the use of the simplified equation in Eq. (3.9) below:

$$J(\omega) = [n * e(1+2n)]^4 \tag{3.9}$$

Photovoltaic (PV) Modelling

The use of the exterior curtain wall skin of a building as a micro photovoltaic panel to carry out ultra-relativistic reaction within the cell has been recommended to deform Bose-Einstein equilibrium photon to numerous HnP^- to transform it into electricity by using a single-diode model PV panel (Fig. 3.2).

Therefore, the PV cell I–V equation within the single diode model describes the photovoltaic model. Within the photovoltaic panel, the equation of V–I relationship can be as under:

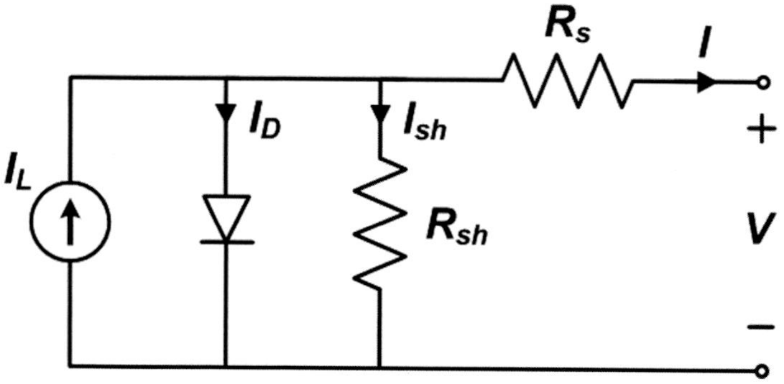

Fig. 3.2 The above model integrates parallel and series circuit connections using two resistors and a diode for PV cell

$$I = I_L - I_O \left\{ \exp^{\left[\frac{q(V+I_{RS})}{AkTc}\right]} - 1 \right\} - \frac{(V + I_{RS})}{R_{Sh}} \tag{3.10}$$

where I_L is the photon creating energy, I_O is the saturated energy within the diode, RS is the resistance within the series, A is the diode inactive function, $k(= 1.38 \times 10^{-23}$ W/ m^2 K) is Boltzmann's constant, $q(= 1.6 \times 10^{-19}$C) is the electron charge amplitude, and T_c is the functional cell temperature. Consequently, within the photovoltaic cells, the I–q connection varies because of the diode energy and/or saturation energy, which may be expressed as follows [18, 19, 32]:

$$I_O = I_{RS} \left(\frac{T_C}{T_{ref}} \right)^3 \exp \left[\frac{qEG\left(\frac{1}{T_{ref}} - \frac{1}{T_C} \right)}{KA} \right] \tag{3.11}$$

I_{RS}, in the above equation, is the saturated current given that the solar irradiance and qEG and functional temperature denote the band-gap energy into the graphene and silicon photovoltaic cell taking into account the perfect, normalized, and normal modes (Fig. 3.3).

In consideration of a photovoltaic module, not including the I–V curve, is I–V curve conjunctions amid all the photovoltaic panel cells. As a result, the I–V equation can be changed as under, to find out the V–R connection:

$$V = -IR_s + K \log \frac{I_{L-I+I_0}}{I_0} \tag{3.12}$$

From Eq. (3.12) V_{mo} and I_{mo} represent the voltage and current within the photovoltaic panel and K stands for a constant $\left(= \frac{AKT}{q}\right)$. Thus the connection amid V_{mo} and I_{mo} will be similar to that of photovoltaic cell I–V relation:

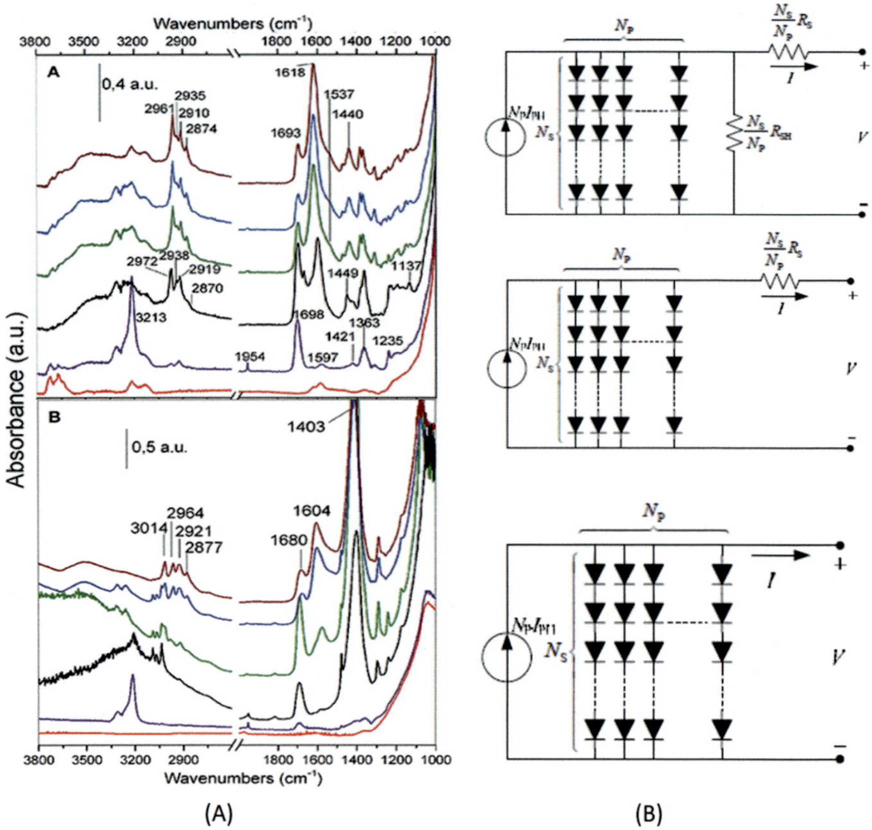

Fig. 3.3 (**a**) The absorption of solar irradiance (a.u.) in relation to the photonic wave per unit area, (**b**) equivalent circuit models of the PV module (a) normal, (b) normalized, and (c) perfect modes

$$V_{\mathrm{mo}} = -I_{\mathrm{mo}}R_{\mathrm{Smo}} + K_{\mathrm{mo}}\log\left(\frac{I_{\mathrm{Lmo}-I_{\mathrm{mo}+I_{\mathrm{Omo}}}}}{I_{\mathrm{Omo}}}\right) \tag{3.13}$$

I_{Lmo} is the photon-generated current, I_{Omo} is the saturated current going into the diode, R_{Smo} is the resistant within the series, and K_{mo} is the fractional constant. As soon as the interconnection of all the non-series (N_{S}) cells within the series is done, the series resistant will be regarded as the sum of each cell series resistant $R_{\mathrm{Smo}} = N_{\mathrm{S}} \times R_{\mathrm{S}}$. The constant factor can be written as follows:

$$V_{\mathrm{mo}} = -I_{\mathrm{mo}}N_{\mathrm{S}}R_{\mathrm{S}} + N_{\mathrm{S}}K\log\left(\frac{I_{\mathrm{L}-I_{\mathrm{mo}+I_{\mathrm{o}}}}}{I_{\mathrm{o}}}\right) \tag{3.14}$$

In the same way, after all N_{P} cells are linked within a parallel mode, the calculation of the current-voltage can be expressed as below, for the parallel link:

$$V_{\text{mo}} = -I_{\text{mo}}\frac{Rs}{Np} + K\log\left(\frac{I_{\text{sh}}\, I_{\text{L}} - I_{\text{mo}+N_{\text{p}}}\, I_{\text{o}}}{N_P I_o}\right) \tag{3.15}$$

Mostly, since the photon-generated current shall rely on the PV panel relativistic temperature states and solar irradiance, the following equation can be utilized to determine the current:

$$I_{\text{L}} = G[I_{\text{SC}} + \text{K}_{\text{I}}(T_{\text{C}} - T_{\text{ref}})] * V_{\text{mo}} \tag{3.16}$$

where G signifies the solar energy in KW/m^2, T_{ref} signifies the functional temperature of photovoltaic panel, K_{I} denotes the PV panel relativistic coefficient factor, and I_{sc} is the photovoltaic current at KW/m^2 and 25 °C.

Results and Discussion

Photon Production Proliferation

Determining arithmetically, the HnP^- production in the photovoltaic panel requires one to first solve the dynamic photon proliferation by incorporating Eqs. (3.7) and (3.8). It is well known that, due to the state of areal unit variable, different dynamic photon proliferation is produced by the photovoltaic panels. $J(\omega)$, unit area, possesses an insistent weak-coupling perimeter, and the approximation rule of Weisskopf-Wigner and/or Markovian master equation is equivalent to the proliferation of photon production. As a result, all proliferation of photon production will take a dynamic photon condition mode (1D, 2D, 3D) within the photovoltaic cell. Finally, minus following the Bose-Einstein photon distribution theory, the production will transform from the current non-equilibrium state. This can clearly be expressed as shown in Table 3.1 [12].

Consequently, to avoid the DOS bifurcation within a 3D photovoltaic cell, a fine cutoff at Ω_{C}, which is of high-level frequency, is employed. Correspondingly, a sharpened cutoff at Ω_{C}, which is also of high level frequency, maintains a positive DOS within 2D and 1D photovoltaic cells. Therefore, while and e_{rfc} (x) functions as an addictive variable, Li_2 (x) acts as a dilogarithm variable. Successively, DOS of different photovoltaic cells represented by $Q_{\text{PC}}(\omega)$ can be determined through calculating eigenfunctions and eigenfrequencies of Maxwell's rules in consideration of the photovoltaic *nano*structure [21–23]. The equivalent DOS for a 1D photovoltaic cell is represented by $Q_{\text{PC}}\,(\omega)\alpha\,\frac{1}{\sqrt{\omega - \omega_{\text{e}}}}\,\theta(\omega - \omega_{\text{e}})$. The variable ω_{e} denotes the existing frequency within the PBE in consideration of DOS, and $\theta(\omega - \omega_{\text{e}})$ denotes the Heaviside step function.

Table 3.1 The photonic structures in different DOS dimensional modes in the PV cell

Photovoltaic (PV)	Unit area $J(\omega)$ for different DOS	Reservoir-induced self-energy correction $\Sigma(\omega)$
1D	$\frac{C}{\pi}\frac{1}{\sqrt{\omega-\omega_e}}\Theta(\omega-\omega_e)$	$-\frac{C}{\sqrt{\omega_e-\omega}}$
2D	$-\eta\left[\ln\left\lvert\frac{\omega-\omega_0}{\omega_0}\right\rvert-1\right]\Theta(\omega-\omega_e)\Theta(\Omega_d-\omega)$	$\eta\left[\mathrm{Li}_2\left(\frac{\Omega_d-\omega_0}{\omega-\omega_0}\right)-\mathrm{Li}_2\left(\frac{\omega_0-\omega_e}{\omega_0-\omega}\right)-\ln\frac{\omega_0-\omega_e}{\Omega_d-\omega_0}\ln\frac{\omega_e-\omega}{\omega_0-\omega}\right]$
3D	$\chi\sqrt{\frac{\omega-\omega_e}{\Omega_C}}\exp\left(-\frac{\omega-\omega_e}{\Omega_C}\right)\Theta(\omega-\omega_e)$	$\chi\left[\pi\sqrt{\frac{\omega_e-\omega}{\Omega_C}}\exp\left(-\frac{\omega-\omega_e}{\Omega_C}\right)e_{\mathrm{rfc}}\sqrt{\frac{\omega_e-\omega}{\Omega_C}}-\sqrt{\pi}\right]$

The photonic structures in this correspond to different $J(\omega)$, as well as self-energy induction at $\Sigma(\omega)$, reservoir. The variables η, χ and C work like joined forces amid the point break and photovoltaic of 3D, 2D, and 1D into the photovoltaic cell

As a result, this density of state is determined to carry out 3D isotropic analysis within photovoltaic cells to project the mistake free qualitative condition of the mode of Weisskopf-Wigner, as well as the photon-photon collision condition within the photovoltaic cell [15, 22, 26]. Consequently, within a 3D photovoltaic cell, the density of state near the photo band energy is affected by anisotropic density of state: $Q_{PC}(\omega)\alpha \frac{1}{\sqrt{\omega-\omega_e}}\theta(\omega-\omega_e)$. In the case of the 1D and 2D photovoltaic cells, the DOS of photon shows a pure logarithm divergence near the photon band energy, which is estimated as $Q_{PC}(\omega)\alpha - \left[\text{In}|\frac{\omega-\omega_0}{\omega_0}|-1\right]\theta(\omega-\omega_e)$, with ω_e being the central peal logarithm point. $J(\omega)$, unit area, is explained as the DOS production field within the photovoltaic cell by $V(\omega)$ in the photon band and photon voltage cell as under [17]:

$$J(\omega)=Q(\omega)|V(\omega)|^2 \tag{3.17}$$

Henceforth, it was assumed through the proliferative photon dynamics and photon band frequency ω_c using $u(t, t_0)$ for the structure of photon in the connection $\langle a(t)\rangle = u(t, t_0)\langle a(t_0)\rangle$.

Therefore,

$$(t, t_0)=\frac{1}{1-\Sigma'(\omega_b)}e^{-i\omega(t-t_0)}+\int_{\omega_e}^{\infty}d\omega\frac{J(\omega)e^{-i\omega(t-t_0)}}{[\omega-\omega_c-\Delta(\omega)]^2+\pi^2J^2(\omega)} \tag{3.18}$$

$\Sigma'(\omega_b)=[\partial\Sigma(\omega)/\partial\omega]_{\omega=\omega_b}$ and $\Sigma(\omega)$ is the reservoir-induced photonic band photon self-energy correction:

$$\Sigma(\omega)=\int_{\omega_e}^{\infty}d\omega'\frac{J(\omega')}{\omega-\omega'} \tag{3.19}$$

The function ω_b in Eq. (3.2) denotes the mode of photonic frequency in the photonic band gap (PBG). PBG ($0<\omega_b<\omega_e$) can be determined through the use of the pole state:

$$\omega_b-\omega_c-\Delta(\omega_b)=0, \text{ with a principal-value integral being}\lesssim\Delta(\omega)$$

$$=\mathcal{P}\left[\int d\omega'\frac{J(\omega')}{\omega-\omega'}\right].$$

Moreover, in consideration of the proliferation magnitude, the comprehensive photonic dynamics have been determined and indicated in Fig. 3.4a for 3D, 2D, and 1D photovoltaic cells with regard to different detuning δ incorporated from the area of photonic band gap to the area of photonic band [16, 23, 41] where $k(t)$, photonic

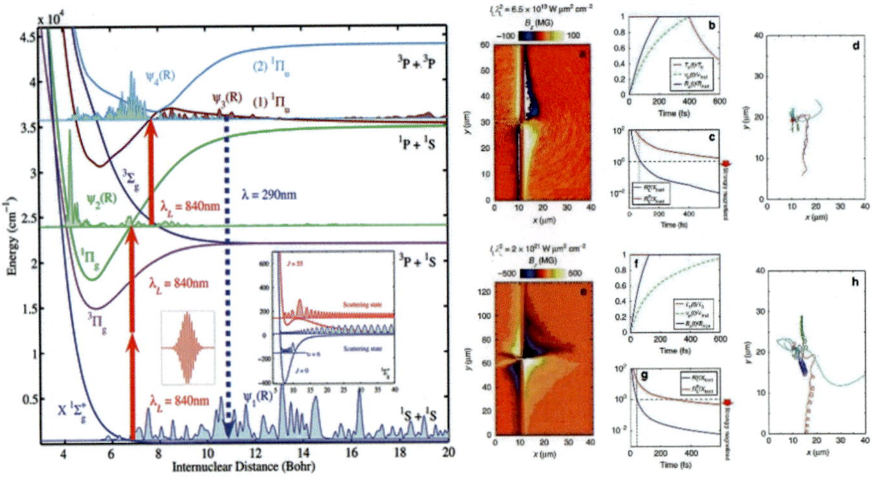

Fig. 3.4 (**a**) Heating photon energy formation against the internuclear distance (Bohr) at different variable frequencies where the photon particle carries out the energy measurements (DQD) by integrating Higgs-Boson quantum field. (**b**) The photon particle dynamics into the particle in simulation of cell (PIC). (Source: M. Nakatsutsumi et al. *Nature Communications*; volume 9, article number: 280 (2018))

dynamic rate, is indicated in Fig. 3.4b disregarding $\delta = 0.1\omega_e$. Based on the calculated result, the rate of producing dynamic photons is high as soon as ω_c crosses to PB area from the photonic band-gap area. Since the range within $u(t, t_0)$ is $1 \geq |u(t, t_0)| \geq 0$, the crossing area can be described as connected to the condition $0.9 \gtrsim |u(t \rightarrow \infty, t_0)| \geq 0$. This relates to $-0.025\omega_e \lesssim \delta \lesssim 0.025\omega_e$, with $k(t)$ as the rate of production in the photonic band gap ($\delta < -0.025\,\omega_e$), as well as within the PBG ($0.025\omega_e \lesssim \delta \lesssim 0.025\omega_e$) vicinity.

Figure 3.5 Dynamic photon proliferation in photovoltaic cells. Figure 3.5a, in consideration of the photon band area, $<a\,(t)> = 5\,u(t, t_0) < a(t_0)>$, Fig. 3.5b is the rate of dynamic photon ($k(t)$), which is plotted for 1D, 2D, and 3D photovoltaic cells [29, 30, 36]

For $\delta \gg 0.025\omega_e$, the production of dynamic photon is nearly exponential, which represents a Markov factor. Since the photonic band frequency in the crossover area is small within the PBE vicinity, this sharply raises the dynamic photon production mode [24, 42, 43]. As a result, this dynamic photon production proliferation proves the ionic photonic condition counts within the PBG vicinity in the photovoltaic cell, where the existing photons are in a non-equilibrium photonic condition.

Dynamic photon proliferation within the photovoltaic panel is afterwards explained in consideration of thermal variations in regard to $v(t, t)$, through the determination of the non-equilibrium photon scattering theorem.

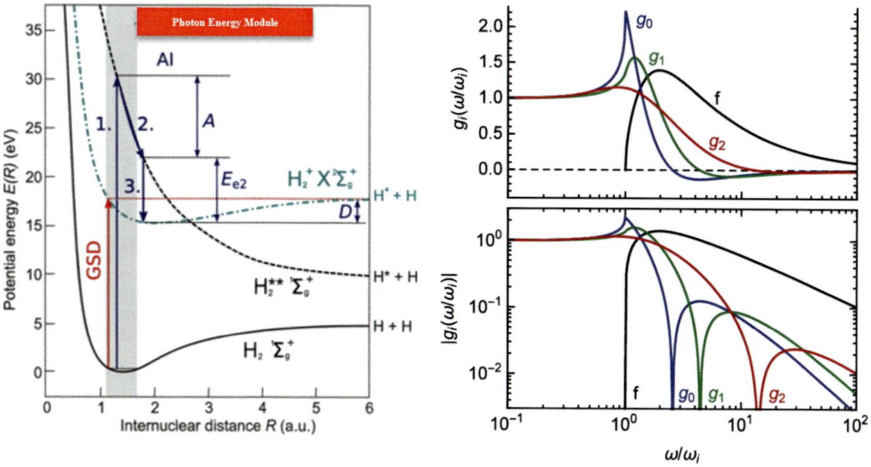

A B

Fig. 3.5 (**a**) *I–V* clarification of the single diode where curtain wall in the standard condition to deform energy considering photonic irradiance. (**b**). Photonic energy factor irradiance where black line is the photon absorptive region. Blue line is the dispersive region for related photon irradiance with static energy and static direction. Green line is the angle average for photonic irradiance isotropic dispersal with static energy. Red line represents photonic thermal isotropic average (Source-Alexandra Dobrynina et al. Demidov Yaroslavl State University, Sovietskaya; 15 December 2014)

$$v(t,t) = \int_{t_0}^{t} dt_1 \int_{t_0}^{t} dt_2 u^*(t_1,t_0)\tilde{g}(t_1,t_2)u(t_2,t_0) \tag{3.20}$$

From the above equation, $\tilde{g}(t_1,t_2) = \int d\omega J(\omega)\bar{n}(\omega,\mathrm{T})e^{-i\omega(t-t')}$ discloses the variations of photonic dynamic induced particularly by the thermal relativistic state, with $\bar{n}(\omega,T) = 1/\left[e^{\hbar\omega/k_BT} - 1\right]$ being the photon production proliferation within the photovoltaic cell at relativistic temperature T, and it is expressed as follows:

$$v(t, t \to \infty) = \int_{\omega_e}^{\infty} d\omega \mathcal{V}(\omega),$$

with

$$\mathcal{V}(\omega) = \bar{n}(\omega,T)[\mathcal{D}_l(\omega) + \mathcal{D}_d(\omega)] \tag{3.21}$$

Equation (3.21) simplified so as to determine $V(\omega) = \bar{n}(\omega,T)\mathcal{D}_d(\omega)$, which represents the non-equilibrium state. Dissipation of Einstein's photon fluctuation under low-temperature states is non-dynamically viable at the photonic band, although the linking photonic structures are measurable, $n(t) = \langle a^{\dagger}(t)a(t)\rangle = |u(t, t_0)|^2 n(t_0)v(t, t)$, with $n(t_0)$ being the prime PB. For this reason, in Fig. 3.3b, the dynamic photon numbers have been plotted versus temperature so as to confirm the non-equilibrium production of proliferated photon, as indicated by the curve that is solid blue. Basically, the distribution of Bose-Einstein does not necessarily apply for non-equilibrium photon generation proliferation within the photonic band-gap religion, because of the extreme relativistic states. More precisely, one has to initially consider the PB as the Fock condition photon number n_0, i. e. $\rho(t_0) = |n_0\rangle\langle n_0|$, in order to convert the distributed Bose-Einstein photons into HnP^{-} proliferated photons. The photon number n_0 is determined as follows:

$$\rho(t) = \sum_{n=0}^{\infty} p_n^{(n_0)}(t)|n_0\rangle\langle n_0| \tag{3.22}$$

$$p_n^{(n_0)}(t) = \frac{[v(t,t)]^n}{[1+v(t,t)]^{n+1}}[1-\Omega(\mathrm{t})]^{n_0} \times \sum_{k=0}^{\min\{n_0 n\}} \binom{n_0}{k}\binom{n}{k}\left[\frac{1}{v(t,t)}\frac{\Omega(\mathrm{t})}{1-\Omega(\mathrm{t})}\right]^k \tag{3.23}$$

In this case, $\Omega(\mathrm{t}) = \frac{|u(t,t_0)|^2}{1+v(t,t)}$ suggesting that a Fock condition photon will transform into different Fock conditions of $n_0|\rangle$ is $p_n^{(n_0)}(t)$. The dissipation of photon proliferation $p_n^{(n_0)}(t)$ within the prime condition $|n_0 = 5\rangle$ as well as the steady condition $p_n^{(n_0)}(t \rightarrow \infty)$ is therefore indicated in Fig. 3.6. The photon generation proliferation therefore will finally reach the thermal non-equilibrium condition with the photonic structure. The distribution of Bose-Einstein photon can be overlooked:

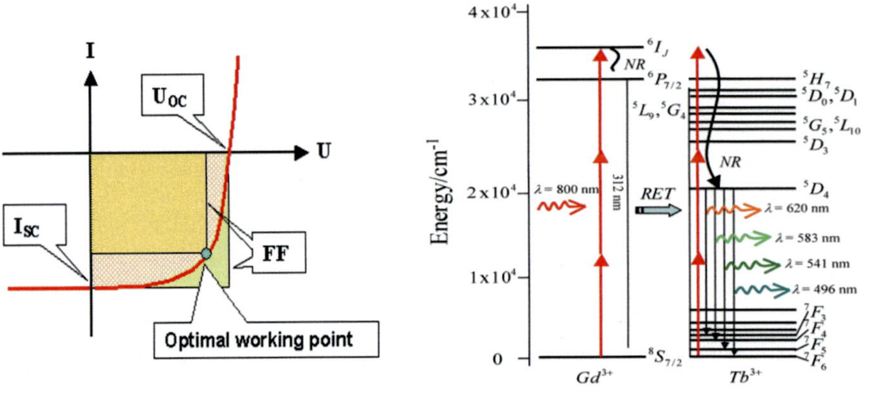

(a) (b)

Fig. 3.6 Represents the solar cell current-voltage distinctive features for conceptual function of (**a**) current production optimum working point, (**b**) the current source current-voltage module nearby the short circuit point, and as a production of energy per unit area within the open-circuit point vicinity

$$p_n^{(n_0)}(t \to \infty) = \frac{[\overline{n}(\omega_c, T)]^n}{[1 + \overline{n}(\omega_c, T)]^{n+1}} \tag{3.24}$$

To confirm this generation of HnP^-, it is suggestively essential to breakdown the Bose-Einstein distribution theory via extreme relativistic comprehensible conditions and work out Eq. (3.2) in consideration of the photon proliferation conditions, which can as well be rewritten as below:

$$\rho(t) = \mathcal{D}[\alpha(t)]\rho_T[v(t,t)]\mathcal{D}^{-1}[\alpha(t)] \tag{3.25}$$

where $\mathcal{D}[\alpha(t)] = \exp\{\alpha(t)a^\dagger - \alpha^*(t)a\}$ represents the displacement driver with $\alpha(t) = u(t, t_0)\alpha_0$ and

$$\rho_T[v(t,t)] = \sum_{n=0}^{\infty} \frac{[v(t,t)^n]}{[1 + v(t,t)]^{n+1}} \mid n\rangle\langle n| \tag{3.26}$$

From the above equation, ρ_T denotes the thermal condition with mean practical number $v(t, t)$, with Eq. (3.11) disclosing that the prime point break cavity condition will transform into a displaced thermal condition [25, 27, 38], which represents the combination of displaced number conditions $\mathcal{D}[\alpha(t)] \mid n\rangle^{37}$. As a result Eq. (3.25) can as well be rewritten as follows:

$$\begin{aligned} \langle m|\rho(t)|n\rangle = J(\omega) &= \mathrm{e}^{-\Omega(t)|\alpha_0|^2} \frac{[\alpha(t)]^m[\alpha^*(t)]^n}{[1 + v(t,t)]^{m+n+1}} \\ &= \sum_{k=0}^{\min\{m,n\}} \frac{\sqrt{m!n!}}{(m-k)!(n-k)!k!} \left[\frac{v(t,t)}{\Omega(t)|\alpha_0|^2}\right]^k \end{aligned} \tag{3.27}$$

Here, the *HnP*- production within the photovoltaic panel ($\langle m|\rho(t)|n\rangle$) will certainly transform into a non-equilibrium state $[\alpha(t)]^m[\alpha^*(t)]^n$ and an extreme relativistic thermal condition $[1 + v(t, t)]^{m+n+1}$. This clearly recommends that the photonic band photon number shall not comply with the Bose-Einstein distribution; instead it produces the photon exponentially $J(\omega) = \sum_{k=0}^{\min\{m,n\}} \frac{\sqrt{m!n!}}{(m-k)!(n-k)!k!} \left[\frac{v(t,t)}{\Omega(t)|\alpha_0|^2}\right]^k = [n * e(1 + 2n)]^4$. Basically, Bose-Einstein photon dissemination breakdown and transformation into numerous photons within the photovoltaic panels are relatively viable through implementing extreme relativistic states and clearly explaining dynamic photon, thus transforming the equilibrium photons into non-equilibrium stable-condition photons in regard to DOS photon activation [35, 37].

Electricity Conversion by a Photovoltaic Panel

Modelling a one-diode photovoltaic panel, comprising of a small semiconductor disk attached by wire to a circuit with a negative and a positive silicon film located beneath a thin glass slice, and attached to graphene, by the use of an exterior curtain wall skin of a building, has been suggested in this study as a way of converting HnP^- into electricity beneath extreme relativistic states. Essentially, the photovoltaic panel will require to have an open-*circuit point*, with maximum voltage and current, open-circuit voltage V_{oc}, and also, the extreme power point can be explained through the use of maximum current as well as calculation of voltage instantaneously upon capturing the non-equilibrium photons [28, 44, 45]. Power that will be supplied by a photovoltaic panel shall therefore have the ability to acquire maximum values at I_{mp} and V_{mp} [31, 40]. Analysing the optimum working point as well as the dark functionality under revealing the photovoltaic panel status will help in confirming the photovoltaic panel ability of such kind of voltage-current flow.

Finally, the total photon production from a single photon is calculated as, $J(\omega) = [n * e(1 + 2n)]^4$ (Eq. 3.9). Therefore, the total number of photons would be 81 photons, which originate from a single photon. Then, I performed an estimate to convert these photons into electricity. The light quanta of a certain type of polarization have a frequency range of ν_r to $\nu_r + d\nu_r$; thus, the maximum solar radiation is achieved from 81 photons at 1.4 eV × 81 = 113.4 eV (one photon is 1.4 eV). Therefore, the total energy would be (1.4 eV = 27.77 mW/m^2 × 81) 2249.37 mW/m^2 eV per hour [31, 33]. The electricity produced at this stage is direct current (DC) that must be converted to alternating current (AC) for use in the building sector and battery system storage. The estimate reveals that in a year with an average of 5 h a day maximum for 365 days, this total energy production is equivalent to 4,105,100.25 kW/year [34, 39, 40]. This means that 11,246.85 kW of energy can be supplied by a 1 m^2 solar panel per day. If an office and commercial building consumption is approximately 3800 kWh/day for a building with a 32 m × 31 m footprint and height of 30 m (10 floors), 0.33 m^2 of the building's exterior skin must be used as solar panels to meet the total energy demand of the building, which is equivalent to 0.00008% of the exterior curtain wall.

Conclusions

Bose-Einstein photon distribution theory is broken down theoretically under ultra-relativistic conditions to create HnP^- and produce clean energy. A series of mathematical calculations have been performed to confirm the production of multiple HnP^- from a single Bose-Einstein dormant photon under extreme relativistic conditions in the PV panel. Mathematical analyses suggest that extreme relativistic conditions break down the Bose-Einstein dormant photons and produce nearly exponential rate photons from an equilibrium to a non-equilibrium stage. The results suggest that the proliferation of these photons is agitated by the extreme relativistic condition to conduct photon-photon interactions in the PV panel to exponentially produce photons. Naturally, these created photons can be captured by the PV circuit system to produce DC current and then convert it into AC current to satisfy the energy demand of a building. The energy production from the PV panel has also been calculated, which revealed that if only a mere 0.00008% of a building exterior curtain wall is used as a *Modern Solar Photovoltaics Energy* panel, it will meet the total energy demand of the building using 100% clean energy. Naturally, implementation of this technology would introduce a new era of science to meet the total energy demand of the building sector and dramatically reduce climate change.

Acknowledgements The author, Md. Faruque Hossain, declares that any findings, predictions, and conclusions described in this article are solely performed by the author and it is confirmed that there is no conflict of interest for publishing this research paper in a suitable journal and/or publisher.

References

1. Armani, D. K., Kippenberg, T. J., Spillane, S. M., & Vahala, K. J. (2003). Ultra-high-Q toroid microcavity on a chip. *Nature, 421*(6926), 925–928.
2. Birnbaum, K. M., Boca, A., Miller, R., Boozer, A. D., Northup, T. E., & Kimble, H. J. (2005). Photon blockade in an optical cavity with one trapped atom. *Nature, 436*(7047), 87–90.
3. Busch, K., Von Freymann, G., Linden, S., Mingaleev, S. F., Tkeshelashvili, L., & Wegener, M. (2007). Periodic nanostructures for photonics. *Physics Reports, 444*(3–6), 101–202.
4. Chang, D. E., Sørensen, A. S., Demler, E. A., & Lukin, M. D. (2007). A single-photon transistor using nanoscale surface plasmons. *Nature Physics, 3*(11), 807–812.
5. Chen, J., Wang, C., Zhang, R., & Xiao, J. (2012). Multiple plasmon-induced transparencies in coupled-resonator systems. *Optics Letters, 37*(24), 5133–5135.
6. Cheng, M., & Song, Y. (2012). Fano resonance analysis in a pair of semiconductor quantum dots coupling to a metal nanowire. *Optics Letters, 37*(5), 978–980.
7. Dayan, B., Parkins, A. S., Aoki, T., Ostby, E. P., Vahala, K. J., & Kimble, H. J. (2008). A photon turnstile dynamically regulated by one atom. *Science, 319*(5866), 1062–1065.
8. Douglas, J. S., Habibian, H., Hung, C., Gorshkov, A. V., Kimble, H. J., & Chang, D. E. (2015). Quantum many-body models with cold atoms coupled to photonic crystals. *Nature Photonics, 9*(5), 326–331.
9. Englund, D., Majumdar, A., Faraon, A., Toishi, M., Stoltz, N., Petroff, P., & Vučković, J. (2010). Resonant excitation of a quantum dot strongly coupled to a photonic crystal nanocavity. *Physical Review Letters, 104*(7), 073904.
10. Gleyzes, S., Kuhr, S., Guerlin, C., Bernu, J., Deleglise, S., Busk Hoff, U., et al. (2007). Quantum jumps of light recording the birth and death of a photon in a cavity. *Nature, 446*(7133), 297–300.
11. Guerlin, C., Bernu, J., Deleglise, S., Sayrin, C., Gleyzes, S., Kuhr, S., et al. (2007). Progressive field-state collapse and quantum non-demolition photon counting. *Nature, 448*(7156), 889–893.
12. Han, Z., & Bozhevolnyi, S. I. (2011). Plasmon-induced transparency with detuned ultracompact Fabry-Perot resonators in integrated plasmonic devices. *Optics Express, 19*(4), 3251–3257.
13. Hossain, M. F. (2016). Solar energy integration into advanced building design for meeting energy demand and environment problem. *International Journal of Energy Research, 40*(9), 1293–1300.

14. Hossain, M. F. (2017a). Design and construction of ultra-relativistic collision PV panel and its application into building sector to mitigate total energy demand. *Journal of Building Engineering, 9*, 147–154.
15. Hossain, M. F. (2017b). Green science: Independent building technology to mitigate energy, environment, and climate change. *Renewable and Sustainable Energy Reviews, 73*, 695–705.
16. Huang, J., Shi, T., Sun, C. P., & Nori, F. (2013). Controlling single-photon transport in waveguides with finite cross section. *Physical Review A, 88*(1), 013836.
17. Joannopoulos, J. D., Villeneuve, P. R., & Fan, S. (1997). Photonic crystals: Putting a new twist on light. *Nature, 386*(6621), 143–149.
18. Kofman, A. G., Kurizki, G., & Sherman, B. (1994). Spontaneous and induced atomic decay in photonic band structures. *Journal of Modern Optics, 41*(2), 353–384.
19. Kolchin, P., Oulton, R. F., & Zhang, X. (2011). Nonlinear quantum optics in a waveguide: Distinct single photons strongly interacting at the single atom level. *Physical Review Letters, 106*(11), 113601.
20. Lang, C., Bozyigit, D., Eichler, C., Steffen, L., Fink, J. M., Abdumalikov, A. A., Jr., et al. (2011). Observation of resonant photon blockade at microwave frequencies using correlation function measurements. *Physical Review Letters, 106*(24), 243601.
21. Lei, C. U., & Zhang, W. (2012). A quantum photonic dissipative transport theory. *Annals of Physics, 327*(5), 1408–1433.
22. Li, Q., Xu, D. Z., Cai, C. Y., & Sun, C. P. (2013). Recoil effects of a motional scatterer on single-photon scattering in one dimension. *Scientific Reports, 3*(1), 1–6.
23. Liao, J., & Law, C. K. (2010). Correlated two-photon transport in a one-dimensional waveguide side-coupled to a nonlinear cavity. *Physical Review A, 82*(5), 053836.
24. Liao, J., & Law, C. K. (2013). Correlated two-photon scattering in cavity optomechanics. *Physical Review A, 87*(4), 043809.
25. Lo, P., Xiong, H., & Zhang, W. (2015). Breakdown of Bose-Einstein distribution in photonic crystals. *Scientific Reports, 5*(1), 1–9.
26. Longo, P., Schmitteckert, P., & Busch, K. (2011). Few-photon transport in low-dimensional systems. *Physical Review A, 83*(6), 063828.
27. Lü, X., Zhang, W., Ashhab, S., Wu, Y., & Nori, F. (2013). Quantum-criticality-induced strong kerr nonlinearities in optomechanical systems. *Scientific Reports, 3*(1), 1–9.
28. Mahan, G. D. (2000). *Many-body physics*. Springer.
29. Martens, C., Longo, P., & Busch, K. (2013). Photon transport in one-dimensional systems coupled to three-level quantum impurities. *New Journal of Physics, 15*(8), 083019.
30. Noda, S., & Baba, T. (2003). Roadmap on photonic crystals Springer Science & Business Media.
31. O'Shea, D., Junge, C., Volz, J., & Rauschenbeutel, A. (2013). Fiber-optical switch controlled by a single atom. *Physical Review Letters, 111*(19), 193601.
32. Reinhard, A., Volz, T., Winger, M., Badolato, A., Hennessy, K. J., Hu, E. L., & Imamoğlu, A. (2012). Strongly correlated photons on a chip. *Nature Photonics, 6*(2), 93–96.
33. Roy, D. (2013). Two-photon scattering of a tightly focused weak light beam from a small atomic ensemble: An optical probe to detect atomic level structures. *Physical Review A, 87*(6), 063819.
34. Saloux, E., Teyssedou, A., & Sorin, M. (2011). Explicit model of photovoltaic panels to determine voltages and currents at the maximum power point. *Solar Energy, 85*(5), 713–722.
35. Shen, J., & Fan, S. (2007). Strongly correlated two-photon transport in a one-dimensional waveguide coupled to a two-level system. *Physical Review Letters, 98*(15), 153003.
36. Sreekumar, S., & Benny, A. (2013). *Maximum power point tracking of photovoltaic system using fuzzy logic controller based boost converter*. Paper presented at the 2013 international conference on current trends in engineering and technology (ICCTET), pp. 275–280.
37. Tame, M. S., McEnery, K. R., Özdemir, Ş. K., Lee, J., Maier, S. A., & Kim, M. S. (2013). Quantum plasmonics. *Nature Physics, 9*(6), 329–340.

38. Tu, M. W., Zhang, W., Jin, J., Entin-Wohlman, O., & Aharony, A. (2012). Transient quantum transport in double-dot Aharonov-Bohm interferometers. *Physical Review B, 86*(11), 115453.
39. Tu, M. W., & Zhang, W. (2008). Non-Markovian decoherence theory for a double-dot charge qubit. *Physical Review B, 78*(23), 235311.
40. Wang, X., Gu, B., Wang, R., & Xu, H. (2003). Decay kinetic properties of atoms in photonic crystals with absolute gaps. *Physical Review Letters, 91*(11), 113904.
41. Xiao, Y., Li, M., Liu, Y., Li, Y., Sun, X., & Gong, Q. (2010). Asymmetric Fano resonance analysis in indirectly coupled microresonators. *Physical Review A, 82*(6), 065804.
42. Yan, W., & Fan, H. (2014). Single-photon quantum router with multiple output ports. *Scientific Reports, 4*(1), 1–6.
43. Yan, W., Huang, J., & Fan, H. (2013). Tunable single-photon frequency conversion in a Sagnac interferometer. *Scientific Reports, 3*(1), 1–6.
44. Zhang, W., Lo, P., Xiong, H., Tu, M. W., & Nori, F. (2012). General non-Markovian dynamics of open quantum systems. *Physical Review Letters, 109*(17), 170402.
45. Zhu, Y., Hu, X., Yang, H., & Gong, Q. (2014). On-chip plasmon-induced transparency based on plasmonic coupled nanocavities. *Scientific Reports, 4*(1), 1–7.

Chapter 4
Modeling of Ultimate Energy to Mitigate Everything

Introduction

Since the imagination power has the extreme influence on mind energy (M_e) which then consequently transform it into body energy (B_e), and then work energy (W_e), this research, a novel mathematical modeling, is being conducted primarily on clarifying the probing of the mind energy (M_e), body energy (B_e), and work energy (W_e) in order to confirm a real physical work by the human body eventually. Simply in this work, a bridge of the mind and body energy has been computed in order to clarify how the mind energy can be decoded from its energy momentum in order to transform it into real activated working energy to do any physical work that is initiated originally from the mind. Naturally, the energy momentum into the human mind is being modeled to release its activated energy considering its nano-point of QED in order to form body state energy [18, 29]. Subsequently, this body state energy is also being modeled to convert it into working state energy by implementing its quantum of Higgs-Boson ($H \rightarrow \gamma\gamma^{-}$) of electromagnetic fields in order to real work physically in general [11, 12]. This energy conversion mechanism from mind to body and then into the work shall indeed be a noble scientific discovery ever to solve any problem in advance.

Methods and Simulation

Computation of Mind Energy ($\mathbf{M_e}$)

The computation of mind energy considering its imagination sphericity and the space parameters of the brain, a mathematical modeling is being conducted to initiate the presence of mind energy in the human brain by expressing following equation:

M. F. Hossain, *Global Sustainability*, https://doi.org/10.1007/978-3-031-34575-3_4

$$\rho \cos(c) = \rho \cos(a)\cos(b) + \rho \sin(a)\sin(b)\cos(C) \tag{4.1}$$

Here *a, b*, and *c* represent the arc lengths in radians corresponding to the sides of a spherical horizon of the mind where *C* represents the angle of the vertex corresponding to the brain of arc with the length *c* [8, 23]. Subsequently, the implementation of its zenith angle Θ, the following terms is being written as the laws of cosines: $C = h = r$; $c = \Theta$; $a = \frac{1}{2}\pi - \phi$; $b = \frac{1}{2}\pi - \delta$ and thus, the equation is being expressed as:

$$\rho \cos(\Theta) = \sin(\phi)\sin(\delta) + \cos(\phi)\cos(\delta)\cos(r) \tag{4.2}$$

To confirm this calculation, a further clarification has been conducted considering a general derivative as below

$$\begin{aligned}\rho \cos(\theta) = {} & \rho \sin(\phi)\sin(\delta)\cos(\beta) + \sin(\delta)\cos(\phi)\sin(\beta)\cos(\gamma) \\ & + \cos(\phi)\cos(\delta)\cos(\beta)\cos(h) - \cos(\delta)\sin(\phi)\sin(\beta)\cos(\gamma)\cos(h) \\ & - \cos(\delta)\sin(\beta)\sin(\gamma)\sin(h)\end{aligned} \tag{4.3}$$

Here, β represents the angle corresponding to the horizon of the mind and γ corresponding to the azimuth angle of the brain and thus the sphere of the imagination and space parameter of the brain, which is originated from the mind energy dynamics, denoted here as R_E, and its mean distance is being denoted here as R_O considering the average distant corresponding of mind-dynamical unit (*m.u.*) and the energy constant is named here S_O [7, 13, 35]. Thus, the mind energy flux onto the imagination of the sphere is calculated as:

$$Q = \begin{cases} S_o \dfrac{R_0^2}{R_E^2} \cos(\theta) & \cos(\theta) > 0 \\ 0 & \cos(\theta) \leq 0 \end{cases} \tag{4.4}$$

Here, the mean Q over a time is the mean value of Q in relation to its rotation within the time of $h = \pi$ to $h = -\pi$, and thus, calculation can be expressed as:

$$Q^{-\text{time}} = -\frac{1}{2\pi}\int_{\pi}^{-\pi} Q dh \tag{4.5}$$

Since h_O is the time-related sphere once Q is the positive, it will be feasible during the mind energy presence in the brain when $\Theta = 1/2\,\pi$ or for h_O considering following equation by solving as:

$$\rho \sin(\phi)\sin(\delta) + \cos(\phi)\cos(\delta)\cos\left(h_{0)}\right) = 0 \tag{4.6}$$

or

$$\rho \cos(h_0) = -\tan(\phi)\tan(\delta) \tag{4.7}$$

Since tan(φ)tan(δ) > 1, the mind energy will confirm its existence at $h = \pi$, so $h_o = \pi$ in a Q^{-} time = 0; $\frac{R_0^2}{R_E^2}$ would remain constant into the mind energy by presenting an imagination sphere that can be expressed by the following integral:

$$\int_{\pi}^{-\pi} Q dh = \int_{h_0}^{-h_0} Q dh = S_0 \frac{R_0^2}{R_E^2} \int_{h_0}^{-h_0} \cos(\theta) dh = S_0 \frac{R_0^2}{R_E^2} \times \left[h \sin(\phi)\sin(\delta) + \cos(\phi)\cos(\delta)\sin(h)\right]_{h=h_0}^{h=-h_0} = -2S_0 \frac{R_0^2}{R_E^2} \times \left[h_0 \sin(\phi)\sin(\delta) + \cos(\phi)\cos(\delta)\sin\left(h_{0)}\right)\right] \tag{4.8}$$

Therefore,

$$Q^{-\text{time}} = \frac{S_0}{\pi} \frac{R_0^2}{R_E^2} \left[h_0 \sin(\phi)\sin(\delta) + \cos(\phi)\cos(\delta)\sin\left(h_{0)}\right)\right] \tag{4.9}$$

Since the θ is representing here as mainstream imagination that describes the orbit of the brain, $\theta = 0$ corresponding to its vernal equinox of δ at its spherical position that can be written as:

$$\delta = \varepsilon \sin(\theta) \tag{4.10}$$

Here ε represents the mainstream longitude of sphere ϖ that is related to the vernal equinox, and thus, the equation can be written as:

$$R_E = \frac{R_0}{1 + e\cos(\theta - \omega)} \tag{4.11}$$

or

$$\frac{R_0}{R_E} = 1 + e\cos(\theta - \omega) \tag{4.12}$$

Given the above clarification of ϖ, ε, s_o, and e from mind-dynamical computation, mind energy L_e can be computed for any latitude φ and θ of the imaginative

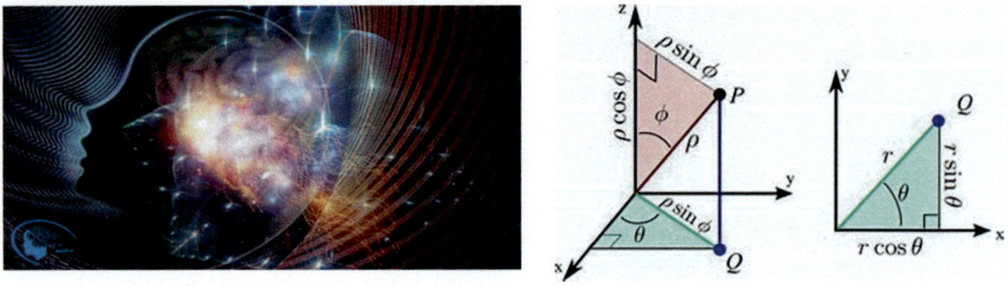

(a) (b)

Fig. 4.1 The distribution of mind energy, (**a**) schematic diagram of the imagination sphericity of the mind (**b**) clarifies the mind energy with respect law of cosines into the brain

sphere (Fig. 4.1). Here, $\theta = 0°$ is representing the vernal equinox, $\theta = \infty$ is representing the sphere of the imagination, and thus, the equation that can be simplified for mind energy activation in the brain can be expressed as:

$$M_e = S_0\left(1 + \cos\left(2\pi \frac{n}{x}\right)\right) \tag{4.13}$$

Here S_o is the energy constant, n is the unit of imagination, x is the imagination, and π is the sphere of the imagination that optimized the generation of mind energy (M_e) in the brain in order to pave the formation of the body energy (B_e).

Computation of Body Energy ($\mathrm{B_e}$)

Since the body energy is a hidden particle of mind energy, the particle of the body energy (B_e) is being clarified considering the quantum electrodynamics (QED) of the body by analyzing its minimal way of new U(1) gauge field considering its vectorial (V) dynamics of energy of its tangential (T) field Z dimension of the body [1, 9, 10]. Simply, here, the new U(1) gauge symmetry and its kinetic energy proliferation corresponding to the body energy field and the *Standard Model Hypercharge Field* are being computed to confirm the determination of striking electron flow from the body energy that will initiate to active the body energy (Fig. 4.2). Here the striking electron of the body energy ($\overrightarrow{De}$) will be scatters at an angle θ with speed of c, which is being expressed as the real fermion of the energy momentum of vectors of constituent particles of the body:

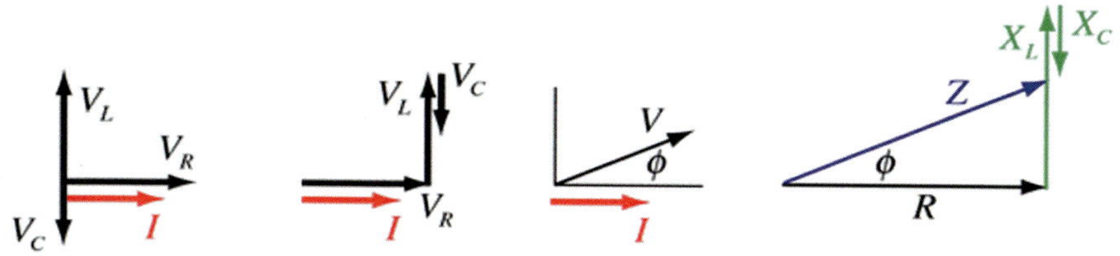

$$V=\sqrt{V_R^2+(V_L-V_C)^2} \qquad Z=\sqrt{R^2+(X_L-X_C)^2}$$

$$\phi=\tan^{-1}\frac{V_L-V_C}{V_R} \qquad \phi=\tan^{-1}\frac{X_L-X_C}{R}$$

Fig. 4.2 The conceptual model of body energy U(1) gauge symmetry corresponding to its vectorial dynamics of tangential striking electron in order to pave it to transform into work energy considering the kinetic energy field parameter of the body

$$\overrightarrow{De}=\left[\frac{\text{Edp}}{c},\ \overrightarrow{\text{Pd}}\right]+\overrightarrow{Xef}\ \left[\frac{Eef}{c},\overrightarrow{Pef}\right] \tag{4.14}$$

Since body energy momentum is conserved and thus, $\Delta P=0$; therefore, the sum of the initial momenta must be equal to the sum of the final momentum. Thus, it can be rewritten the sum of the vectors as:

$$\overrightarrow{De}+\overrightarrow{Xei}=\overrightarrow{Xp}+\overrightarrow{Xef} \tag{4.15}$$

Since the isolated the four momentums of the released electron are being calculated and thus the equation can be rewritten as:

$$\left(\overrightarrow{\overrightarrow{De}} + \overrightarrow{Xei} - \overrightarrow{Xp}\right)^2 = \overrightarrow{Xef}^2 \tag{4.16}$$

Multiplying the energy momentum of the vectors together the cross term here it can be rewritten as:

$$2\,\overrightarrow{Xd}\,.\overrightarrow{Xei} = 2 * \left[\frac{E_d}{c}, \overrightarrow{P_d}\right].\left[m_e c, \vec{0}\right] = 2E_d m_e \tag{4.17}$$

Subsequently, multiplying together the momentums of the initially stationary electron considering the generation of body energy, the equation can be rewritten as:

$$m_d^2c^2 - m_e^2c^2 + -2m_eE_p - 2\left(\frac{E_dE_p}{c^2} - \frac{P_dE_p\cos(\theta)}{c}\right) + 2E_dm_e = m_e^2c^2 \tag{4.18}$$

which can be simplified as:

$$m_d^2c^2 - 2\left(m_eE_p + \left(\frac{E_dE_p}{c^2} - \frac{P_dE_p}{c}\cos(\theta)\right) - E_dm_e\right) = 0 \tag{4.19}$$

Here, to confirm the energy generation from body of the momentums, the vector energy momentum has been calculated by conducting the following mathematical probe as:

$$2\left(m_eE_p + \left(\frac{E_dE_p}{c^2}\right) - \frac{P_dE_p}{c}\cos(\theta)\;\right) = m_d^2c^2 - 2E_dm_e \tag{4.20}$$

Here, the expression for energy generation is related to the vectorial momentum of body energy; thus, the limit that $m_d \to 0$ can be expressed as the case of scattering in order to transform it into work energy. Since the $m_d \to 0$, the $P_d \to E_c$ and thus the body energy generation can be expressed by adding the differential scattering cross section in order to transform it into the work energy by the configuration of this body energy transform dynamics mechanism as the following equation:

$$De = \frac{\partial \sigma_{kn}}{\partial \Omega} = \frac{r_0^2}{2} \frac{\left(P_d^2c^2 + m_d^2c^4\right)}{\left(\frac{-\left(\frac{m_d^2c^2}{2m_e} - E_d\right)}{\left(1+\left(\frac{E_d}{m_ec^2} - \frac{P_d}{m_e^c}\cos(\theta)\right)\right)}\right)^2} \left[\left(\frac{\left(P_d^2c^2 + m_d^2c^4\right)^{1/2}}{\frac{-\left(\frac{m_d^2c^2}{2m_e} - E_d\right)}{\left(1+\left(\frac{E_d}{m_ec^2} - \frac{P_d}{m_ec}\cos(\theta)\right)\right)}}\right) + \frac{\frac{-\left(\frac{m_d^2}{2m_e} - E_d\right)}{\left(1+\left(\frac{E_d}{m_ec^2} - \frac{P_d}{m_ec}\cos(\theta)\right)\right)}}{\left(P^2C^2 + m_d^2C^4\right)^{1/2}} - \sin(\theta)^2\right] \tag{4.21}$$

Computation of Work Energy (W_m)

Since the work energy is the force particles of the body energy, nanoscopic numbers of body particles are being calculated here in relation to its force generation by implementing Higgs-Boson ($H \to \gamma\gamma^{-}$) electromagnetic fields of the body [2, 6, 38]. Since a standard particle and its anti-particle have the same mass, its electric charge will have opposite quantum numbers in order to confirm that proton has a positive charge and anti-proton has a negative charge [15, 33, 36]. Simply, the energy content within the work energy will be released from the body energy as a mass energy of $F = ma$ considering its standard matter (H^*) on its free quantum fields (n) as an activated state of energy (Fig. 4.3).

Subsequently, the quantum field of this work energy is to be the components of all free quantum momentum of body; thus, its components (t, x) are being clarified into a vector $x = (ct, x)$ considering its energy momentum of $\eta_{\mu v} =$ diag $(\pm1, \mp1, \mp1, \mp1)$ of Klein-Gordon equations, expressed by the following table Tables 4.1 and 4.2:

Here, the Klein-Gordon equation permits two variables where one variable is positive and another is negative, which can be obtained by describing a relativistic wavefunction of the work energy as:

$$\left[\nabla^2 - \frac{m^2c^2}{\hbar^2}\right]\psi(r) = 0 \tag{4.22}$$

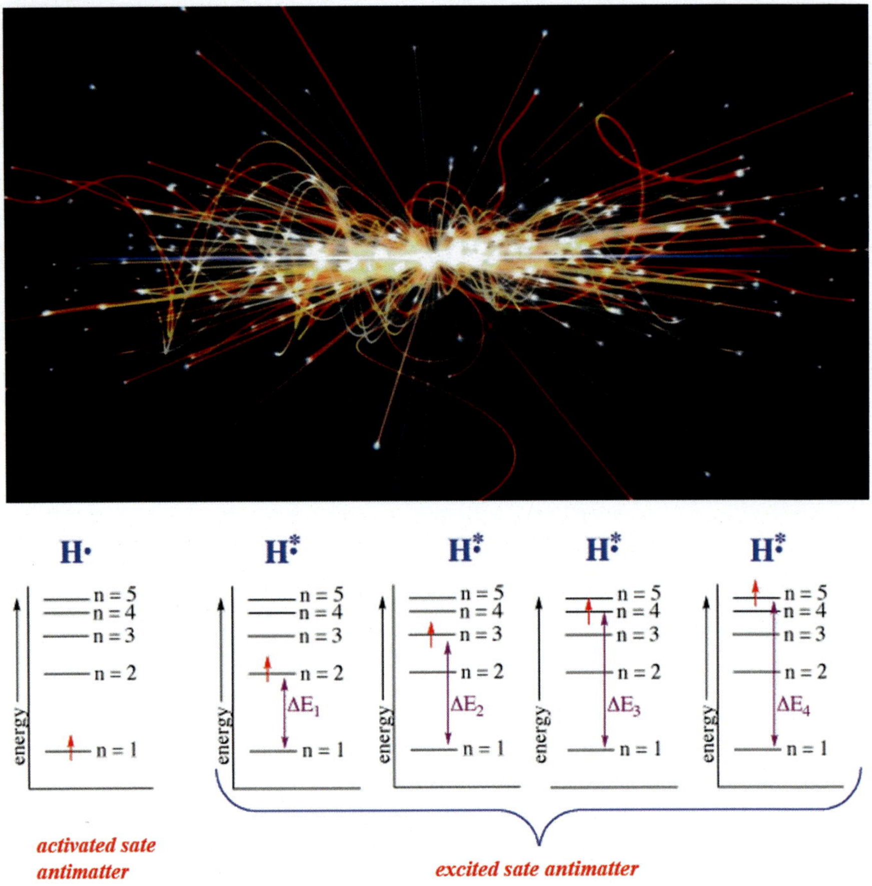

Fig. 4.3 The concept of work energy distribution (H^*) due to the quantum mechanical functions of the human body considering free quantum fields (n) corresponding to its vectorial state

Since the Klein-Gordon function is a standard unit, $(c + m^2)\psi(x) = 0$,with the metric signature of $\eta_{\mu\nu} =$ diag $(\pm 1, -1, -1, -1)$, it is solved by Fourier transformation as:

$$\psi(x) = \int \frac{d^4 p}{(2\pi)^4} e^{-ip.x} \psi(p) \tag{4.23}$$

and using spherical dimension of body the complex exponentials described here as:

$$p^2 = \left(p^0\right)^2 - p^2 = m^2 \tag{4.24}$$

This result will confirm the momenta of the work energy's positive and negative energy dynamics as:

Table 4.1 Klein-Gordon equation in normal units with metric of $\eta_{\mu\nu} = \text{diag}\ (\pm 1, \mp 1, \mp 1, \mp 1)$

	Position space $x = (ct, x)$	Fourier transformation $\omega = \frac{E}{\hbar},\ k = p/\hbar$	Momentum space $p = \left(\frac{E}{c}, p\right)$
Body and work	$\left(\frac{1}{c^2}\frac{\partial^2}{\partial t^2} - \nabla^2 + \frac{m^2c^2}{\hbar^2}\right)\psi(t,x) = 0$	$\psi(t,x) = \int \frac{d\omega}{2\pi\hbar} \int \frac{d^3k}{(2\pi\hbar)^3} e^{\mp i(\omega t - k.x)} \psi(\omega, k)$	$E^2 = p^2c^2 + m^2c^4$
Vector form	$(c + \mu^2)\psi(x) = 0,\ \mu = mc/\hbar$	$\psi(x) = \int \frac{d^4p}{(2\pi\hbar)^4} e^{-ip.\frac{x}{\hbar}} \psi(p)$	$p^2 = \pm m^2c^2$

Here, $\pm\eta^{\mu\nu}\partial_\mu\partial_\nu$ is representing the d'Alembert function and ∇^2 is representing the Laplace variable, while c is representing the light speed, and ħ is the Planck constant, while $c = \hbar = 1$

Table 4.2 Klein-Gordon equation in natural unit metric of $\eta_{\mu\nu} = diag\ (\pm1, \mp1, \mp1, \mp1)$

	Position space	Fourier transformation $\omega = E,\ k = p$	Momentum space $p = (E, p)$
Body and Work	$\left(\partial_t^2 - \nabla^2 + m^2\right)\psi(t,x) = 0$	$\psi(t,x) = \int \frac{dw}{2\pi} \int \frac{d^3k}{(2\pi)^3} e^{\mp i(wt - k.x)} \psi(w,k)$	$E^2 = p^2 + m^2$
Vector form	$(^{c+} m^2)\psi(x) = 0$	$\psi(x) = \int \frac{d^4p}{(2\pi)^4} e^{-ip.x} \psi(p)$	$p^2 = \pm m^2$

$$p^0 = \pm E(p) \text{ where } E(p) = \sqrt{p^2 + m^2}. \tag{4.25}$$

Using a new set of constant $C_{(p)}$,the solution will become

$$\psi(x) = \int \frac{d^4p}{(2\pi)^4} e^{ip.x} C(P)\delta(P^0)^2 - E(P)^2) \tag{4.26}$$

It is a standard practice to segregate the positive and negative energy dynamics while working with the positive energy only as:

$$\psi(x) = \int \frac{d^4p}{(2\pi)^4} \delta\left(\left(p^0\right)^2 - E(p)^2\right)\left(A(p)e^{-ip^0x^0 + ip^ix^i} + B(p)e^{+ip^0x^0 + ip^ix^i}\right)\theta\left(p^0\right) \tag{4.27}$$

$$= \int \frac{d^4p}{(2\pi)^4} \delta\left(\left(p^0\right)^2 - E(p)^2\right)\left(A(P)e^{-ip^0x^0 + ip^ix^i} + B(-P)e^{+ip^0x^0 - ip^ix^i}\right)\theta\left(p^0\right)$$

$$\rightarrow \int \frac{d^4p}{(2\pi)^4} \delta\left(\left(p^0\right)^2 - E(p)^2\right)\left(A(p)e^{-ip.x} + B(P)e^{+ip.x}\right)\theta\left(p^0\right) \tag{4.28}$$

Here, $B(p) \rightarrow B(-p)$ can be considered p^0- in order to pick up the positive frequency from its delta as:

$$\psi(x) = \int \frac{d^4p}{(2\pi)^4} \frac{\delta(p^0 - E(p))}{2E(p)} \left(A(p)e^{-ip.x} + B(p)e^{+ip.x}\right)\theta\left(p^0\right) \tag{4.29}$$

$$= \int \frac{d^3p}{(2\pi)^3} \frac{1}{2E(p)} \left(A(p)e^{-ip.x} + B(p)e^{+ip.x}\right)\Big|_{p^0 = +E(p)} \tag{4.30}$$

Since it is a general solution to the Klein-Gordon equation, Lorentz invariant quantities of $p.\ x = p_\mu x^\mu$ is the final variable to solve the Klein-Gordon equation [23, 36, 37]. Subsequently, the Lorentz invariance will release the ½$E(p)$ factor into the coefficient $A(p)$and $B(p)$, and thus, the work energy of a free particle can be written as:

$$\frac{p^2}{2m} = E \tag{4.31}$$

By quantizing this, it can confirm the non-relativistic Schrödinger equation for a free particle energy as:

$$\frac{\widehat{p}^2}{2m}\psi = \widehat{E}\psi \tag{4.32}$$

where $\widehat{p} = -i\hbar\nabla$ is representing the momentum function (∇ is the delta variable) and and $\widehat{E} = i\hbar\frac{\partial}{\partial t}$ is the energy dynamic.

Since the Schrödinger equation is a non-relativistic invariant, it is standard to use its free particle as the momentum energy as:

$$\sqrt{p^2c^2 + m^2c^4} = E \tag{4.33}$$

Here, implementing the quantum momentum of the free particle, the energy yields can be written as:

$$\sqrt{(-i\hbar\nabla)^2c^2 + m^2c^4\psi} = i\hbar\frac{\partial}{\partial t}\psi \tag{4.34}$$

Subsequently, the electromagnetic fields of the work energy yield can be further clarified by the Klein and Gordon equation of e identity, i.e., $p^2c^2 + m^2c^4 = E^2$, which will give:

$$\left((-i\hbar\nabla)^2c^2 + m^2c^4\right)\psi = \left(i\hbar\frac{\partial}{\partial t}\right)^2\psi \tag{4.35}$$

which simplifies to

$$-\hbar^2c^2\nabla^2\psi + m^2c^4\psi = \hbar^2\frac{\partial^2}{\partial t^2}\psi \tag{4.36}$$

Rearranging terms yields

$$\frac{1}{c^2}\frac{\partial^2}{\partial t^2}\psi - \nabla^2\psi + \frac{m^2c^2}{\hbar^2}\psi = 0 \tag{4.37}$$

Rewriting these functions considering the inverse metric $(-c^2, 1, 1, 1)$ in relation to Einstein's summational variable, it can be written as:

$$-\eta^{\mu v}\partial_{\mu}\partial_{v}\psi=\sum_{\mu=0}^{\mu=3}\sum_{v=0}^{v=3}-\eta^{\mu v}\partial_{\mu}\partial_{v}\psi=\frac{1}{c^2}\partial_0^2\psi-\sum_{v=1}^{v=3}\partial_{v}\partial_{v}\psi=\frac{1}{c^2}\frac{\partial^2}{\partial t^2}\psi-\nabla^2\psi \tag{4.38}$$

Therefore, the Klein-Gordon equation can be expressed as a covariant function, which is the abbreviation of the form of $(c + \mu^2)\psi = 0$, where $\mu = \frac{mc}{\hbar}$ and $c = \frac{1}{c^2}\frac{\partial^2}{\partial t^2} - \nabla^2$ and, thus, the Klein-Gordon equation can be described as a field of work energy dynamics $V(\psi)$ as:

$$c\psi+\frac{\partial V}{\partial\psi}=0 \tag{4.39}$$

Here, the gauge U(1) symmetry is a complex field $\varphi(x)\epsilon\mathbb{C}$ that can satisfy the Klein-Gordon as:

$$\partial_{\mu}j^{\mu}(x)=0, j^{\mu}(x)=\frac{e}{2m}(\varphi^*(x)\partial^{\mu}\varphi(x)-\varphi(x)\partial^{\mu}\varphi^*(x)) \tag{4.40}$$

Subsequently, proof Klein-Gordon equation using algebraic manipulations from to form a complex field $\varphi(x)$ of mass m can be expressed as a covariant notation as follows:

$$(c+\mu^2)\varphi(x)=0 \tag{4.41}$$

And its complex conjugate

$$(c+\mu^2)\varphi^*(x)=0 \tag{4.42}$$

Then, multiplying by the vectorial variables of $\varphi^*(x)$ and $\varphi(x)$, it can be expressed as:

$$\varphi^*(c+\mu^2)\varphi=0 \tag{4.43}$$

$$\varphi(c+\mu^2)\varphi^*=0 \tag{4.44}$$

Subtracting the former from the latter, it can be obtained

$$\varphi^*c\varphi-\varphi c\varphi^*=0 \tag{4.45}$$

$$\varphi^*\partial_{\mu}\partial^{\mu}\varphi-\varphi\partial_{\mu}\partial^{\mu}\varphi^*=0 \tag{4.46}$$

Then it will confirm

$$\partial_{\mu}(\varphi^*\partial^{\mu}\varphi)=\partial_{\mu}\varphi^*\partial^{\mu}\varphi+\varphi^*\partial_{\mu}\partial^{\mu}\varphi \tag{4.47}$$

From which it can be obtained the conservation law for the Klein-Gordon field:

$$\partial_\mu j^\mu(x) = 0, \qquad j^\mu(x) \equiv \varphi^*(x)\partial^\mu\varphi(x) - \varphi(x)\partial^\mu\varphi^*(x) \tag{4.48}$$

Here, implementing the vectorial variable to the Klein-Gordon field, the scalar field derived work energy can be calculated as:

$$T^{\mu\nu} = \frac{\hbar^2}{m}\left(\eta^{\mu\alpha}\eta^{\nu\beta} + \eta^{\mu\beta}\eta^{\nu\alpha} - \eta^{\mu\nu}\eta^{\alpha\beta}\right)\partial_\alpha\overline{\psi}\partial_\beta\psi - \eta^{\mu\nu}mc^2\overline{\psi}\psi \tag{4.49}$$

Here, the energy component T^{00} will confirm the positive frequency related to the particles of work energy as $\psi(x,t) = \varnothing(x,t)e^{-\frac{i}{\hbar}mc^2t}$, where $\varnothing(x,t) = u_E(x)e^{-\frac{i}{\hbar}Et}$. Thus, defining the work energy of body $E' = E - mc^2 = \sqrt{m^2c^4 + c^2p^2} - mc^2 \approx \frac{p^2}{2m}, E' \ll mc^2$ in the non-relativistic limit $v{\sim}p <\quad< c$, and hence $\left|i\hbar\frac{\partial\varnothing}{\partial t}\right| = E'\varnothing \ll mc^2\varnothing$, thus, this yield of the derivative ψ can be written as:

$$\frac{\partial\psi}{\partial t} = \left(-i\frac{mc^2}{\hbar}\o; + \frac{\partial\o}{\partial t}\right)e^{-\frac{i}{\hbar}mc^2t} \approx -i\frac{mc^2}{\hbar}\o e^{-\frac{i}{\hbar}mc^2t}$$

$$\frac{\partial^2\psi}{\partial t^2} \approx -\left(i\frac{2mc^2}{\hbar}\frac{\partial\varnothing}{\partial t} + \left(\frac{mc^2}{\hbar}\right)^2\varnothing\right)e^{-\frac{i}{\hbar}mc^2t} \tag{4.50}$$

Substituting it with the free Klein-Gordon equation, $c^{-2}\partial_t^2\psi = \nabla^2\psi - m^2\psi$, yield $-\frac{1}{c^2}\left(i\frac{2mc^2}{\hbar}\frac{\partial\varnothing}{\partial t} + \left(\frac{mc^2}{\hbar}\right)^2\varnothing\right)e^{-\frac{i}{\hbar}mc^2t} \approx \left(\nabla^2 - \left(\frac{mc^2}{\hbar}\right)^2\right)\varnothing e^{-\frac{i}{\hbar}mc^2t}$ can be simplified as:

$$-i\hbar\frac{\partial\varnothing}{\partial t} = -\frac{\hbar^2}{2m}\nabla^2\varnothing \tag{4.51}$$

Since free Klein-Gordon equation is a *classical* Schrödinger field, thus it is further analyzed by quantum field of Klein-Gordon of the body [3, 32]. Subsequently, the quantum field of Klein-Gordon equation is being analyzed to confirm the symmetry of the work energy function φ is in the local U(1) gauge of the body of $\varphi \rightarrow \varphi' =$ exp $(i\theta)\varphi$ where $\theta(t,x)$ is a local vectorial angle, which transforms the functional phase as exp$(i\theta) =$ cos θ + i sin θ, where derivative ∂_μ can be replaced by gauge-covariant variable $D_\mu = \partial_\mu - ieA_\mu$, while the gauge fields transform as $eA_\mu \rightarrow eA'_\mu + \partial_\mu\theta$, which confirm the activation of work energy in the human body as:

$$D_\mu D^\mu\varphi = -(\partial_t - ieA_0)^2\varphi + (\partial_i - ieA_i)^2\varphi = m^2\varphi \tag{4.52}$$

Results and Discussion

Computation of Mind Energy (M_e)

To computationally confirm the formation of imagination power in the human brain, the vectorial momentum of its steady weak coupling energy field has been clarified in order to determine its mind energy (M_e). Here, mind energy is thus considered as the quantum field of the density of states (DOS) imagination power proliferation of $J(\omega)$, which, in fact, deform it into a mind energy magnitude of $V(\omega)$ [19, 21, 25]. Thus, mind energy formation follows the Weisskopf-Wigner theory is therefore, all the generated from imagination power and it pass through all dimensional modes (1D, 2D, and 3D) of the brain (Table 4.3).

Subsequently, all dimensions of 1D, 2D, and 3D of the brain are being further clarified by implanting frequency cutoff to avoid the bifurcation of the DOS to pave the path for mind energy formation [30, 34]. Thus, the DOS analyzed at various dimensions of the imagination has been calculated, here named as $\varrho_{PC}(\omega)$, which is being confirmed by the clarifying of mind energy frequency formation of Maxwell's theory of nanostructure corresponding to the imagination power and expressed by $\varrho_{PC}(\omega) \propto \frac{1}{\sqrt{\omega - \omega_e}}\Theta(\omega - \omega_e)$, where $\Theta(\omega - \omega_e)$ where ω_e is denoted as the frequency of the DOS [1, 31].

Then, the density of state (DOS) has been implemented to confirm the qualitative form of the non-Weisskopf-Wigner module that represents the mind energy, calculated by the dimensional clarification of the imagination power [4, 8, 28]. This DOS is thereafter determined considering the electromagnetic field of the brain, which is determined by $\varrho_{PC}(\omega) \propto \frac{1}{\sqrt{\omega - \omega_e}}\Theta(\omega - \omega_e)$. Simply, here, the mind energy of DOS reveals a perfect algorithmic function close to the PBE, which is $\varrho_{PC}(\omega) \propto -[\ln|(\omega - \omega_0)/\omega_0| - 1]\Theta(\omega - \omega_e)$, where ω_e defines the prime tip point of the DOS distribution in the mind.

Subsequently, this distributed DOS imagination will transform it into the mind energy state that will confirm the total deliberation of energy dynamics of the mind corresponding to its quantum fields, and, thus, defined above, $J(\omega)$ will confirm the quantum field of mind energy magnitudes $V(\omega)$ corresponding to the imaginational power as:

$$J(\omega) = \varrho(\omega)|V(\omega)|^2 \tag{4.53}$$

Therefore, considering here, the energy frequency of ω_c and the generated mind energy dynamic of $\langle a(t)\rangle = u(t, t_0)\langle a(t_0)\rangle$, the function $u(t, t_0)$ has been described as the mind energy structure as $u(t, t_0)$ by clarifying the integral-differential equation as:

Table 4.3 Mind energy dynamics considering its densities of states (DOS) in the brain. The unit areas $J(\omega)$ and the generation of self-energy inductions in the mind $\Sigma(\omega)$, demonstrated by the mind energy corresponding to its variables of C, η, and χ as coupled forces among three-dimensional (1D, 2D, and 3D) point break of energy band (Lo. et. al. 2015: Sci. Rep.)

Energy particle	Unit area $J(\omega)$ for different DOS*	Reservoir-induced self-energy induction $\Sigma(\omega)$*
1D	$\frac{C-1}{\pi}\frac{1}{\sqrt{\omega-\omega_e}}\Theta(\omega-\omega_e)$	$-\frac{C-1}{\sqrt{\omega_e-\omega}}$
2D	$C-\eta\left[ln\left\lvert\frac{\omega-\omega_0}{\omega_0}\right\rvert-1\right]\Theta(\omega-\omega_e)\Theta(\Omega_d-\omega)$	$C-\eta\left[Li_2\left(\frac{\Omega_d-\omega_0}{\omega-\omega_0}\right)-Li_2\left(\frac{\omega_0-\omega_e}{\omega_0-\omega}\right)-ln\frac{\omega_0-\omega_e}{\Omega_d-\omega_0}ln\frac{\omega_e-\omega}{\omega_0-\omega}\right]$
3D	$\chi\sqrt{\frac{\omega-\omega_e}{\Omega_C}}exp\left(-\frac{\omega-\omega_e}{\Omega_C}\right)\Theta(\omega-\omega_e)$	$\chi\left[\pi\sqrt{\frac{\omega_e-\omega}{\Omega_C}}exp\left(-\frac{\omega-\omega_e}{\Omega_C}\right)erfc\sqrt{\frac{\omega_e-\omega}{\Omega_C}}-\sqrt{\pi}\right]$

$$u(t, t_0) = \frac{1}{1 - \Sigma'(\omega_b)} e^{-i\omega(t - t_0)} + \int_{\omega_e}^{\infty} d\omega \frac{J(\omega) e^{-i\omega(t - t_0)}}{[\omega - \omega_c - \Delta(\omega)]^2 + \pi^2 J^2(\omega)}, \quad (4.54)$$

where $\Sigma'(\omega_b) = [\partial\Sigma(\omega)/\partial\omega]_{\omega = \omega_b}$ and $\Sigma(\omega)$ confirm the mind energy induction into the reservoir,

$$\Sigma(\omega) = \int_{\omega_e}^{\infty} d\omega' \frac{J(\omega')}{\omega - \omega'} \quad . \quad (4.55)$$

Here, the frequency ω_b represents the mind energy frequency mode ($0 < \omega_b < \omega_e$),calculated as $\omega_b - \omega_c - \Delta(\omega_b) = 0$, where $\lesssim\Delta(\omega) = \mathcal{P}\left[\int d\omega' \frac{J(\omega')}{\omega - \omega'}\right]$ is a primary-valued integral of the mind. In details, it can be explained that energy dynamics (*eV*) in the brain is considered as the Fock mind energy state n_0, *i. e.* , $\rho(t_0) = |n_0\rangle\langle n_0|$, which is obtained computationally from the real-time quantum field of its radial (*r*) dimension of the imagination at a time *t* and expressed as:

$$\rho(t) = \sum_{n=0}^{\infty} \mathcal{P}_n^{(n_0)}(t)|n_0\rangle\langle n_0| \quad (4.56)$$

$$\mathcal{P}_n^{(n_0)}(t) = \frac{[v(t,t)]^n}{[1 + v(t,t)]^{n+1}} [1 - \Omega(\mathrm{t})]^{n_0} \text{ x } \sum_{k=0}^{min\{n_0, n\}} \binom{n_0}{k}\binom{n}{k} \left[\frac{1}{v(t,t)} \frac{\Omega(\mathrm{t})}{1 - \Omega(\mathrm{t})}\right]^k \quad (4.57)$$

where $\Omega(\mathrm{t}) = \frac{|u(t, t_0)|^2}{1 + v(t,t)}$. Thus, this result suggested that the Fock state mind energy is indeed generated into dynamic states of the imagination $\mathcal{P}_n^{(n_0)}(t)$ of $|n_0\rangle$ where the proliferation of energy deliberation $\mathcal{P}_n^{(n_0)}(t)$ is in the prime state $|n_0 = 5\rangle$ and in the steady-state limits $\mathcal{P}_n^{(n_0)}(t \to \infty)$, which will eventually reach at a non-equilibrium mind energy in order to transform it into body state energy.

Computation of Body Energy (B$_e$)

To transform mind energy into body energy, the local symmetry of Higgs-Boson quantum field of the body has been simulated considering Abelian local symmetries by applying Higgs-Boson electromagnetic field of the body [5, 16]. Simply the momentum of this energy dynamic will interrupt the gauge-field symmetries of the body, and thus, the Goldstone scalar field will act as the longitude modes of the Higgs-Boson field [14, 33]. Therefore, the local symmetries of quantum field of the body will be broken down as particle of T^{α} related to the gauge field of $A_{\mu}^{\alpha}(x)$, and thus, the Higgs-Boson quantum fields will then start to generate local U(1) phase symmetries to create body energy [11, 16]. So, this transformation process can be consisting of the scalar fields Φ (x) of energetically charged q paired with the electromagnetic field $A^{\mu}(x)$ of the body energy that could be written as:

$$\mathcal{L} = -\frac{1}{4} F_{\mu\nu}F^{\mu\nu} + D_{\mu}\Phi^* D^{\mu}\Phi - V(\Phi^*\Phi), \tag{4.58}$$

where

$$\begin{aligned} D_{\mu}\Phi(x) &= \partial_{\mu}\Phi(x) + \mathrm{iq}A_{\mu}(x)\Phi\,(x) \\ D_{\mu}\Phi^*(x) &= \partial_{\mu}\Phi^*(\mathrm{x}) - \mathrm{iq}A_{\mu}(x)\Phi^*(x), \end{aligned} \tag{4.59}$$

and

$$V(\Phi^*\Phi) = \frac{\lambda}{2}\,(\Phi^*\Phi)^2 + m^2(\Phi^*\Phi) \tag{4.60}$$

Here $\lambda > 0$ but $m^2 < 0$; therefore, $\Phi = 0$ is a local peak factor of the scalar field, and the precise form of quantum dynamics is $\Phi = \frac{v}{\sqrt{2}} * e^{i\theta}$ that can be expressed as:

$$v = \sqrt{\frac{-2m^2}{\lambda}} \text{ for any real } \theta \tag{4.61}$$

Consequently, here, the scalar quantum dynamics Φ (x) will form a nonzero value of $\langle\Phi\rangle \neq 0$, which will form the local U(1) symmetries into electromagnetic field of body energy. Therefore, in this local U(1) symmetries, the quantum dynamics of Φ (x) will be an activated scalar field of the body energy that can be expressed as:

$$\Phi\,(x) = \frac{1}{\sqrt{2}}\,\Phi_r(x) * e^{i\Theta(x)}, \text{real } \Phi_r(\mathrm{x}) > 0, \text{real } \Phi(x). \tag{4.62}$$

Since the scalar field of body energy is a dynamic momentum of the quantum Φ (x) $= 0$, it will meet the principle of $\langle\Phi\rangle \neq 0$ in order to be considered as the instinctual function of the body energy. Thus, considering the momentum field of $\phi_r(x)$ and $\Theta(x)$, the scalar field related to the radial field of ϕ_r of the body energy can be simplified:

$$V(\phi) = \frac{\lambda}{8}\,(\phi_r^2 - v^2)^2 + \text{const.} \tag{4.63}$$

Subsequently, the momentum field can be shifted by applying the variable scalars, $\Phi_r(x) = v + \sigma(x)$,and expressed as:

$$\phi_r^2 - v^2 = (v+\sigma)^2 - v^2 = 2v\sigma + \sigma^2 \tag{4.67}$$

$$V = \frac{\lambda}{8}\left(2v\sigma - \sigma^2\right)^2 = \frac{\lambda v^2}{2} * \sigma^2 + \frac{\lambda v}{2} * \sigma^3 + \frac{\lambda}{8} * \sigma^4. \tag{4.68}$$

Thus, the functional derivative $D_\mu \phi$ will become

$$\begin{aligned} D_\mu \phi &= \frac{1}{\sqrt{2}}\left(\partial_\mu\left(\phi_r e^{i\Theta}\right) + iqA_\mu{}^* \phi_r e^{i\Theta}\right) = \frac{e^{i\Theta}}{\sqrt{2}}\left(\partial_\mu \phi_r + \phi_r{}^* i\partial_\mu \Theta + \phi_r{}^* iqA_\mu\right) \\ \left|D_\mu \phi\right|^2 &= \frac{1}{2}\left|\partial_\mu \phi_r + \phi_r{}^* i\partial_\mu \Theta + \phi_r{}^* iqA_\mu\right|^2 \\ &= \frac{1}{2}\left(\partial_\mu \phi_r\right) + \frac{\phi_r^{2*}}{2}\left(\partial_\mu \Theta qA_\mu\right)^2 \end{aligned} \tag{4.69}$$

$$= \frac{1}{2}\left(\partial_\mu \sigma\right)^2 + \frac{(v+\sigma)^2}{2} * \left(\partial_\mu \Theta + qA_\mu\right)^2. \tag{4.70}$$

The Lagrangian is then given by

$$\mathcal{L} = \frac{1}{2}\left(\partial_\mu \sigma\right)^2 - v(\sigma) - \frac{1}{4} F_{\mu\nu}F^{\mu\nu} + \frac{(v+\sigma)^2}{2} * \left(\partial_\mu \Theta + qA_\mu\right)^2 \tag{4.71}$$

Simply, to conduct the heat ($\mathcal{L}_{\text{heat}}$) generation through electromagnetic field of the body energy considering this *Lagrangian*, it has been further expanded $\mathcal{L}_{\text{heat}}$ as the energy dynamics of the electromagnetic field of the body energy that can be expressed as:

$$\mathcal{L}_{\text{heat}} = \frac{1}{2}\left(\partial_\mu \sigma\right)^2 - \frac{\lambda v^2}{2} * \sigma^2 - \frac{1}{4} F_{\mu\nu}F^{\mu\nu} + \frac{v^2}{2} * \left(qA_\mu + \partial_\mu \Theta\right)^2. \tag{4.72}$$

Subsequently, it will initiate to generate work energy into the quantum field of the body energy considering the scalar particles of λv^2 corresponding to its energy spectrum of the human body [17, 20, 26].

Computation of Work Energy ($\mathbf{W_e}$)

Finally, the electromagnetic field of the body energy is being modeled by Higgs-Boson quantum dynamics considering its local symmetry of U(1) analysis in order to from gauge-variable QED, in terms of gauge particles $\varnothing' \rightarrow e^{i\alpha(x)}\varnothing$ to transform body energy into work energy. This mechanism is being confirmed by the following variable derivatives considering the specific transformational laws of the scalar field, written as:

$$\partial_\mu \rightarrow D_\mu = \partial_\mu = ieA_\mu \qquad \text{[covariant derivatives]}$$
$$A'_\mu = A_\mu + \frac{1}{e}\ \partial_\mu \alpha \qquad [A_\mu \text{ derivatives}] \tag{4.73}$$

Here, the local U(1) gauge-invariant Lagrangian is being considered as the perplex scalar field that is expressed by

$$\mathcal{L} = (D^\mu)^\dagger \left(D_\mu \varnothing\right) - \frac{1}{4}\ F_{\mu\nu}F^{\mu\nu} - V(\varnothing). \tag{4.74}$$

The term $\frac{1}{4}\ F_{\mu\nu}F^{\mu\nu}$ is the kinetic motion of the gauge field for considering thermal energy and $V(\varnothing)$ is denoted the kinetic term written as $V(\varnothing^*\varnothing) = \mu^2(\varnothing^*\varnothing) + \lambda\ (\varnothing^*\varnothing)^2$.

Thus, the equation of the Lagrangian $\mathcal{L}$ is being considered as the quantum field of scalar particle ϕ_1 and ϕ_2 and a mass μ of the body energy; thus, here, $\mu^2 < 0$ will confirm an infinite number of quanta to satisfy the equation $\phi_1^2 + \phi_2^2 = -\mu^2/\lambda = v^2$. So, the quantum field can be clarified as $\phi_0 = \frac{1}{\sqrt{2}}\ [(\upsilon + \eta) + i\xi]$, and then the derivative of the Lagrangian can be expressed as kinetic mode of:

$$\begin{aligned}\mathcal{L}\text{kin}(\eta, \xi) &= (D^\mu\phi)^\dagger (D^\mu\phi) \\ &= (\partial^\mu + ieA^\mu)\phi^*\ \left(\partial_\mu - ieA_\mu\right)\ \phi\end{aligned} \tag{4.75}$$

Here, $V(\eta, \xi) = \lambda\ \upsilon^2\eta^2$ is the final term to the second order, and thus, the full Lagrangian can be expressed as:

$$\mathcal{L}_{\text{kin}}(\eta, \xi) = \frac{1}{2}\ \left(\partial_\mu\eta\right)^2 - \lambda\upsilon^2\eta^2 + \frac{1}{2}\ \left(\partial_\mu\xi\right)^2 - \frac{1}{4}\ F_{\mu\nu}F^{\mu\nu} + \frac{1}{2}\ e^2\upsilon^2A_\mu^2 - e\upsilon A_\mu(\partial^\mu\xi) \tag{4.76}$$

Here, η represents the mass, ξ represents the massless, μ represents the mass for the quanta, and thus A_μ is defined as the term $\partial_\mu\alpha$, as is the function of the quantum field [23, 27]. Naturally, A_μ and ϕ can be changed spontaneously, so the equation can be rewritten as a Lagrangian scalar to confirm the formation of the work energy particles within the quantum field of the body:

$$\begin{aligned}\mathcal{L}_{\text{scalar}} &= (D^\mu\phi)^\dagger (D^\mu\phi) - V\left(\phi^\dagger\phi\right) \\ &= \ (\partial^\mu + ieA^\mu)\frac{1}{\sqrt{2}}(v + h)\ \left(\partial_\mu - ieA_\mu\right)\frac{1}{\sqrt{2}}(v + h) - V\ \left(\phi^\dagger\phi\right)\end{aligned} \tag{4.77}$$

$$= \frac{1}{2}\left(\partial_\mu h\right)^2 + \frac{1}{2}\, e^2 A_\mu^2\, (v+h)^2 - \lambda v^2 h^2 - \lambda v h^3 - \frac{1}{4}\,\lambda h^4 + \frac{1}{4}\,\lambda h^4 \tag{4.78}$$

Here, the Lagrangian scalar thus revealed that the Higgs-Boson quantum field can certainly be initiated to form work energy in the body. To confirm to determine this work energy generation in the body, a further calculation considering its isotropic distributed kinetic energy has been conducted with respect to the angle θ from the dimensional axis of the body and the differential density of energy $\in$ and written as:

$$\mathrm{dn} = \frac{1}{2}\mathrm{n}\,(\in)\,\sin\theta\mathrm{d}\in \mathrm{d}\theta. \tag{4.79}$$

Here, the speed of high-energy frequency is being implemented as $c\ (1\text{–}cos\theta)$ considering the release of energy by expressing the following equation:

$$\frac{d\tau_{\text{abs}}}{dx} = \int\!\!\int \frac{1}{2}\sigma n\,(\in)\,(1-\cos\theta)\,\sin\theta\mathrm{d}\in \mathrm{d}\theta. \tag{4.80}$$

Rewriting these variables as integral over s instead of θ, by (4.79) and (4.80), it has been determined as:

$$\frac{d\tau_{\text{abs}}}{dx} = \pi r_0^2 \left(\frac{m^2c^4}{E}\right)^2 \int_{\frac{m^2c^4}{E}}^{\infty} \in^{-2} n(\in)\,\overline{\phi}\,[s_0\,(\in)]\,de, \tag{4.81}$$

where

$$\overline{\phi}\,[s_0\,(\in)] = \int_1^{s_0(\in)} s\overline{\sigma}\,(s)ds, \overline{\sigma}(s) = \frac{2\sigma(s)}{\pi r_0^2}. \tag{4.82}$$

Thus, this result confirms the representation of the work energy generation in that is readily allow the calculation of $\overline{\phi}\,[s_0]$ to determine the network energy formation of s_0 [22, 24, 37] in the body. Therefore, the net functional asymptotic formula is expressed as follows to confirm the work energy formation:

$$\overline{\phi}[s_0] = 2s_0(\ln 4s_0\text{–}2) + \ln 4s_0(\ln 4s_0\text{–}2)\text{–}\frac{(\pi^2-9)}{3} + s_0^{-1}\left(ln\,4s_0 + \frac{9}{8}\right) \tag{4.83}$$

$$+ \ldots (s_0 >> 1);$$

$$\overline{\phi}\left[s_{0]} = \left(\frac{2}{3}\right)(S_0-1)^{\frac{3}{2}} + \left(\frac{5}{3}\right)(S_0-1)^{\frac{5}{2}}\text{–}\left(\frac{1507}{420}\right)(S_0-1)^{\frac{7}{2}} + \ldots (s_0\text{–}1 << 1). \right. \tag{4.84}$$

The function $\frac{\overline{\phi}[s_0]}{(s_0-1)}$ is revealed as $1 < s_0 < 10$; at larger s_0, it confirms a standard algorithm function of s_0. Thus, the energy spectra of the work energy can be written as $n(\in) \propto \in^m$, and thus, the calculation of the work energy in terms energy spectra in the body can be expressed as:

$$\begin{aligned} n(\in) &= 0, \quad \in \; < \; \in_0 \\ &= C_{\in}^{-\alpha} \; or \; D_{\in}^{\beta}, \quad \in_0 \; < \; \in \; < \; \in_m \\ &= 0, \quad \in \; > \; \in_m \end{aligned} \tag{4.85}$$

Then, it can be transformed as:

$$\left(\frac{d\tau_{\text{abs}}}{dx}\right)_{\alpha} = \pi r_0^2 C\left(\frac{m^2c^4}{E}\right)^{1-\alpha}$$

$$\times \begin{cases} 0 \quad , & E < E_m \\ [F_\alpha(1) – \mathrm{F}_\alpha(\sigma_m)], & E_m < E < E_0 \\ [F_\alpha(\sigma_0) – \mathrm{F}_\alpha(\sigma_m)], & E > E_0; \end{cases} \tag{4.86}$$

$$\left(\frac{d\tau_{\text{abs}}}{dx}\right)_{\beta} = \pi r_0^2 D\left(\frac{m^2c^4}{E}\right)^{1+\beta}$$

$$\times \begin{cases} 0 \quad , & E < E_m \\ \left[F_\beta(\sigma_m)\right) \quad , & E_m < E < E_0 \\ \left[F_\beta(\sigma_m) – \mathrm{F}_\beta(\sigma_0)\right], & E > E_0. \end{cases} \tag{4.87}$$

In these functional variables, the work energy spectra on the body can be properly defined by asymptotic formula. Thus, the term Γ_γ^{LPM} defines the energy generation in relation to the energy dynamics as:

$$\Gamma_\gamma \equiv \frac{dn_\gamma}{dVdt}. \tag{4.88}$$

Here, the contributions $\Gamma_\gamma^{\text{LPM}}$ and the rate of work energy generation are being confirmed as $O(\alpha_{\text{EM}}\ \alpha_s)$. Thus, it has been confirmed by implementing the contributed work energy generation $\Gamma_\gamma^{\text{LPM}}$ to energy-physical reaction μ of the body by expressing the following equation:

$$\frac{d\Gamma_\gamma^{LPM}}{d^3k} = \frac{d_F q_s^2 \alpha_{EM}}{4\pi^2 k} \int_{-\infty}^{\infty} \frac{dp_{\parallel}}{2\pi} \int \frac{d^2\mathrm{p}_{\perp}}{(2\pi)^2} A\left(p_{\parallel}, k\right) \; \mathrm{Re}\left\{2\mathrm{P}_{\perp} \cdot f\left(\mathrm{p}_{\perp}; p_{\parallel}, k\right)\right\} \tag{4.89}$$

where d_F is the variable state of the work energy particles [N_c in SU(N_c)] and q_s is the Abelian charge of the work energy quark, $k \equiv |k|$, and thus, the kinetic functional mode A(p||, k) is being expressed by:

$$A\left(p_{\|},k\right) \equiv \left\{ \begin{array}{ll} \dfrac{n_b\left(k+p_{\|}\right)\left[1+n_b\left(p_{\|}\right)\right]}{2p_{\|}\left(p_{\|}+k\right)} & \text{scalars} \\ \dfrac{n_f\left(k+p_{\|}\right)\left[1-n_f\left(p_{\|}\right)\right]}{2\left[p_{\|}\left(p_{\|}+k\right)\right]^2}\left[{p_{\|}}^2+\left(p_{\|}+k\right)^2\right], & \text{fermions} \end{array} \right\} \tag{4.90}$$

with

$$n_b(p) \equiv \frac{1}{\exp[\beta(p-\mu)]-1}, \quad n_f(p) \equiv \frac{1}{\exp[\beta(p-\mu)]+1} \tag{4.91}$$

The function $f(p\perp;\, p\|,\, k)$ is then integrated to resolve the below equation that suggested that the work energy proliferation in the body is very much possible, which can be expressed as the following derivative of energy generation:

$$\begin{aligned} 2p_{\perp} = i\delta E\, \mathrm{f}\left(p_{\perp};p_{\|},k\right) + \frac{\pi}{2}C_F g_s^2 m_D^2 \int \frac{d^2q_{\perp}}{(2\pi)^2}\frac{dq_{\|}}{2\pi}\frac{dq^0}{2\pi}2\pi\delta\left(q^0-q_{\|}\right) \text{ x } \frac{T}{|q|} \\ \times \left[\frac{2}{\left|q^2-\Pi_L(Q)\right|^2} + \frac{\left[1-\left(q^0/\left|\mathrm{q}_{\|}\right|\right)^2\right]^2}{\left|(q^0)^2-\mathrm{q}^2-\Pi_T(Q)\right|^2}\right] \\ \times \left[f\left(p_{\perp};p_{\|},k\right) - f\left(q+p_{\perp};p_{\|},k\right)\right] \end{aligned} \tag{4.92}$$

Simply, this work energy generation is derived from the explicit forms of body energy corresponding to its given energy function of $f(p\perp;\, p\|,\, k)$ to transform it into body energy to work energy in the human body in order to work physically.

Conclusions

The conduction of mind energy (M_e), body energy (B_e), and work energy (W_e) are being modeled to confirm that mind energy has the enough hidden imagination power in the human brain to transform it into body state energy and then work state energy, which is in fact a final transformational energy for imagination power. It is because the results of this computational modeling suggested that the availability of mind energy is very much doable in human brain that can be transformed into body energy by implementing the quantum electrodynamics (QED) of the human mind energy. Subsequently, this body energy can also be reformed into the work energy by integrating Higgs-Boson [BR ($H \rightarrow \gamma\gamma^{-}$)] quantum dynamics of the human body in order to work physically. Simply the transformation mechanism of mind energy into work energy eventually through the process of body energy shall indeed be the most innovative scientific discovery to release extreme energy to mitigate everything in the daily life.

Acknowledgments The author, Md. Faruque Hossain, declares that any findings, predictions, and conclusions described in this article are solely performed by the author and it is confirmed that there is no conflict of interest for publishing this research paper in a suitable journal and/or publisher.

References

1. Berges, J., & Mesterhazy, D. (2012). Introduction to the nonequilibrium functional renormalization group. *Nuclear Physics B-Proceedings Supplements, 228*, 37–60.
2. Bertozzi, E. (2010). Hunting the ghosts of a 'strictly quantum field': The Klein–Gordon equation. European Journal of Physics, 31(6), 1499.
3. Birnbaum, K. M., Boca, A., Miller, R., Boozer, A. D., Northup, T. E., & Kimble, H. J. (2005). Photon blockade in an optical cavity with one trapped atom. *Nature, 436*(7047), 87–90.
4. Boettcher, I., Pawlowski, J. M., & Diehl, S. (2012b). Ultracold atoms and the functional renormalization group. *Nuclear Physics B-Proceedings Supplements, 228*, 63–135.
5. Broz, M., Contreras, J. G., & Takaki, J. T. (2020a). A generator of forward neutrons for ultra-peripheral collisions: nOOn. *Computer Physics Communications, 253*, 107181.
6. Chang, D. E., Sørensen, A. S., Demler, E. A., & Lukin, M. D. (2007). A single-photon transistor using nanoscale surface plasmons. *Nature Physics, 3*(11), 807–812.
7. Chen, G., Chen, S., Li, C., & Chen, Y. (2013). Examining non-locality and quantum coherent dynamics induced by a common reservoir. *Scientific Reports, 3*(1), 1–6.
8. Chen, J., Wang, C., Zhang, R., & Xiao, J. (2012). Multiple plasmon-induced transparencies in coupled-resonator systems. *Optics Letters, 37*(24), 5133–5135.
9. Cheng, M., & Song, Y. (2012). Fano resonance analysis in a pair of semiconductor quantum dots coupling to a metal nanowire. *Optics Letters, 37*(5), 978–980.
10. Douglas, J. S., Habibian, H., Hung, C., Gorshkov, A. V., Kimble, H. J., & Chang, D. E. (2015). Quantum many-body models with cold atoms coupled to photonic crystals. *Nature Photonics, 9*(5), 326–331.
11. Dupuis, N., Canet, L., Eichhorn, A., Metzner, W., Pawlowski, J. M., Tissier, M., & Wschebor, N. (2021a). The nonperturbative functional renormalization group and its applications. *Physics Reports, 910*, 1–114.
12. Eichler, J., & Stöhlker, T. (2007). Radiative electron capture in relativistic ion–atom collisions and the photoelectric effect in hydrogen-like high-Z systems. *Physics Reports, 439*(1-2), 1–99.
13. Englund, D., Majumdar, A., Faraon, A., Toishi, M., Stoltz, N., Petroff, P., & Vučković, J. (2010). Resonant excitation of a quantum dot strongly coupled to a photonic crystal nanocavity. *Physical Review Letters, 104*(7), 073904.
14. Fernández, J., & Martín, F. (2009). Electron and ion angular distributions in resonant dissociative photoionization of H2 and D2 using linearly polarized light. *New Journal of Physics, 11*(4), 043020.

15. Gazi, V., & Passino, K. M. (2005). Stability of a one-dimensional discrete-time asynchronous swarm. *IEEE Transactions on Systems, Man, and Cybernetics, Part B (Cybernetics), 35*(4), 834–841.
16. Hencken, K., Baur, G., & Trautmann, D. (2006). Transverse momentum distribution of vector mesons produced in ultraperipheral relativistic heavy ion collisions. *Physical Review Letters, 96*(1), 012303.
17. Jahnke, V., Luna, A., Patino, L., & Trancanelli, D. (2014). More on thermal probes of a strongly coupled anisotropic plasma. *Journal of High Energy Physics, 2014*(1), 1–40.
18. Jiaqi, L., Yi, Z., Chengkai, T., & Xingxing, Z. (2019). INS aided high dynamic single-satellite position algorithm. Paper presented at the 2019 IEEE International Conference on Signal Processing, Communications and Computing (ICSPCC), 1–5
19. Johnson, B. R., Reed, M. D., Houck, A. A., Schuster, D. I., Bishop, L. S., Ginossar, E., . . . Girvin, S. M. (2010). Quantum non-demolition detection of single microwave photons in a circuit. Nature Physics, 6(9), 663-667.
20. Md. Faruque Hossain (2022). Ultraviolet Germicidal Irradiation (UVGI) Application in Building Design to Terminate Pathogens Naturally. Material Today Sustainability. DOI: https://doi.org/10.1016/j.mtsust.2022.100161. (Springer).
21. Md. Faruque Hossain (2021). Sustainable Building Technology: Thermal Control of Solar Energy to Cool and Heat the Building Naturally. Environment, Development and Sustainability. DOI: https://doi.org/10.1007/s10668-020-01212-z. (Springer).
22. Md. Faruque Hossain (2020). Modeling of Global Temperature Control. Environment, Development and Sustainability. DOI: https://doi.org/10.1007/s10668-020-00924-6. (Springer).
23. Md. Faruque Hossain (2018). Green Science: Advanced Building Design Technology to Mitigate Energy and Environment. Renewable and Sustainable Energy Reviews. 81 (2), 3051-3060. (Elsevier).
24. Md. Faruque Hossain (2018). Photon energy amplification for the design of a micro-PV panel. International Journal of Energy Research. DOI: https://doi.org/10.1002/er.4118. (Wiley).
25. Md. Faruque Hossain (2018). Design and Construction of Ultra-Relativistic Collision PV Panel and Its Application into Building Sector to Mitigate Total Energy Demand. Journal of Building Engineering. 9, 147-154. (Elsevier).
26. Md. Faruque Hossain (2017). Green Science: Independent Building Technology to Mitigate Energy, Environment, and Climate Change. Renewable and Sustainable Energy Reviews. 73; 695-705. (Elsevier).
27. Md. Faruque Hossain (2016). Solar Energy Integration into Advanced Building Design for Meeting Energy Demand. International Journal of Energy Research. 40, 1293-1300. (Wiley).
28. Md. Faruque Hossain (2016). Theory of Global Cooling. Energy, Sustainability, and Society. 6: 24. (Springer).
29. Naghiloo, M., Foroozani, N., Tan, D., Jadbabaie, A., & Murch, K. W. (2016). Mapping quantum state dynamics in spontaneous emission. *Nature Communications, 7*(1), 1–7.
30. Reinhard, P., & Nazarewicz, W. (2021). Nuclear charge densities in spherical and deformed nuclei: Toward precise calculations of charge radii. *Physical Review C, 103*(5), 054310.
31. Shalm, L. K., Hamel, D. R., Yan, Z., Simon, C., Resch, K. J., & Jennewein, T. (2013). Three-photon energy–time entanglement. *Nature Physics, 9*(1), 19–22.
32. Stehle, C., Zimmermann, C., & Slama, S. (2014). Cooperative coupling of ultracold atoms and surface plasmons. *Nature Physics, 10*(12), 937–942.
33. Szafron, R., & Czarnecki, A. (2016). High-energy electrons from the muon decay in orbit: Radiative corrections. *Physics Letters B, 753*, 61–64.
34. Tame, M. S., McEnery, K. R., Özdemir, Ş. K., Lee, J., Maier, S. A., & Kim, M. S. (2013). Quantum plasmonics. *Nature Physics, 9*(6), 329–340.
35. Ting, T. (2004). The polarization vector and secular equation for surface waves in an anisotropic elastic half-space. *International Journal of Solids and Structures, 41*(8), 2065–2083.
36. Tu, M. W., & Zhang, W. (2008). Non-markovian decoherence theory for a double-dot charge qubit. *Physical Review B, 78*(23), 235311.

37. Wang, C., Wang, W., & Chen, Z. (2017a). Single-satellite positioning algorithm based on direction-finding. Paper presented at the 2017 Progress in Electromagnetics Research Symposium-Spring (PIERS), 2533-2538.
38. Yan, J. Z., & Chang, S. (2018). The contingent effects of political strategies on firm performance: A political network perspective. *Strategic Management Journal, 39*(8), 2152–2177.

Part III
Sustainable Housing and Building Technology

Chapter 5
Remodeling of Photon Structure to Design High-Performance Solar Cell

Introduction

Energy from solar irradiance has always been interesting, but the conversion rate has never been satisfactory to implement this renewable energy massively to meet the global energy demand [2, 6, 12]. In the research, an approach is, thus, initiated to form thousands of photons from a single photon to ultimately accelerate the solar energy harvesting rate a thousand times higher than currently available in any solar cell system throughout the world. Here, *Higgs-Boson* BR($H \rightarrow \gamma\gamma^{-}$) quantum field can play a vital role once it is employed under the high-temperature condition to transform a single photon into pair photons and then further form thousands of photons with the assistance of two feed semiconductors and three-diode superconductors into the high-performance solar cell. In recent years, some model of ultra-relativistic reaction has been studied by G. Baur et al. (2007), J. Eichle et al. (2007), Najjari A. B et al. (2009), and Li et al. (2013), which suggested that pair photon production is feasible via ultra-relativistic collision of two bare ions of a single photon once it is implemented by thermal non-equilibrium condition [3, 5, 10]. These findings are interesting since it can produce pair electrons from a single one, and none of these studies demonstrates the detailed mechanism of the pair photon production and then the paired photon's conversion mechanism into multiple photons by the implementation of a two-layer semiconductor and then be scattered further by a three-diode silicon superconductor to produce a tremendous number of photons. To overcome this knowledge gap, in this research, the Higgs-Boson mechanism has been implemented to create a quantum field under the high-temperature condition of the solar cell in order to form a symmetry-breaking mechanism for the interactions of electrons. Consequently, this breaking symmetry will trigger the Higgs mechanism to cause the Bosons to interact with its bare ions to generate energy through the electroweak symmetry-breaking mechanism of the electrons, and thus, these electrons will carry a unit negative electron charge, whereas two electrons can interact with each other via the electromagnetic force; it forms the electron-

M. F. Hossain, *Global Sustainability*, https://doi.org/10.1007/978-3-031-34575-3_5

positron pair photons by inducing relativistic ion-ion (e^+e^-) collisions [31, 34, 38]. Subsequently, produced pair photon can be introduced into two feed semiconductors and then three diode superconductors for scattering the photon quantum electrodynamics to confirm the proliferation of thousands of photon energy to confirm the utilization of this energy commercially to meet the global energy demand totally.

Methods and Materials

Higgs-Boson Quantum Field and Pair Photon Production

To generate a local Higgs quantum field into the silicon solar cell upon penetration of the photon irradiance in the solar cell, the local *Abelian* symmetries of this solar cell have been analyzed considering its gauge field symmetry, while the Goldstone scalar is being considered as the longitudinal mode of the vector of the solar irradiance [4, 7, 14]. Consequently, here all local symmetry has been impulsively unified using MATLAB software into particle of T^α in order to form resultant gauge field of $A_\mu^\alpha\ (x)$ of the solar cell where the Higgs quantum field will be commencing work at a local symmetry of U (1) [1, 8, 9]. Therefore, this model shall encompass an intricate scalar field $\Phi\ (x)$ of having an electric charge (q) to fix the EM field $A^\mu(x)$ of the solar energy, which can be represented by Lagrangian as follows:

$$\mathcal{L} = -\frac{1}{4}F_{\mu\nu}F^{\mu\nu} + D_\mu\Phi^* D^\mu\Phi - V(\Phi^*\Phi) \tag{5.1}$$

where

$$\begin{aligned} D_\mu\Phi(x) &= \partial_\mu\Phi(x) + iqA_\mu(x)\Phi\ (x) \\ D_\mu\Phi^*(x) &= \partial_\mu\Phi^*(x) - iqA_\mu(x)\Phi^*(x) \end{aligned} \tag{5.2}$$

and

$$V(\Phi^*\Phi) = \frac{\lambda}{2}(\Phi^*\Phi)^2 + m^2(\Phi^*\Phi) \tag{5.3}$$

Considering $\lambda > 0$ but $m^2 < 0$, then $\Phi = 0$ and is the scalar potential's local maximum is being clarified as the minima creating a degenerate circle $\Phi = \frac{v}{\sqrt{2}} * e^{i\theta}$,

$$v = \sqrt{\frac{-2m^2}{\lambda}}, \text{anyreal}\theta \tag{5.4}$$

As a result, a non-zero vacuum expectation value $\langle \Phi \rangle \neq 0$ is developed in the scalar field Φ, impulsively creating the magnetic field's U (1) symmetry. Such a breakdown would result in a stemming of a massless Goldstone scalar from the complex field's Φ (x) phase. Therefore, the local U(1) symmetry, the Φ (x) phase is not simply the expectation value's $\langle \Phi \rangle$ phase, but also represents the dynamical Φ (x) field's x-dependent phase. This mechanism has thus been confirmed by using polar coordinates in the scalar field as:

$$\Phi(\mathrm{x}) = \frac{1}{\sqrt{2}} \Phi_r(x) * e^{i\Theta(x)}, \text{real } \Phi_r(x) > 0, \text{real } \Phi(x) \tag{5.5}$$

where Φ $(x) = 0$, the redefinition of the field, meaning it is never been considered of having$\langle \Phi \rangle \neq 0$, although it is perfect to consider $\Phi\langle x \rangle \neq 0$ as an expected value, and thus, the ideal fields $\phi_r(x)$ and $\Theta(x)$ are the scalar potential that depended singly on the radial field ϕ_r that can be expressed as:

$$V(\phi) = \frac{\lambda}{8} \left(\phi_r^2 - v^2\right)^2 + \text{const}, \tag{5.6}$$

or considering the radial field shifted by its VEV, $\Phi_r(x) = v + \sigma(x)$,

$$\phi_r^2 - v^2 = (v + \sigma)^2 - v^2 = 2v\sigma + \sigma^2 \tag{5.7}$$

$$V = \frac{\lambda}{8} \left(2v\sigma - \sigma^2\right)^2 = \frac{\lambda v^2}{2} * \sigma^2 + \frac{\lambda v}{2} * \sigma^3 + \frac{\lambda}{8} * \sigma^4 \tag{5.8}$$

Altogether the covariant derivative $D_\mu \phi$ is as follows:

$$D_\mu \phi = \frac{1}{\sqrt{2}} \left(\partial_\mu \left(\phi_r e^{i\Theta}\right) + iqA_\mu{}^* \phi_r e^{i\Theta}\right) = \frac{e^{i\Theta}}{\sqrt{2}} \left(\partial_\mu \phi_r + \phi_r{}^* i\partial_\mu \Theta + \phi_r{}^* iqA_\mu\right)$$

$$\left|D_\mu \phi\right|^2 = \frac{1}{2} \left|\partial_\mu \phi_r + \phi_r{}^* i\partial_\mu \Theta + \phi_r{}^* iqA_\mu\right|^2 \tag{5.9}$$

$$= \frac{1}{2} \left(\partial_\mu \phi_r\right) + \frac{\phi_r^{2*}}{2} \left(\partial_\mu \Theta qA_\mu\right)^2$$

$$= \frac{1}{2} \left(\partial_\mu \sigma\right)^2 + \frac{(v + \sigma)^2}{2} * \left(\partial_\mu \Theta + qA_\mu\right)^2 \tag{5.10}$$

Altogether,

$$\mathcal{L} = \frac{1}{2} \left(\partial_\mu \sigma\right)^2 - v(\sigma) - \frac{1}{4} F_{\mu\nu} F^{\mu\nu} + \frac{(v + \sigma)^2}{2} * \left(\partial_\mu \Theta + qA_\mu\right)^2 \tag{5.11}$$

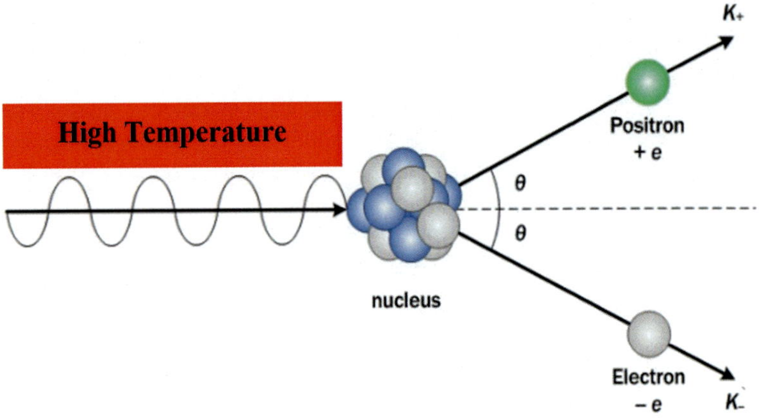

Fig. 5.1 The pair photon production (e^+e^-) from a single photon due to the Higgs-Boson quantum field activation upon high-temperature condition into the solar cell

To estimate Lagrangian's magnetic field properties, the expansion has been analyzed in powers of the fields (similar to their derivatives) and concentrates on the quadratic section defining the free particles as:

$$\mathcal{L}_{\text{free}} = \frac{1}{2}\left(\partial_\mu \sigma\right)^2 - \frac{\lambda v^2}{2} * \sigma^2 - \frac{1}{4} F_{\mu\nu} F^{\mu\nu} + \frac{v^2}{2} * \left(q A_\mu + \partial_\mu \Theta\right)^2 \tag{5.12}$$

Here, the *Lagrangian* ($\mathcal{L}_{\text{free}}$) function is definitely indicative of a real scalar particle whose positive mass^2 = λv^2 linking to the A_μ (x) and the Θ (x) fields to commence and generate high temperature inside the quantum field in order to pave form pair photons from single one [2, 11, 40]. Here this Lagrangian function, subsequently, with Higgs-Boson quantum field is being acquired high-temperature condition because of solar irradiance penetration into the solar cell, the photon energy will get excited causing collision with its bare ions inside the quantum field results in, pair photon will be formed from a single photon (Fig. 5.1).

In essence, a comprehensive mathematical computation concerning accuracy, the solar electromagnetic radiation, and features of a current–voltage correlation has been undertaken to verify the conversion of a single photon into a photon pair [12, 25, 26]. Therefore, the photovoltaic radiation mechanism's momentum is computed as $2mc^2$, with m representing an electron's mass [15, 43] and photon pair generation into the quantum field is being calculated using the following equation:

$$\sigma = \frac{1}{2}\pi r_0^2 \left(1 - \beta^2\right)\left[\left(3 - \beta^4\right) \ln \frac{1+\beta}{1-\beta} - 2\beta\left(2 - \beta^2\right)\right] \tag{5.13}$$

with $r_0 = \frac{e^2}{mc^2}$ indicating the classical radius, while βc shows the electrons' momentum velocity. Therefore, the coordination of β and *E (high-energy photon)*, $\in$ *(low-energy photon)*, and θ (*angle with respect to the x axis*) may be computed taking into consideration the two photons' momentum vectors and may be stated as follows:

$$2 \in E\,(1\text{–cost}) = 4E_e^2, \tag{5.14}$$

with E_{el} indicating the electron momentum into the PV panel and can be additionally rewritten as

$$s = \left(\frac{E_e}{mc^2}\right)^2 = \left(\frac{\in E}{2m^2c^4}\right)(1 - \cos\theta), \tag{5.15}$$

so

$$\beta = \mid p_e \mid \frac{c}{E_e} = \left(1 - \frac{1}{s}\right)^{\frac{1}{2}} \tag{5.16}$$

Also, it may be expressed as

$$s = s_0 z, \tag{5.17}$$

$$s_0 = \frac{\in E}{m^2c^4}, z = \frac{1}{2}\,(1 - \cos\theta).$$

This is a clear indication that the production of photon pair ($\sigma_{\gamma\gamma}$) does occur take place where, $s > 1$ in the relativistic condition, with the threshold condition for a head-on ($\theta = \pi$, $z = 1$) photon collision being $\in E = m^2c^4$, thus, the photon par production can be represented as follows:

$$\sigma_{\gamma\gamma} \approx \frac{3}{8}\,\sigma T\left(\frac{mc^2}{hw}\right)^2\left(ln\,\frac{2hw}{mc^2} - 1\right) \tag{5.18}$$

Three-Diode Photon Scattering

To verify the production of multiple photons from photon pairs, the adequate energy densities have been calculated considering the interaction of Higgs-Boson quantum field into the photon flux densities in the solar cell [24, 27, 33]. Due to photonic particle collision, the pair photon's physical interaction is being created, and thus, the photo-dissociation will act to generate thousands of photons from the pair photons through the three-diode scattering process (Fig. 5.2).

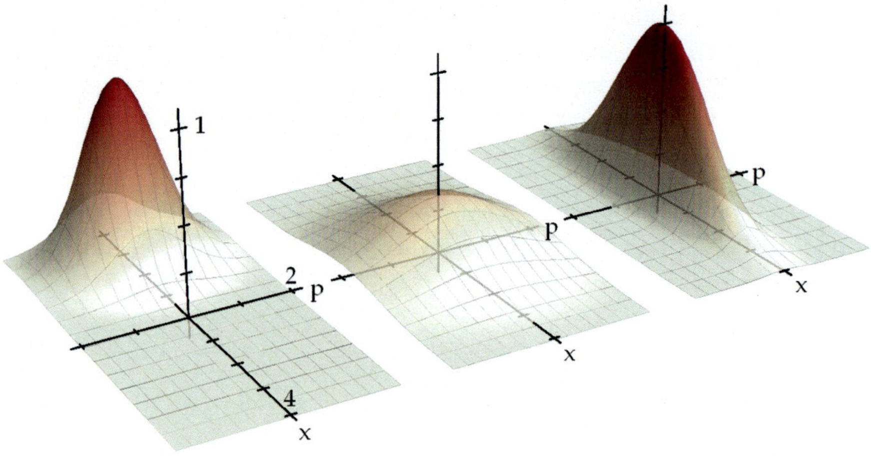

Fig. 5.2 The photonic particle motion directs the pair photon transformation into the photo-dissociation process through the three-diode scattering to generate thousands of photons

Here, the three-diode scattering process is being considered in determining the photon dynamics at the *nano*scale via photonic waveguides integrated in the superconductor circuit [13, 16, 18]. Subsequently, the electron dynamics for constant states of photon generation at atomic spectra taking into solar cell consideration contour maps can be exemplified as the following Hamiltonian:

$$H = \sum \omega_{ci} a_i^{\dagger} a_i + \sum_K \omega_k b_k^{\dagger} b_k + \sum_{ik} \left(V_{ik} a^{\dagger} b_k + V_{ik}^{*} b_k^{\dagger} a_i \right) \tag{5.19}$$

with $a_i \left(a_i^{\dagger} \right)$ indicating the photonic waveguide mode, $b_k \left(b_k^{\dagger} \right)$ indicating the photo-dynamic modes' driver of the photon structure, while the coefficients V_{ik} indicate the photonic mode's magnitude within the photon structure.

Taking into consideration the primary photonic structure in formation of equilibrium, here, the entire photonic reservoir structure has been integrated taking into consideration the solar cell's excited coherent state photons, and thus, the subsequent functional equation, here, can be mathematically expressed as:

$$\begin{aligned} \rho(t) = & -i\left[H_c'(t)\rho(t)\right] + \sum_{ij} \left\{ k_{ij}(t) \left[2a_j \rho(t) a_i^{\dagger} - a_i^{\dagger} a_j \rho(t) - \rho(t) a_i^{\dagger} a_j \right] \right. \\ & \left. + k_{ij}(t) \left[a_i^{\dagger} \rho(t) a_j + a_j \rho(t) a_i^{\dagger} - a_i^{\dagger} a_j \rho(t) - \rho(t) a_j a_i^{\dagger} \right] \right\} \end{aligned} \tag{5.20}$$

where $\rho(t)$ shows the photon's attenuated density in photonic wave, $H_c'(t) = \sum_{ij} \omega_{cij}' (t) a_i^{\dagger} a_j$ indicates the photonic re-standardized Hamiltonian relating to the photonic wave frequencies $\omega'_{cii}(t) = \omega'_{ci}(t)$, as well as the function initiated induced couplings of photons among the particles $\omega'_{cij}(t)$.

Here, the factors featured as photonic dynamics include $\kappa_{ij}(t)$ and $\tilde{\kappa}_{ij}(t)$ and the non-perturbative principle has been considered in purely solving the time-dependent factor, $\kappa_{ij}(t)$ and$\tilde{\kappa}_{ij}(t)$, and the re-standardized frequency, ω'_{cij} [17, 24, 37]. Considering the photon reservoir, the expression for Hamiltonian thus becomes $H_I = \sum_k \lambda_k x q_k$, with $q_k$and x indicating the location of secondary breakpoint and primary unique point reservoirs. Therefore, taking into account the photon's quantum dynamics, the Hamiltonian's entire reservoir of photons is modified as $H_I = \sum_k V_k \left(a^\dagger b_k + b_k^\dagger a + a^\dagger b_k^\dagger + ab_k\right)$ to validate the photonic dynamics' magnitude within the solar cell [11, 48]. This means that the dynamics of photons are being defined by the dissipated photon factors κ (t) and $\tilde{\kappa}(t)$ (Eq. 5.20's sub-indices, j) that can be computed by use of the following equation:

$$\omega'_c(t) = -Im[u\ (t, t_0)/u\ (t, t_0)] \tag{5.21}$$

$$k\ (t) = -\ Re\ [u\ (t, t_0)/u\ (t, t_0)] \tag{5.22}$$

$$\widehat{k}(t) = \dot{v}(t, t) + 2v\ (t, t)\ k(t). \tag{5.23}$$

with $u\ (t, t_0)$ being the photonic area, while $v\ (t, t)$ indicates the photon dynamics resulting from the induced reservoir. $v\ (t, t)$ is further classified by considering the subsequent non-equilibrium dynamic of the integral-differential expression as follows:

$$\dot{u}(t, t_0) = -i\omega_c u\ (t, t_0) - \int_{t_0}^{t} dt' g(t - t')\ u\ (t', t_0) \tag{5.24}$$

$$v\ (t, t) = \int_{t_0}^{t} dt \int_{0}^{t} dt_2\ u^*(t_1, t_0)\widehat{g}(t_1 - t_2)\ u\ (t_2, t_0) \tag{5.25}$$

with v_c being considered as the photonic waveguide's prime frequency, which is the integral functions in Eqs. 5.24 and 5.25 calculated as the total photon generation by the non-equilibrium state of the photonic structure per unit area $J\ (\varepsilon)$ as:$(t - t') = \int d\omega J(\omega) e^{-i\omega(t-t')}$ and $\tilde{g}(t - t') = \int d\omega J(\omega)\overline{n}(\omega, T)e^{-i\omega(t-t')}$, where $\overline{n}(\omega, T) = 1/\left[e^{\hbar\omega/k_B T} - 1\right]$ indicates the initial solar cell's photon dynamics at temperature T, and the classification of *unit* area $J(\omega)$ is being functioned based on the density of states (DOS) $\varrho(\omega)$ photon generation in the solar cell at the magnitude of V_k, which can be expressed as:

$$J\ (\omega) = \sum_k |V_k|^2\ \delta(\omega - \omega_k) = \varrho\ (\omega)\ |\ V\ (\omega)|^2 = \left[e * n(\gamma + \gamma)^2\right]^3 \tag{5.26}$$

Therefore, the last part summarizes the photon generation proliferation is based on the photon dynamic state of the quantum field in that $V_k \rightarrow V(\omega)$ and i of V_{ik} in Eq. (5.26) can be quantified at the single-diode mode and thus the overall photon generation$J(\omega)$ can be simplified as expression:

$$J\,(\omega) = \left[e * n(\gamma + \gamma)^2\right]^3 \tag{5.27}$$

with $n=E=hf$ being the primary quantity of photons in the equilibrium stage, while $e=1$ represents the solar irradiance constant where there is zero interruption, γ is the pair photon at electron state, while $J\,(\omega)$ is the total number of photons being generated.

Results and Discussion

Higgs-Boson Quantum Field and Pair Photon Production

Since the Higgs-Boson quantum field acquired temperature as a result of solar irradiance penetration into the solar cell, the local U (1) gauge invariant (QED), here, permits the addition of a mass term for gauge particle under $\varnothing' \rightarrow e^{i\alpha(x)}\varnothing$ to clarify a covariant derivative photon particle that has a unique rule of transformation of scalar that is being expressed as:

$$\begin{aligned} &\partial_\mu \rightarrow D_\mu = \partial_\mu = ieA_\mu && [\text{covariant derivatives}] \\ &A'_\mu = A_\mu + \frac{1}{e}\partial_\mu \alpha && [A_\mu \text{ derivatives}] \end{aligned} \tag{5.28}$$

Hence, the complex scalar field's local U(1) gauge-invariant *Lagrangian* can be expressed as follows:

$$\mathcal{L} = (D^\mu)^\dagger (D_\mu \varnothing) - \frac{1}{4} F_{\mu\nu}F^{\mu\nu} - V(\varnothing) \tag{5.29}$$

The expression $\frac{1}{4}F_{\mu\nu}F^{\mu\nu}$is the kinetic motion for the gauge field (photon), while $V(\varnothing)$ is the extra term in the *Lagrangian* considered as:

$$V(\varnothing^*\varnothing) = \mu^2(\varnothing^*\varnothing) + \lambda(\varnothing^*\varnothing)^2 \tag{5.30}$$

As a result, initiation of *Lagrangian* ($\mathcal{L}$) under perturbations has been clarified considering the quantum field of the massive scalar particles $\phi_1$and $\phi_2$in addition to mass μ under this condition, $\mu^2 < 0$ with each satisfied variable of $\phi_1^2 + \phi_2^2 = -\mu^2/\lambda = v^2$ plus the *Lagrangian* via the covariant derivatives of $\eta and \xi$ in order to define the quantum field as $\phi_0 = \frac{1}{\sqrt{2}}[(v + \eta) + i\xi]$ to confirm the kinetic term as follows:

$$\mathcal{L}_{\text{kin}}(\eta,\xi)=\frac{1}{2}\left(\partial_{\mu}\eta\right)^{2}-\lambda v^{2}\eta^{2}+\frac{1}{2}\left(\partial_{\mu}\xi\right)^{2}-\frac{1}{4}F_{\mu\nu}F^{\mu\nu}+\frac{1}{2}e^{2}v^{2}A_{\mu}^{2}$$
$$-evA_{\mu}(\partial^{\mu}\xi) \tag{5.31}$$

Here, massive η, mass less ξ(as earlier on) and equally a mass term for the quantum while $A_{\mu}$is fixed up to expression $\partial_{\mu}\alpha$ as evident activated quantum field. Overall, A_{μ} and ϕ change at the same time; therefore it can be redefined so that the photon particle spectrum can be incorporated within the quantum field via the following expression:

$$\begin{aligned}\mathcal{L}_{\text{scalar}}&=(D^{\mu}\phi)^{\dagger}(D^{\mu}\phi)-V\left(\phi^{\dagger}\phi\right)\\&=\left(\partial^{\mu}+ieA^{\mu}\right)\frac{1}{\sqrt{2}}(v+h)\left(\partial_{\mu}-ieA_{\mu}\right)\frac{1}{\sqrt{2}}(v+h)-V\left(\phi^{\dagger}\phi\right)\end{aligned} \tag{5.32}$$

$$=\frac{1}{2}\left(\partial_{\mu}h\right)^{2}+\frac{1}{2}e^{2}A_{\mu}^{2}(v+h)^{2}-\lambda v^{2}h^{2}-\lambda vh^{3}-\frac{1}{4}\lambda h^{4}+\frac{1}{4}\lambda h^{4} \tag{5.33}$$

Therefore, the term expansion in *Lagrangian* as linked to the scalar field indicates how Higgs-Boson quantum field is ready to start photon-photon connections with its quantum field [1, 27]. Simply, with the solar irradiance having the heated Higgs-Boson quantum field, therefore, computing for solar irradiance absorption per unit time $\frac{d\tau_{abs}}{dx}$ can be considered in determining the photon pair generation [19, 45, 50]. In this expression, τ_{abs} is indicative of the photonic energy's optical depth passing through electron spectrum n ($\in$) and n ($\in$) is indicative of photon generation per unit volume regarding deliverable energy per unit time [22, 23, 47]. This means the differential cone movements' isotropic distribution of the photons can be computed for by taking into consideration the angle θ and within θ and $\theta + d\theta$ is $\frac{1}{2}\sin\theta d\theta$. Therefore, the energy's $\in$ and angle's θ differential photon density is expressed as follows:

$$dn=\frac{1}{2}n\left(\in\right)\,\sin\theta d\in d\theta. \tag{5.34}$$

As a result, the high-energy photon's functional speeds are being computed taking into consideration the directional form of c (1 – cosθ), with the absorption per unit area computed as follows:

$$\frac{d\tau_{\text{abs}}}{dx}=\iint\frac{1}{2}\sigma n\left(\in\right)(1-\cos\theta)\,\sin\theta d\in d\theta. \tag{5.35}$$

By having the functions modified into an integration over s rather than θ by (34) and (35), then the following can be obtained:

$$\frac{d\tau_{\text{abs}}}{dx}=\pi r_{0}^{2}\left(\frac{m^{2}c^{4}}{E}\right)^{2}\int_{\frac{m^{2}c^{4}}{E}}^{\infty}\in^{-2}n(\in)\,\overline{\phi}\left[s_{0}\left(\in\right)\right]de, \tag{5.36}$$

where

$$\overline{\phi}\,[s_0\,(\in)] = \int_1^{s_0(\in)} s\overline{\sigma}\,(s)ds, \overline{\sigma}(s) = \frac{2\sigma(s)}{\pi r_0^2}. \tag{5.37}$$

The outcome has been considered a dimensional variable $\overline{\phi}$ as well as dimensionless cross section$\overline{\sigma}$. Computation for variable $\overline{\phi}[s_0]$ is being clarified based on a comprehensive solar cell for $1 < s_0 < 10$ in order to determine a reliable functional asymptotic computation for $\overline{\phi}$ by expressing as to consider the variable thermal state energy spectrum, y_1/D in the solar cell in relation to the various frequencies, f(Hz) of the quantum fields, $k_1\eta$ fundamental, higher, and peak modes (Fig. 5.3):

$$\begin{aligned}\overline{\phi}\,[s_0] = (\,y_1/D + f(\text{Hz}) + k_1\eta)\,\frac{1+\beta_0^2}{1-\beta_0^2}\ln\omega_0 - + \beta_0^2\,\ln\omega_0 - \ln^2\omega_0 - \frac{4\beta_0}{1-\beta_0^2}\\ + 2\beta_0 + 4\,\ln\omega_0\,\ln\,(\omega_0+1) - L(\omega_0) + x_1/D = 2) + x_1/D = 6)\\ + x_1/D = 10)\end{aligned} \tag{5.38}$$

where

$$\beta_0^2 = \frac{1-1}{s_0}, \omega_0 = \frac{(1+\beta_0)}{(1-\beta_0)} \text{ and } L(\omega_0) = \int_1^{\omega_0} \omega^{-1}\ln\,(\omega+1)d\omega.$$

Thus, the finally integral can be expressed as follows:

$$(\omega+1) = \omega\left(\frac{1+1}{\omega}\right), L(\omega_0) = \frac{1}{2}\ln^2\omega_0 + L'\,(\omega_0) \tag{5.39}$$

where

$$\begin{aligned} L'\,(\omega_0) &= \int_1^{\omega_0} \omega^{-1}\ln\,\left(1+\frac{1}{\omega}\right)d\omega,\\ &= \frac{\pi^2}{12} - \sum_{n=1}^{\infty}(-1)n^{-1}n^{-2}\omega_0^{-n}.\end{aligned}$$

Since the light absorption function for the spectrum to adhere to the features of the solar cell, thus, the photon spectra shall have a high-energy wave in consideration of the form's spectrum as:

$$n(\in) = D \in^{\beta},\ \in\ <\ \in_m, \qquad \beta \leq 0 \tag{5.40}$$

$$= 0,\ \in\ >\ \in_m. \tag{5.41}$$

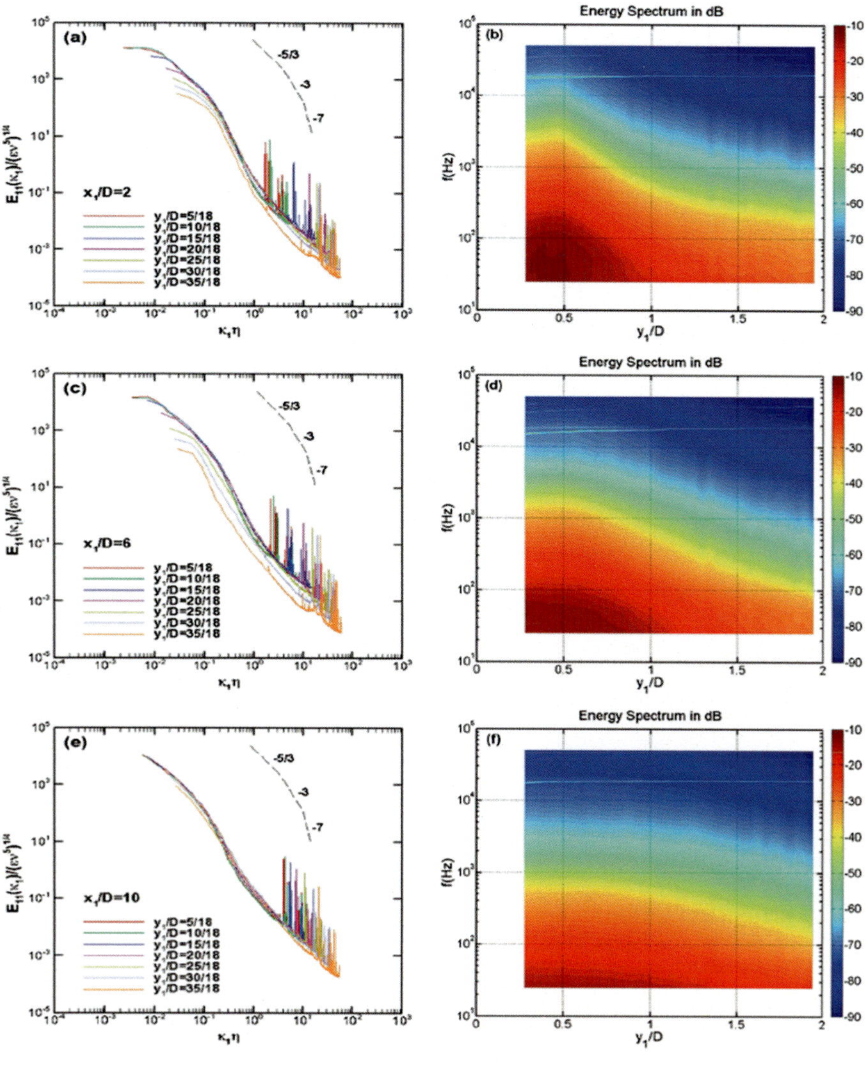

Fig. 5.3 (**a–b**) Graph relating to function $\overline{\phi}\,[s_0]$ indicates the energy generation through the various energy spectrum and frequencies being concurrently integrated into the quantum field's fundamental mode ($x_1/D = 2$), (**c–d**) higher mode ($x_1/D = 6$), and (**e–f**) peak mode (($x_1/D = 10$). The error bars signify plot measurements' standard error

For this spectrum, it can be determined as:

$$\frac{d\tau_{abs}}{dx} = \pi r_0^2 D\left(\frac{m^2c^4}{E}\right)^{1+\beta} \times \begin{cases} 0, & E < E_m \\ F_\beta(\sigma_m), E > E_m, \end{cases} \tag{5.42}$$

where

$$\sigma_m = \frac{E}{E_m} = \frac{\in_m E}{m^2c^4}, \tag{5.43}$$

$$F_\beta(\sigma_m) = \int_1^{\sigma m} s_0^{\beta-2}\, \overline{\phi}[s_0]ds_0. \tag{5.44}$$

Additionally, through Eqs. (5.43) and (5.44), it is possible to attain the asymptotic form photon agitation as to form pair photon:

$$\begin{aligned} &\beta = 0 : F_\beta(\sigma_m) \rightarrow A_\beta + \ln^2\sigma_m – 4\ \ln\sigma_m + \ldots, \\ &\beta \neq 0 : F_\beta(\sigma_m) \rightarrow A_\beta + 2\beta^{-1}\sigma_m^\beta(\ln 4\sigma_m – \beta^{-1} – 2) + \ldots, \sigma_m > 10 \end{aligned} \tag{5.45}$$

$$\text{all } \beta : F_\beta\ (\sigma_m) \rightarrow \left(\frac{4}{15}\right)(\sigma_m - 1)^{\frac{5}{2}} + \left[\frac{2(2\beta+1)}{21}\right](\sigma_m - 1)^{\frac{7}{2}} + \ldots, \sigma_m – 1 \ll 1 \tag{5.46}$$

Here, $\sigma_m^{-\beta}F_\beta(\sigma_m)$ is being taken into consideration pair terms equivalent to the spectra for both positive and negative indices:

$$n(\in) = 0,\ \in\ <\ \in_0 \tag{5.47}$$

$$= C_\in^{-\alpha} \text{ or } D_\in^\beta,\ \in_0 <\ \in\ <\ \in_m \tag{5.48}$$

$$= 0,\ \in\ >\ \in_m \tag{5.49}$$

followed by calculating,

$$\left(\frac{d\tau_{abs}}{dx}\right)_\alpha = \pi r_0^2 C\left(\frac{m^2c^4}{E}\right)^{1-\alpha} \times \begin{cases} 0 & , \quad E < E_m \\ [\mathrm{F}_\alpha(1) – \mathrm{F}_\alpha(\sigma_m)], & E_m < E < E_0 \\ [\mathrm{F}_\alpha(\sigma_0) – \mathrm{F}_\alpha(\sigma_m)], & E > E_0; \end{cases} \tag{5.50}$$

$$\left(\frac{d\tau_{abs}}{dx}\right)_\beta = \pi r_0^2 D\left(\frac{m^2c^4}{E}\right)^{1+\beta} \times \begin{cases} 0 & , \quad E < E_m \\ [\mathrm{F}_\beta(\sigma_\mathrm{m}) & , \quad E_m < E < E_0 \\ [\mathrm{F}_\beta(\sigma_\mathrm{m}) – \mathrm{F}_\beta(\sigma_0)], & E > E_0. \end{cases} \tag{5.51}$$

In this case, the asymptotic analysis of photon spectrum, and by further clarifying, Γ_{γ}^{LPM}confirm the contribution of the photon to emitting irradiance at a rate of per unit volume in accordance with bremsstrahlung as well as inelastic pair photon production technique as:

$$\Gamma_{\gamma} \equiv \frac{dn_{\gamma}}{dVdt} \tag{5.52}$$

The input of Γ_{γ}^{LPM} is thus added into the photon generation technique to validate the rate of O(α_{EM} αs). The assumption considered here is that the findings from the polarized rate of emission Γ_{γ}^{LPM} within the thermodynamically regulated plasma surface's equilibrium at temperature T plus the photophysical reaction is μ; therefore, it can be expressed in detail as follows:

$$\frac{d\Gamma_{\gamma}^{LPM}}{d^3k} = \frac{d_F q_s^2 \alpha_{EM}}{4\pi^2 k} \int_{-\infty}^{\infty} \frac{dp_{\parallel}}{2\pi} \int \frac{d^2 \mathrm{p}_{\perp}}{(2\pi)^2} A\left(p_{\parallel}, k\right) \mathrm{Re}\left\{2P_{\perp} \cdot \mathrm{f}\left(p_{\perp}; p_{\parallel}, k\right)\right\} \tag{5.53}$$

with d_{F} being the functional particle [N_c in SU(N_c)] and q_s being the Abelian charge of a particle, k ≡ |k|, and the kinetic function $A(p\|, k)$ of the emitted particle

$$A\left(p_{\parallel}, k\right) \equiv \begin{cases} \dfrac{n_b\left(k+p_{\parallel}\right)\left[1+n_b\left(p_{\parallel}\right)\right]}{2p_{\parallel}\left(p_{\parallel}+k\right)}, & \text{scalars} \\ \dfrac{n_f\left(k+p_{\parallel}\right)\left[1-n_f\left(p_{\parallel}\right)\right]}{2\left[p_{\parallel}\left(p_{\parallel}+k\right)\right]^2}\left[{p_{\parallel}}^2+\left(p_{\parallel}+k\right)^2\right], & \text{fermions} \end{cases} \tag{5.54}$$

with

$$n_b(p) \equiv \frac{1}{\exp[\beta(p-\mu)]-1}, \quad n_f(p) \equiv \frac{1}{\exp[\beta(p-\mu)]+1} \tag{5.55}$$

With calculation $f(p\perp;\, p\|,\, k)$ and having been incorporated into Eq. (5.55), the solution to the subsequent linear integral equation will validate the pair photon production as:

$$2p_{\perp} = i\delta E\, f\left(p_{\perp}; p_{\parallel}, k\right) + \frac{\pi}{2} C_F g_s^2 m_D^2 \int \frac{d^2 q_{\perp}}{(2\pi)^2} \frac{dq_{\parallel}}{2\pi} \frac{dq^0}{2\pi} 2\pi\delta\left(q^0 - q_{\parallel}\right) \times \frac{T}{|\mathrm{q}|}$$
$$\times \left[\frac{2}{|\mathrm{q}^2 - \Pi_{\mathrm{L}}(\mathrm{Q})|^2} + \frac{\left[1-\left(q^0/\left|q_{\parallel}\right|\right)^2\right]^2}{\left|(q^0)^2 - q^2 - \Pi_T(\mathrm{Q})\right|^2}\right] \times \left[f\left(p_{\perp}; p_{\parallel}, k\right) - f\left(q + p_{\perp}; p_{\parallel}, k\right)\right] \tag{5.56}$$

where C_F is a quadratic particle [$C_F = (N_c^{\,2}-1)/2\mathrm{Nc} = 4/3$ in QCD], m_D is the leading-order Debye mass, while δE is the variation in quasi-particle energies, which has equally been computed taking into consideration the photon emission and the condition of thermodynamic temperature equilibrium as:

$$\delta E \equiv k^0 + E_{\mathrm{p}}\mathrm{sign}\left(p_{\|}\right) - E_{\mathrm{p+k}}\mathrm{sign}\left(p_{\|} + k\right) \tag{5.57}$$

By considering the SU(N) gauge theory having N_f Dirac fermions and N_s complex scalars in the basic representation, the following equation gives the

$$m_D^2 = \frac{1}{6}\left(2N + N_s + N_f\right)g^2T^2 + \frac{N_f}{2\pi^2}g^2\mu^2 \tag{5.58}$$

During the final integration (5.58), so as to validate the precision of energy's emission rate in the region $p\| > 0$, and thus, it has been determined the distribution of $n(k+p\|)[1\pm n(p\|)]$ comprising of $A(p\|, k)$, and this confirms the generation of pair photons by applying the subsequent equation:

$$n_b(-p) = -\left[1 + \overline{n}_b(p)\right], \qquad n_f(-p) = \left[1 - \overline{n}_f(p)\right], \tag{5.59}$$

where $n(p) \equiv 1/[e\beta(p+\mu) \mp 1]$ is the suitable anti-particle distribution function; in this interval, the factor $A(p", k)$, can be restated as follows:

$$A\left(p_{\|}, k\right) \equiv \begin{cases} \dfrac{n_b\left(k - \left|p_{\|}\right|\right)\overline{n}_b\left(\left|p_{\|}\right|\right)}{2\left|p_{\|}\right|\left(k - \left|p_{\|}\right|\right)}, & \text{scalars} \\ \dfrac{n_f\left(k - \left|p_{\|}\right|\right)\overline{n}_f\left(\left|p_{\|}\right|\right)}{2\left[\left|p_{\|}\right|\left(k - \left|p_{\|}\right|\right)\right]^2}\left[p_{\|}^{\,2} + \left(k - \left|p_{\|}\right|\right)^2\right], & \text{fermions} \end{cases} \tag{5.60}$$

Therefore, the overt energy Ep form of a photon particle with momentum $|p|$ is expressed as follows:

$$E_{\mathrm{p}} = \sqrt{p^2 + m_\infty^2} \simeq |p| + \frac{m_\infty^2}{2|\mathrm{p}|} \simeq \left|p_{\|}\right| + \frac{p_\perp^2 + m_\infty^2}{2\left|p_{\|}\right|} \tag{5.61}$$

with the asymptotic thermal "mass" being

$$m_\infty^2 = \frac{C_f g^2 T^2}{4} \tag{5.62}$$

Replacing the overt form of Ep into the definition (62), the following equation is arrived at:

$$\delta E = \left[\frac{\mathrm{p}_\perp^2 + m_\infty^2}{2}\right]\left[\frac{k}{p_\parallel\left(k + p_\parallel\right)}\right] \tag{5.63}$$

Lastly, for the result's simplification purposes, it has been considered to compute for the particle's scalar quantities, which is actually the photodynamic of photon momenta $k = O\ (T)$ with $|K^2| \leq g^2T^2$. Therefore, utilizing this finding of multiple photon generation from the single one, the photon momentum's emission rate has been calculated at $|k| \gg g^4T$ scattering rate in g^{-1} per unit volume to leading order in e^2, illustrated as follows:

$$\sigma = W_{\mu\nu}(K) = \int d^4xe^{-iKx}\langle j_\mu(x)j_\nu(0)\rangle = (\gamma + \gamma)^2 \tag{5.64}$$

with K being the photon 4-momenta with 1-momenta $\mathbf{k}$ and positive energy $k^0 = |k| \equiv k$, while $W_{\mu\nu}\ (K)= \sigma=(\gamma\ +\ \gamma)^2$ is the generation of pair photons in the two-semiconductor parameter$(2\ +\ 2)^2)$ to confirm net production of a total of 16 photons.

Three-Diode Photon Scattering

To mathematically compute for the thousands of photon generation from 16 photons in the quantum field in the three-diode photon scattering superconductor, the dynamic photon proliferation flux density (kT), considering the variable the unit area $J(\omega)$ as well as the excited quantum field has been calculated (Fig. 5.4). Subsequently, the computation of collision state of the photon-photon collision in the solar cell considering the anisotropic DOS:$\varrho_{PC}(\omega) \propto \frac{1}{\sqrt{\omega - \omega_e}}\Theta(\omega - \omega_e)$ has been clarified based on the vector of electromagnetic field of the solar cell estimated to be $\varrho_{PC}(\omega) \propto -[\ln|(\omega - \omega_0)/\omega_0| - 1]\Theta(\omega - \omega_e)$, with ω_e indicating the peak logarithm's central point with the photon generation $J(\omega)$ as:

$$J(\omega) = \varrho(\omega)|V(\omega)|^2 \tag{5.65}$$

Here, proliferative photon dynamics have been considered by using the function $u(t, t_0)$ for photon structure as follows:

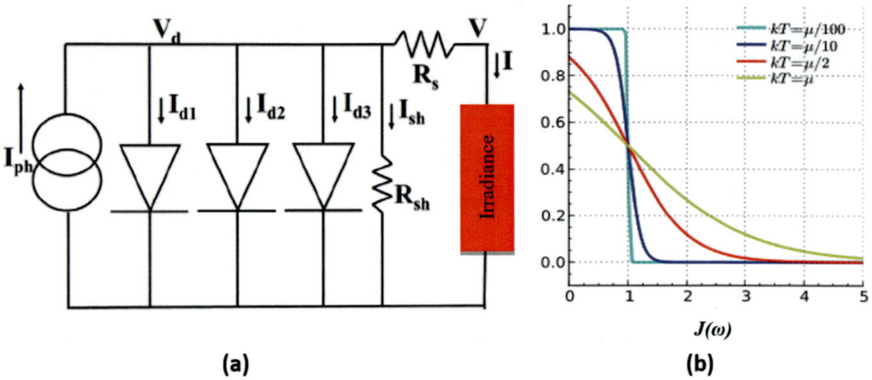

Fig. 5.4 (**a**) Photon scattering mechanism involving three-diode circuit. (**b**) The photon proliferation flux density parameter (kilo tesla-kT) at various colors where red color is related to I_{d2} diode, deep blue is related to I_{d1} diode, while pale green is related to I_{d3} diode, and the pale green is the standard line for all three diodes

$$u(t, t_0) = \frac{1}{1 - \Sigma'(\omega_b)} e^{-i\omega(t - t_0)} + \int_{\omega_e}^{\infty} d\omega \frac{J(\omega) e^{-i\omega(t - t_0)}}{[\omega - \omega_c - \Delta(\omega)]^2 + \pi^2 J^2(\omega)} \tag{5.66}$$

Here, the comprehensive photonic dynamics, taking into consideration the proliferation level, has been computed for $|u(t, t_0)|$ for the quantum field based on various detuning δ in relation to the condition $|u(t \to \infty, t_0)| \geq 0$ of the solar cell [20, 28, 29]. Thus, the dynamic photon generation proliferation validates the ionic photonic states counts in the solar cell with the photons of a non-equilibrium state where the quantum field's photon proliferation dynamics is being taken into consideration for the thermal variation of the photon correlation function $v(t, t)$ in order to estimate the photon scattering as:

$$v(t, t) = \int_{t_0}^{t} dt_1 \int_{t_0}^{t} dt_2 u^*(t_1, t_0) \tilde{g}(t_1, t_2) u(t_2, t_0) \tag{5.68}$$

Hence, simplifying the equation, revealing that the photonic dynamic deviations induced by the Higgs-Boson quantum field of $\overline{n}(\omega, T) = 1/\left[e^{\hbar\omega/k_B T} - 1\right]$ where photon generation proliferation in the solar cell at temperature T can be expressed as

$$v(t, t \to \infty) = \int_{\omega_e}^{\infty} d\omega \mathcal{V}(\omega)$$

with

$$\mathcal{V}(\omega)=\bar{n}(\omega,T)[\mathcal{D}_l(\omega)+\mathcal{D}_d(\omega)] \tag{5.69}$$

Here, the equation confirms the estimate of the non-equilibrium condition: $\mathcal{V}(\omega)=\bar{n}(\omega,T)\mathcal{D}_d(\omega)$ which is linking to the photonic structures of $n(t) = \langle a^{\dagger}(t)\, a(t)\rangle = |u(t,t_0)|^2 n(t_0)v(t,t)$, with $n(t_0)$, the primary state photon number n_0, *i. e.* , $\rho(t_0) = |n_0\rangle\langle n_0|$ via the real-time quantum, taking into consideration the state of photon generation at time t as the following equation:

$$\rho(t)=\sum_{n=0}^{\infty}\mathcal{P}_n^{(n_0)}(t)|n_0\rangle\langle n_0| \tag{5.70}$$

$$\mathcal{P}_n^{(n_0)}(t)=\frac{[v(t,t)]^n}{[1+v(t,t)]^{n+1}}[1-\Omega(\mathrm{t})]^{n_0}\mathrm{x}\sum_{k=0}^{\min\{n_0,n\}}\binom{n_0}{k}\binom{n}{k}\left[\frac{1}{v(t,t)}\frac{\Omega(\mathrm{t})}{1-\Omega(\mathrm{t})}\right]^k \tag{5.71}$$

Hence, $\Omega(t)=\frac{|u(t,t_0)|^2}{1+v(t,t)}$ is the result of the indicative photon dynamics that are being transformed into magnitudal dynamic of states $|n_0\rangle is\ \mathcal{P}_n^{(n_0)}(t)$ [21, 30, 42]. Subsequently, the photon dissipation proliferation, $\mathcal{P}_n^{(n_0)}(t)$, in the initial state $|n_0=5\rangle$ and steady-state limit, $\mathcal{P}_n^{(n_0)}(t\to\infty)$, has therefore been eventually attained at the thermal non-equilibrium state of

$$\mathcal{P}_n^{(n_0)}(t\to\infty)=\frac{[\bar{n}(\omega_c,T)]^n}{[1+\bar{n}(\omega_c,T)]^{n+1}} \tag{5.72}$$

This is the confirmation of examining the large amount of photon generation, and thus, the photon distribution within the quantum field is being finally calculated as:

$$\langle m|\rho(t)|n\rangle=J(\omega)=e^{-\Omega(t)|\alpha_0|^2}\frac{[\alpha(t)]^m[\alpha^*(t)]^n}{[1+v(t,t)]^{m+n+1}}=\sum_{k=0}^{\min\{m,n\}} \times\frac{\sqrt{m!n!}}{(m-k)!(n-k)!k!}\left[\frac{v(t,t)}{\Omega(t)|\alpha_0|^2}\right]^k \tag{5.75}$$

Here, the generation of total photons in the quantum area ($\langle m|\,\rho(t)|\,n\rangle$) defines the great relativistic thermal state $[1 + v(t,t)]^{m\,+\,n\,+\,1}$ and non-equilibrium condition $[\alpha(t)]^m[\alpha^*(t)]^n$ of the photons, which counts the net number as $J(\omega)=\sum_{k=0}^{\min\{m,n\}}\times$ $\frac{\sqrt{m!n!}}{(m-k)!(n-k)!k!}\left[\frac{v(t,t)}{\Omega(t)|\alpha_0|^2}\right]^k=\left[e*n(\gamma+\gamma)^2\right]^3=\left[1*1(2+2)^2\right]^3$ that is equal to 4,092 photons.

Electricity Transformation

To implement these thousands of photons into electricity energy, this solar cell is being connected both in series and parallel using a photon energy (current), two resistors, and diodes and expressed by the I-V relationship by using the following equation:

$$I = IL - IO\left\{ \exp\left[\frac{q(V+I_{RS})}{AkTc}\right] - 1\right\} - \frac{(V + I_{RS})}{R_{Sh}} \tag{5.76}$$

I_L indicates the photon-generating current, IO indicates the diode's saturated current, R_s indicates resistance in a series, A indicates the passive function of the diode, k (= 1.38×10^{-23} W/m^{2}K) indicates Boltzmann's constant, q (=1.6×10^{-19}C) indicates an electron's charge amplitude, while T_C indicates the functional cell temperature [32, 44, 49]. Consequently, the PV cells' I-q relationship fluctuates based on the saturation current and/or diode current, which can be denoted as follows:

$$I_O = I_{RS}\left(\frac{T_C}{T_{\text{ref}}}\right)^3 \exp\left[\frac{qEG\left(\frac{1}{T_{\text{ref}}} - \frac{1}{T_C}\right)}{KA}\right] \tag{5.77}$$

Here, I_{RS} indicates the saturation current in view of the solar irradiance and functional temperature, while qEG indicates the band-gap energy into the silicon solar cell in view of the normal, normalized, and perfect modes (Fig. 5.5).

Taking these solar cell modules, the I-V equation, apart from the I-V curve, there is a concurrence of I-V curves for all of (a) normal, (b) normalized, and (c) perfect modes, which means that the equation can be expressed in the following way in order to establish the V-R relationship:

$$V = -IR_s + K\log\left[\frac{I_L - I + I_O}{I_O}\right] \tag{5.78}$$

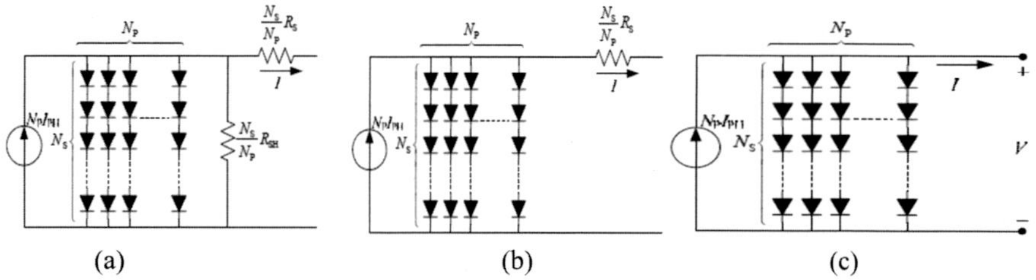

Fig. 5.5 Equivalent solar cell connected to the module of circuit of (**a**) normal, (**b**) normalized, and (**c**) perfect modes

with K being a constant $\left(=\frac{\text{AkT}}{q}\right)$, while I_{mo} and V_{mo} being the PV panel's current and voltage. As a result, the connection between I_{mo} and V_{mo} remains identical to the solar cell I-V correlation:

$$V_{\text{mo}} = -I_{\text{mo}} R_{S\text{mo}} + K_{\text{mo}} \log\left(\frac{I_{L\text{mo}} - I_{\text{mo}} + I_{O\text{mo}}}{I_{O\text{mo}}}\right) \tag{5.79}$$

with $I_{L\text{mo}}$ being the photon-generated current, $I_{O\text{mo}}$ being the diode's saturated current, $R_{s\text{mo}}$ being the resistance in series, and K_{mo} being the factorial constant. After interconnecting all non-series cells in series, the subsequent step involved measuring and series resistance as the summation of the series resistance of each cell, $R_{s\text{mo}} = N_S \times R_S$; hence the constant factor can be expressed as $K_{\text{mo}} = N_S \times K$ [35, 36, 48]. Given a specific level of current flowing into cells connected in series, the current flow is to be similar in each component, i.e., $I_{o\text{mo}} = I_o$ and $I_{L\text{mo}} = I_L$ [41, 47, 48]. Therefore, the following expression represents the connected cells' module I_{mo}-V_{mo} equation for the N_S series:

$$V_{\text{mo}} = -I_{\text{mo}} N_S R_S + N_S K \log\left(\frac{I_L - I_{mo} + I_o}{I_o}\right) \tag{5.80}$$

In the same way, the current–voltage computations may be modified for the parallel connection after connecting all N_P cells in parallel and may be expressed in the following manner:

$$V_{\text{mo}} = -I_{\text{mo}} \frac{Rs}{Np} + K \log\left(\frac{N_{sh} I_L - I_{\text{mo}} + N_p I_o}{N_p I_o}\right) \tag{5.81}$$

Since the photon-derived current will initially rely on the solar cell's relative temperature conditions and solar irradiance, the following expression can be considered in calculating the current:

$$I_L = G\left[I_{SC} + K_I\left(T_C - T_{\text{ref}}\right)\right] * V_{\text{mo}} \tag{5.82}$$

where I_{sc} represents PV current at 25°C and KW/m^2, K_I represents the relativistic PV panel coefficient factor, T_{ref} represents the PV panel's functional temperature, and G represents the solar energy in mW/m^2 [39, 46, 49].

Finally, the total photon production of photons from a single one is calculated as $J(\omega) = [e * n(\gamma + \gamma)^2]^3 = [1 * 1(2 + 2)^2]^3$, with a sum of 4,092 photons. Since the each photon quantum is a certain type of polarization of the frequency range of ν_r to $\nu_r + d\nu_r$, the solar energy is calculated as 6,728.8 eV [1.4 eV × 4,092] where one photon contains 1.4 eV equivalent to 27.77 mW/ m^2/h [33, 46]). Therefore, the total energy would be (1.4 eV = 27.77 mW/m^2 × 4,092) =159,088.78 mW/m^2 · eV per hour [14, 34, 42]. Simply, usage of only 1cm^2 Higgs-Boson quantum field-assisted solar cell will produce 1,590.9 mW/h, which is a tremendous source of energy that is benign to the environment.

Conclusions

Designing a high-performance silicon solar cell to transform single photon into thousands of photons indeed would be a new field of science to meet the total energy demand in the world. Simply, it is a just a photophysical reaction mechanism remodeling by implementing Higgs-Boson quantum field under high temperature to create pair photon e^+e^-, which is further photo-dissociated by two surfaced semiconductors and then further scattered by three-diode superconductor to create a total of 4,092 photons from a single photon. The calculation confirms that only a mere of 1 cm^2 solar cell can produce 1,590.9 MW/h, which indeed would be a cutting-edge science to employ the photon energy much more efficiently to meet the global energy demand dramatically.

Acknowledgments I sincerely thank Professor MA Karim for plagiarism checking for this paper by *iThenticate* software. This research was supported by Green Globe Technology, Inc. under the grant of RD-2022-04 for building a better environment for the planet mother Earth. Any findings, predictions, and conclusions described in this article are solely those of the authors, who confirm that the article has no conflicts of interest for publication in a suitable journal.

References

1. Ruiz, A. (2014). Partial recovery of a potential from backscattering data, Communications in Partial Differential Equations. *Springer Tracts in Modern Physics.*
2. Alammari, and A. Iqbal. (2016). Modelling and simulation of single- and triple-junction solar cells using MATLAB/SIMULINK. *International Journal of Ambient Energy.*
3. Reinhard, A. (2011). Strongly correlated photons on a chip. *Nature Photonics.*
4. Armani, D. K., Kippenberg, T. J., Spillane, S. M., & Vahala, K. J. (2003). Ultra-high-Q toroid microcavity on a chip. *Nature, 421*, 925.
5. Birnbaum, K. M., et al. (2005). Photon blockade in an optical cavity with one trapped atom. *Nature, 436*, 87–90.
6. Busch, K., von Freymann, G., Linden, S., Mingaleev, S. F., Tkeshelashvili, L., & Wegener, M. (2007). Periodic nanostructures for photonics. *Phys. Rep., 444*, 101.
7. Chang, D. E., Sørensen, A. S., Demler, E. A., & Lukin, M. D. (2007). A single-photon transistor using nanoscale surface plasmons. *Nature Physics., 3*, 807–812.
8. Chen, J., Wang, C., Zhang, R., & Xiao, J. (2012). Multiple plasmon-induced transparencies in coupled-resonator systems. *Opt. Lett., 37*, 5133–5135.
9. Cheng, M. T., & Song, Y. Y. (2012). Fano resonance analysis in a pair of semiconductor quantum dots coupling to a metal nanowire. *Opt. Lett., 37*, 978–980.
10. Dayan, B., et al. (2008). A photon turnstile dynamically regulated by one atom. *Science, 319*, 1062–1065.
11. Douglas, J. S., Habibian, H., Hung, C.-L., Gorshkov, A. V., Kimble, H. J., & Chang, D. E. (2015). Quantum many-body models with cold atoms coupled to photonic crystals. *Nature Photonics.*
12. Englund, D., et al. (2010). Resonant excitation of a quantum dot strongly coupled to a photonic crystal nanocavity. *Phys. Rev. Lett., 104*, 073904.
13. Baur, G., Hencken, K., & Trautmann, D. (2007). Revisiting unitarity corrections for electromagnetic processes in collisions of relativistic nuclei. *Phys. Rep., 453*, 1.

14. Gleyzes, S., et al. (2007). Quantum jumps of light recording the birth and death of a photon in a cavity. *Nature, 446*, 297.
15. Guerlin, C., et al. (2007). Progressive field-state collapse and quantum non-demolition photon counting. *Nature, 448*, 889.
16. Gupta, N., Singh, S. P., Dubey, S. P., & Palwalia, D. K. (2011). Fuzzy logic controlled three-phase three-wired shunt active power filter for power quality improvement. *International Review of Electrical Engineering (IREE), 6*(3), 1118–1129.
17. Han, Z., & Bozhevolnyi, S. I. (2011). Plasmon-induced transparency with detuned ultracompact Fabry-Pérot resonators in integrated plasmonic devices. *Opt. Express, 19*, 3251–3257.
18. Hossain, M. F. (2017). Green science: independent building technology to mitigate energy, environment, and climate change. *Renewable and Sustainable Energy Reviews*. https://doi.org/10.1016/j.rser.2017.01.136 (Elsevier).
19. Hossain, M. F. (2017). Design and construction of ultra-relativistic collision PV panel and its application into building sector to mitigate total energy demand. *Journal of Building Engineering*. https://doi.org/10.1016/j.jobe.2016.12.005 (Elsevier).
20. Huang, J. F., Shi, T., Sun, C. P., & Nori, F. (2013). Controlling single-photon transport in waveguides with finite cross section. *Physical Review A, 88*, 013836.
21. Huang, Y., Min, C., & Veronis, G. (2011). Subwavelength slow-light waveguides based on a plasmonic analogue of electromagnetically induced transparency. *Applied Physics Letter, 99*, 143117.
22. Eichler, J., & Sthölker, T. (2007). Radiative electron capture in relativistic ion-atom collisions and the photoelectric effect in hydrogen-like high-Z systems. *Physics Report, 439*, 1.
23. Hencken, K. (2006). Transverse momentum distribution of vector mesons produced in ultraperipheral relativistic heavy ion collisions. *Physical Review Letters*.
24. Kofman, A. G., Kurizki, G., & Sherman, B. (1994). Spontaneous and Induced Atomic Decay in Photonic Band Structures. *J. Mod. Opt., 41*, 353.
25. Kolchin, P., Oulton, R. F., & Zhang, X. (2011). Nonlinear quantum optics in a waveguide: Distinct single photons strongly interacting at the single atom level. *Phys. Rev. Lett., 106*, 113601.
26. Lang, C., et al. (2011). Observation of resonant photon blockade at microwave frequencies using correlation function measurements. *Phys. Rev. Lett., 106*, 243601.
27. Liao, J. Q., & Law, C. K. (2010). Correlated two-photon transport in a one-dimensional waveguide side-coupled to a nonlinear cavity. *Phys. Rev. A, 82*, 053836.
28. Li, Q., Xu, D. Z., Cai, C. Y., & Sun, C. P. (2013). Recoil effects of a motional scatterer on single-photon scattering in one dimension. *Scientific Reports*.
29. Liao, J. Q., & Law, C. K. (2013). Correlated two-photon scattering in cavity optomechanics. *Phys. Rev. A, 87*, 043809.
30. Lei, C. U., & Zhang, W. M. (2012). A quantum photonic dissipative transport theory. *Ann. Phys., 327*, 1408.
31. Yang, L., Wang, S., Zeng, Q., Zhang, Z., Pei, T., Li, Y., & Peng, L.-M. (2011). Efficient photovoltage multiplication in carbon nanotubes. *Nature Photonics*, 672–676.
32. Lo, Ping-Yuan, Heng-Na Xiong, and Wei-Min Zhang. "Breakdown of Bose-Einstein Distribution in Photonic Crystals.", Scientific Reports, 2015.
33. Longo, P., Schmitteckert, P., & Busch, K. (2011). Few-photon transport in low-dimensional systems. *Phys. Rev. A, 83*, 063828.
34. Lü, Xin-You, Wei-Min Zhang, Sahel Ashhab, Ying Wu, and Franco Nori. "Quantum-criticality-induced strong Kerr nonlinearities in optomechanical systems.", Scientific Reports, 2013.
35. Najjari, A. B., Voitkiv, A., & Artemyev, A. S. (2009). Simultaneous electron capture and bound-free pair production in relativistic collisions of heavy nuclei with atoms. *Phys. Rev. A, 80*, 012701.
36. Artemyev, N., Jentschura, U. D., Serbo, V. G., & Surzhykov, A. (2012). Strong Electromagnetic Field EFFECTS in Ultra-Relativistic Heavy-Ion Collisions Eur. *Phys. J. C, 72*, 1935.

37. O'Shea, D., Junge, C., Volz, J., & Rauschenbeutel, A. (2013). Fiber-optical switch controlled by a single atom. *Phys. Rev. Lett., 111*, 193601.
38. Klein, S. A. (1975). Calculation of flat-plate collector loss coefficients. *Solar Energy, 17*, 79–80.
39. Valluri, S. R., Becker, U., Grün, N., & Scheid, W. (1984). Relativistic Collisions of Highly-Charged Ions. *Journal of Physics B: Atomic, Molecular Physics., 17*, 4359.
40. Gould, R. J. (1967). Pair production in photon-photon collisions. *Physical Review*.
41. Roy, D. (2013). Two-photon scattering of a tightly focused weak light beam from a small atomic ensemble: An optical probe to detect atomic level structures. *Phys. Rev. A, 87*, 063819.
42. Sayrin, C., et al. (2011). Real-time quantum feedback prepares and stabilizes photon number states. *Nature, 477*, 73.
43. Shen, J. T., & Fan, S. (2007). Strongly correlated two-photon transport in a one-dimensional waveguide coupled to a two-level system. *Phys. Rev. Lett., 98*, 153003.
44. Shi, T., Fan, S., & Sun, C. P. (2011). Two-photon transport in a waveguide coupled to a cavity in a two-level system. *Phys. Rev. A, 84*, 063803.
45. Tame, M. S., McEnery, K. R., Özdemir, Ş. K., Lee, J., Maier, S. A., & Kim, M. S. (2013). Quantum plasmonics. *Nature Physics*.
46. Tu, M. W. Y., & Zhang, W. M. (2008). Non-Markovian decoherence theory for a double-dot charge qubit. *Physics Review B, 78*, 235311.
47. Xiao, Y. F., et al. (2010). Asymmetric Fano resonance analysis in indirectly coupled microresonators. *Phys. Rev. A, 82*, 065804.
48. Yan, Wei-Bin, Jin-Feng Huang, and Heng Fan. "Tunable single-photon frequency conversion in a Sagnac interferometer.", Scientific Reports, 2013.
49. Yan, Wei-Bin, and Heng Fan. "Single-photon quantum router with multiple output ports.", Scientific Reports, 2014.
50. Zhang, W. M., Lo, P. Y., Xiong, H. N., Tu, M. W. Y., & Nori, F. (2012). General Non-Markovian Dynamics of Open Quantum Systems. *Phys. Rev. Lett., 109*, 170402.

Chapter 6
Splitting Photon into Pair Photons to Design a High-Performance Printable Solar Panel

Introduction

Radiant energy from the sun has always been intriguing, but the conversion rate of this renewable energy through the solar cell has never been satisfactory to implement this power commercially to satisfy the global energy needs (3, 6, 120). Although several excellent works on perovskite solar cells (PSC) that developed into a promising branch of renewable energy conversion have been conducted to increase up to 22.1% efficiency during the last several years, stabilizing them under the operating condition without compromising optimal optoelectronic properties remains challenging, and thus, this achievement is not satisfactory enough to trap solar energy massively for commercial use throughout the world [1]. Thus, it has presented an insurmountable challenge to date for commercially used solar energy globally with an attractive efficiency rate [2, 3]. Interestingly, recent studies revealed that pair-photon formation from a single one is doable via photon-photon collision of a single photon once implemented by high thermal conditions but difficult to demonstrate the potential applications of this energy commercially [4–6]. However, these extraordinary works show us a new path to understanding the possible inconsistency of proton-electron quantum physics deeply that we had barriers in the past several years on how to break down the photon in pair photons successfully for the use of commercial applications [7, 8]. Therefore, in this research, an approach to form pair photons by breaking down the photon-photon collisions of its photo-electron by the induction of gamma ray inside the printable solar cell has been conducted to accelerate the solar energy harvesting rate two times more than the currently available maximum of 22.1% rate to use this energy commercially throughout the world [9, 10]. Therefore, the development of this high-performance printable solar cell, the strategy involves printing base solar cell would enable using solution-based photopolymer coating methods to pave to release of gammy rays assist to break the photon particles to form pair photons via photon-photon collision and release energy with the efficiency of 44.2% [2, 11]. Simply, finding this new

M. F. Hossain, *Global Sustainability*, https://doi.org/10.1007/978-3-031-34575-3_6

solar energy efficiency rate, the deluge of the energy pouring down on us from the sun would be quite outstanding research in the photophysical field to transform solar energy much more efficiently that has ever been thought to meet the global energy crisis dramatically [12, 13].

Materials and Methods

In this research, firstly, the printable solar cell is coated with a thin photopolymer to allow the cell to absorb elevated temperature from the sun to pave to release gammy ray inside the silicon solar cell due to the high-temperature condition [14, 15]. Subsequently, the release of gamma ray inside the solar cell is being clarified using *MATLAB* software to facilitate to conduct the intense photophysical reactions to split the photon into pair photons. Then, a model of a high-performance printable solar cell is designed considering the application of this released pair-photon energy by conducting the following steps.

Simply, in the first step, a fully cross-linked photopolymer (*polymer, chloro-trifluoro-ethylene vinyl ether fluoropolymer binder and dimethacrylic perfluoropolyether oligomer*) coating is used on the thin film of the solar cell to absorb elevated temperature inside the cells by the process of intense absorption of luminescent of the sunlight [16, 17]. Since the solar silicon cell is being covered by a fully cross-linked coating photopolymer, it will absorb high-intensity luminophore (i.e., *high absorption coefficient and fluorescence quantum yield*) due to the aggregation-induced quenching of the solar irradiance by creating charge career dynamics of photons and release gammy ray [18, 19] (Fig. 6.1).

Naturally, the charge carrier dynamics of these photons are being clarified by investigating the emission features using steady-state and time-resolved photoluminescence (PL) spectroscopy corresponding to band-to-band recombination photonic irradiance, which is being connected to a *Lumiso Pooltest 6 Photometer (LP6)*.

Then in the second step, *splitting a single photon into pair-photon* production is clarified by introducing *SENTINEL™ 880 Series Gamma-Ray* device to determine

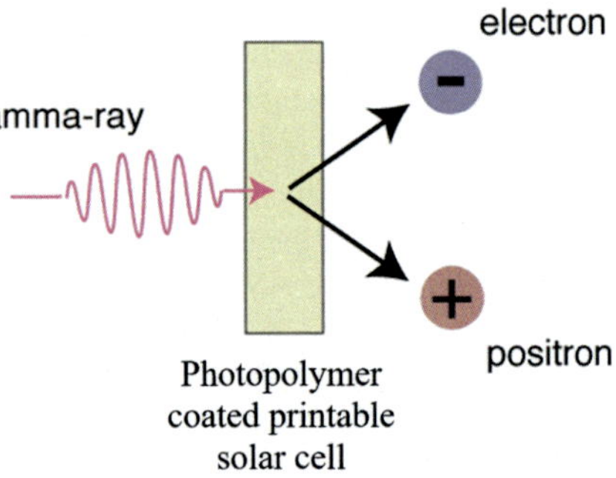

Fig. 6.1 A conceptual diagram of the process of electron-positron pair-photon production by the induction of gamma ray (λ) to develop a high-performance printable solar cell

gamma-ray presence within the solar cell [20, 21]. Here, the *SENTINEL™ 880 Series Gamma-Ray* device is also being connected to the *PerkinElmer® Lambda 25 UV/Vis Spectrometer* to determine the pair-photon production inside the solar cells. Since the preservation of solar irradiance and momentum are the primary l constraints on the action of the breakdown of the photons, all preserved quantum numbers are being kept at the sum to zero by controlling *SENTINEL™ 880 Series Gamma-Ray* device connecting to the *MATLAB* programming software [1, 22]. Subsequently, the production of pair photons due to the photon-photon interactions is being clarified in correlation to the photon energy emission in the solar cell considering its photonic dynamics (*photon to electron and positron, basic kinematics, and energy transfer*) to determine the required wavelength (λ) and frequencies (f) to release high-efficiency solar energy.

Finally, a design of a high-efficiency printable solar cell has been modeled considering the formation of pair photons from a single photon in the silicon solar cell by implementing the required photonic wavelength (λ) and frequency (f), which can release solar energy two times more respecting currently available of maximum solar energy efficiently rate 22.1%.

Results and Discussion

The *monitoring of photon to electron and positron* has been conducted upon the connection of SENTINEL™ 880 Series Gamma-Ray on the printable silicon solar cell to confirm the identification of the breakdown of the photon to release photonic energy due to the gamma-ray induction into the solar cell [9, 11]. Subsequently, this photonic energy release state has been monitored to determine the clarification of this pair-photon production mode due to the photon-photon collision when the photons are getting near to their atomic nucleus to form electron-positron pair. Subsequently, the pair-photon energy is being measured as the converted particle mass considering Einstein's equation, $E = m\,c^2$, where E represents energy, m represents mass, and c represents the speed of light. Subsequently, this pair photon is having much higher energy than the sum of the mass energies of an electron and positron of a single photon; therefore, the electron-positron pair photons are being clarified within the solar cell by the following basic kinematics of the pair photon due to the photon-photon interaction [15, 23].

Here, the *monitoring basic kinematics* of the interaction of the photon-photon is being monitored by using four vector notations to determine the preservation of energy-momentum before and after the interaction occurs within the photonic particles to characterize the formation of two electrons where one is negative and the other is positive (positron), which is originated through photonic electromagnetic energy traveling via its atomic nucleus. Since pair-photon formation is a direct preservation of radiant energy to matter, the direct conservation of radiant energy is being monitored by using *PerkinElmer® Lambda 25 UV/Vis Spectrometer*. Subsequently, the electromagnetic energy of the photon is also being monitored to confirm that it has a minimum mass of two electrons since the mass m of a single electron is equal to 0.51 million electron volts (MeV) of energy E as computed from the equation of Albert Einstein, $E = mc^2$ thus, producing two electrons, is, therefore, the photon energy is equal to 1.02 MeV. Therefore, once pair-photon formation occurs, it will be transformed easily into the motion of the electron-positron pair photon as $\mathrm{E} = (E_k^{pp})_{tr}$, which is being detected by using a cloud chamber of *MATLAB* software to calculate the energy transfer from the pair photons [4, 13].

Consequently, the *energy transfer* to electron and positron from the pair-photon formation is being calculated as:

$$\left(E_k^{pp}\right)_{tr} = hv - 2m_e c^2 \tag{6.1}$$

where h represents Planck's constant, v represents the frequency of the photon, and the $2m_e c^2$ represents the combined rest mass of the electron-positron. Since the electron and positron is being released with pair production, the average energy transformation to each is being calculated as:

$$\left(E_k^{pp}\right)_{tr} = \frac{1}{2}\left(hv - 2m_e c^2\right) \tag{6.2}$$

Solving these two equations, it can get solar energy required to produce pair photons in the silicon solar cell vicinity:

$$hv - 2m_e c^2 = \frac{1}{2}\left(hv - 2m_e c^2\right) \tag{6.3}$$

$$hv = 2m_e c^2, \text{here } hv = E \tag{6.4}$$

$$\text{So}, E = 2mc^2 \tag{6.5}$$

Finally considering Eq. 6.5, a *design of a high-efficiency solar cell* is being modeled by determining the required wavelength (λ) and frequency (f) of a solar cell. Here, a photon turns into a particle by the implementation of gamma ray as described above due to the electron and a positron reaction, and thus, energy generation is being created through the masses of the two particles (2.7). This released energy is thus calculated by the notable equation of Einstein $E = mc^2$, where m represents the mass of the particle and c represents the speed of light (3×10^8 ms^{-1}). Here, the two particles are each other's antiparticle due to the pair-photon production, so, they have identical masses, and thus, the net energy is being calculated by the pair photons as follows:

$$E = 2mc^2 \tag{6.6}$$

$$\text{Since } E = hf = \frac{hc}{\lambda} \tag{6.7}$$

Here, $h = 6.626 \times 10^{-34}$ J.s, Planck's constant, $f =$ frequency of the photon/electromagnetic radiation, $c = 3 \times 10^{8}$ ms^{-1}, speed of light, and $\lambda =$ wavelength of photon/electromagnetic radiation.

Since the mass of an electron is 9.11×10^{-31} kg and the $c = 3 \times 10^{8}$ ms^{-1}, the pair photons are being released energy as

$$E = 2mc^2 \tag{6.8}$$

$$\text{So}, E = 2^* \left[9.11 \times 10^{-31}\ \text{kg} \times \left(3 \times 10^{8}\ \text{m/s}\right)^2\right] \tag{6.9}$$

$$= 2^* 8.2 \times 10^{-14}\ \text{j} \tag{6.10}$$

$$= 2^* 8.2 \times 10^{-14}\ \text{j}\ \left(1\ \text{eV}/1.602 \times 10^{-19}\ \text{j}\right) \tag{6.11}$$

$$= 1{,}024{,}000\ \text{eV} \tag{6.12}$$

To release this amount of energy, it needs to model the high-performance solar cell that is capable of having the following wavelength (λ) and frequencies (f):

$$\lambda = \frac{hc}{2E} \tag{6.13}$$

$$\mathbf{So}, \boldsymbol{\lambda} = \left[\left(\mathbf{6.626 \times 10^{-34}\ J.Sec}\right) \times \left(\mathbf{3x10^{8}\ m/s}\right)\right] / \left[\mathbf{2} \times \left(\mathbf{8.2 \times 10^{-14}\ j}\right)\right] \tag{6.14}$$

$$= \mathbf{1.21 \times 10^{-12}\ m} \tag{6.15}$$

$$\text{and the } E = hf \tag{6.16}$$

$$E = 16.4 \times 10^{-14}\ \text{J} = \left(6.626 \times 10^{-34}\right) \times f \tag{6.17}$$

$$\text{So}, f = 2.5 \times 10^{20}\ \text{Hz} \tag{6.18}$$

Naturally, by these above calculative results from Eqs. 6.14 and 6.18, the design of a high-performance printable solar cell has been conducted by implementing a required photonic wavelength (λ) of 1.21×10^{-12} m and frequency (f) of 2.5×10^{20} Hz to produce energy 1,024,000 eV/photon (Eq. 6.12).

Considering the standard average per m^2 area of the solar energy release of 3.6×10^{20} photons/second/m^2 on a bright sunny day, here, the total photon energy calculation would be equivalent to [(1,024,000 eV) *(3.6×10^{20})] = 3.6864×10^{26} eV/m^2 [12, 20]. Since 1 eV is equal to $1.6021773 \times 10^{-19}$ J, 3.6864×10^{27} eV/m^2 would be equivalent to 59,062,663.99 J/m^2. Since 1 joule is equal to $2.777777778 \times 10^{-7}$ kWh/m^2, 59,062,663.99 J/m^2 would be equivalent to 16.4 kWh/m^2. Considering the 44.2% efficiency of this high-performance solar cell, the net energy generation can be achieved at 7.253 kWh/m^2 cell, which indeed would be an excellent finding in photo-physics and energy science research to trap solar energy much more efficiently for the application of this energy commercially.

Simply, the applications of this photonic breakdown model to produce pair photons from a single one will indeed open a new door to photophysical science to design high-performance printable solar cells, which will release energy much more efficiently to mitigate the global energy crisis dramatically.

Conclusions

The research of this work confirms that the photopolymer-coated printable solar panel is capable of releasing gamma ray to break the photon particle and create pair photon $e^{+}e^{-}$ to release energy two times more with an efficiency of 44.2% requiring solar cell needs to have a photonic wavelength (λ) of 1.21×10^{-12}m, and frequency (f) of 2.5×10^{20} Hz to release energy 1,024,000 eV which is equivalent to energy 7.253 kWh/m^2. Simply, application of the photo-physics in designing high-performance solar cells would indeed be an excellent finding among the physicists and energy scientists in contributing to applying this science commercially to mitigate the global energy crisis completely.

Acknowledgments Thanks to MA Karim for their valuable comments on this manuscript. The authors declare no competing interests as defined by Springer or other interests that might be perceived to influence the results and/or discussion reported in this paper.

References

1. Javadi, A., et al. (2015). Single-photon non-linear optics with a quantum dot in a waveguide. *Nature Communications, 6*, 8655.
2. Schneider, A., Efrati, A., Alon, S., Sohmer, M., & Etgar, L. (2020). Green energy by recoverable triple-oxide mesostructured perovskite photovoltaics. *Proceedings of the National Academy of Sciences of the United States of America, 117*, 31010–31017.
3. Hossain, M. F. (2018). Photon energy amplification for the design of a micro PV panel. *International Journal of Energy Research, 42*, 3861–3876.
4. Cartwright, J. (2010). Photons meet with three-way split. *Nature*. https://doi.org/10.1038/news.2010.381
5. Zhang, H., et al. (2021). Multimodal host–guest complexation for efficient and stable perovskite photovoltaics. *Nature Communications, 12*, 3383.
6. Dobrynina, A., Kartavtsev, A., & Raffelt, G. (2015). Photon-photon dispersion of TeV gamma rays and its role for photon-ALP conversion. *Physical Review D, 91*, 083003.
7. Langford, N. K., et al. (2011). Efficient quantum computing using coherent photon conversion. *Nature, 478*, 360–363.
8. Kaneda, F., & Kwiat, P. G. High-efficiency single-photon generation via large-scale active time multiplexing. *Science Advances, 5*, eaaw8586.
9. Najjari, B., Voitkiv, A. B., Artemyev, A., & Surzhykov, A. (2009). Simultaneous electron capture and bound-free pair production in relativistic collisions of heavy nuclei with atoms. *Physical Review A, 80*, 012701.
10. Ouedraogo, N. A. N., Yan, H., Han, C. B., & Zhang, Y. (2021). Influence of fluorinated components on perovskite solar cells performance and stability. *Small, 17*, 2004081.

11. Fischer, K. A., et al. (2017). Signatures of two-photon pulses from a quantum two-level system. *Nature Physics, 13*, 649–654.
12. Gleyzes, S., et al. (2007). Quantum jumps of light recording the birth and death of a photon in a cavity. *Nature, 446*, 297–300.
13. Grinin, A., et al. (2020). Two-photon frequency comb spectroscopy of atomic hydrogen. *Science, 370*, 1061–1066.
14. Hencken, K., Baur, G., & Trautmann, D. (2006). Transverse momentum distribution of vector mesons produced in ultraperipheral relativistic heavy ion collisions. *Physical Review Letters, 96*, 012303.
15. Hossain, M. F. (2021). Sustainable building technology: Thermal control of solar energy to cool and heat the building naturally. *Environment, Development and Sustainability, 23*, 13304–13323.
16. Patra, S., et al. (2020). Proton-electron mass ratio from laser spectroscopy of HD+ at the part-per-trillion level. *Science, 369*, 1238–1241.
17. Guerlin, C., et al. (2007). Progressive field-state collapse and quantum non-demolition photon counting. *Nature, 448*, 889–893.
18. Martín, F., et al. (2007). Single photon-induced symmetry breaking of H2 dissociation. *Science, 315*, 629–633.
19. Battersby, S. (2019). The solar cell of the future. *Proceedings of the National Academy of Sciences of the United States of America, 116*, 7–10.
20. Hübel, H., et al. (2010). Direct generation of photon triplets using cascaded photon-pair sources. *Nature, 466*, 601–603.
21. Sheng, X., et al. (2014). Printing-based assembly of quadruple-junction four-terminal micro-scale solar cells and their use in high-efficiency modules. *Nature Materials, 13*, 593–598.
22. Dayan, B., et al. (2008). A photon turnstile dynamically regulated by one atom. *Science, 319*, 1062–1065.
23. Silverstone, J. W., et al. (2014). On-chip quantum interference between silicon photon-pair sources. *Nature Photonics, 8*, 104–108.

Chapter 7
Thermal Control of Solar Energy by Advanced Building Design to Cool and Heat the Premises Naturally

Introduction

The conventional heating and cooling of a building is causing serious environmental and atmospheric impacts. Traditional heating technology consumes fossil fuels and releases CO_2, causing climate change, which triggers a deadly natural disaster on the Earth and makes its environment vulnerable. On the other hand, conventional cooling technology releases Chlorofluorocarbons (CFCs), creating holes in the ozone layers. The ozone layer lies between 9.3 and 18.6 miles (15 and 30 km) above the Earth's surface that act as a blanket to block most of the sun's high-frequency ultraviolet rays. Due to the creation of holes in the ozone layer, UV rays are easily penetrating the surface of the Earth causing deadly skin cancer to humans and causing serious reproductive problems in all mammals [1, 2]. Although there have been numerous studies in the past on clean energy technology and climate change, conventional heating and cooling has been performed [3–6], but no study has been done on the natural cooling and heating system in the building sectors. To avoid greenhouse gases and CFC emissions, in this study, I proposed a natural cooling and heating technology by using Bose-Einstein photon distribution mechanism and Higgs boson quantum activation to decode photons (solar energy) in the states of cooling and heating. To decode this photon, cooling photon emission panel consisting of *nano* point breaks and waveguides have been proposed using helium in a portion of the exterior curtain wall [7, 8]. Therefore, the quantum dynamics of photons will be cooled by the quantum electrodynamics (QED) waveguides using photon band edges (PBEs) by photon emission from the sun to cool the building naturally [9]. Then this cooling state photon can be transformed into a heating state photon by employing quantum Higgs boson BR ($H \rightarrow \gamma\gamma^{-}$) to create an electromagnetic field with the help of two-diode semiconductors to naturally heat the building [10–12]. This cooling and heating transformation process will indeed be a new field of science to mitigate energy and protect the environment and the ozone layer.

M. F. Hossain, *Global Sustainability*, https://doi.org/10.1007/978-3-031-34575-3_7

Methods and Simulation

Cooling Mechanism

To decode activated photon into cooling state one, photon emission networks of *nano* point breaks, waveguides, and helium-assisted curtain wall will create point defects into the photon emission panel [13, 14]. This would provide a mechanism for incorporation of photonic band-gap (PBG) waveguide defect arrays into the curtain wall [15, 16]. Therefore, the quantum dynamics of photons will be decoded by point defects and PBG waveguides under helium cooling conditions. Consequently, the solar state photon will be cooled down within this regime. To calculate this detailed formation of cooling state of photon conversion from the sun, I have used MATLAB software to calculate detailed mathematical analysis. Since the *nano* point break through helium waveguides is embedded in a curtain wall, I have treated it as waveguides for reservoirs of photons to satisfy purely electron dynamics for cooling states of photon by considering contour maps (Fig. 7.1), which can be expressed by Hamiltonian as [9, 17, 18]:

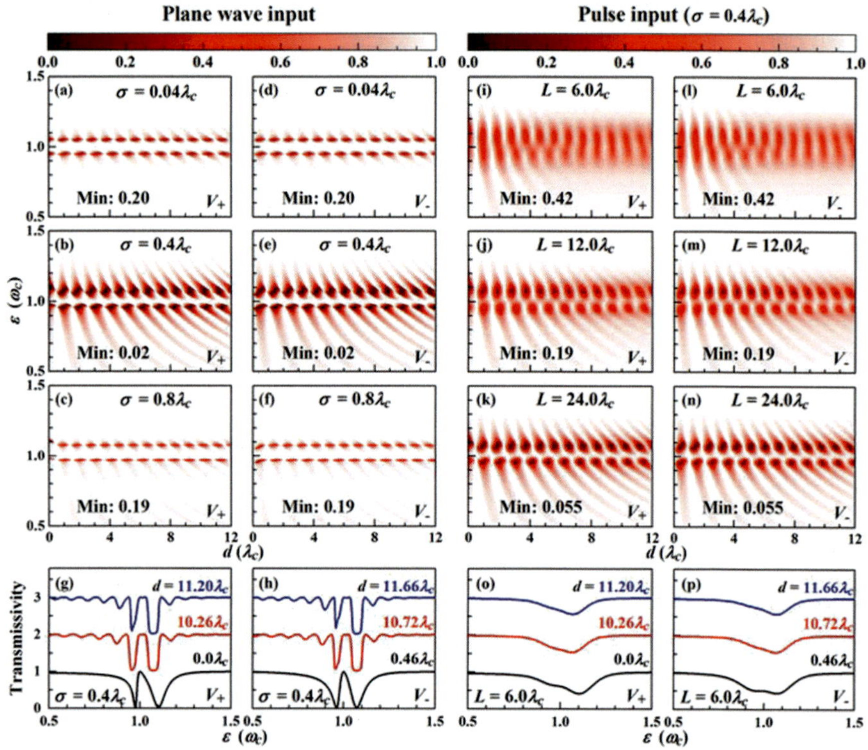

Fig. 7.1 Contour maps of the transmissivity of the single photon (**a–f**) plane wave and (**i–n**) pulse as functions of d and ε. Transmission spectra for the (**g, h**) plane wave and (**o, p**) pulse for different values of d, offset in steps of 1 denoted in the figures

$$H = \sum \omega_{ci} a_i^\dagger a_i + \sum_K \omega_k b_k^\dagger b_k + \sum_{ik} \left(V_{ik} a^\dagger b_k + V_{ik}^* b_k^\dagger a_i \right) \quad (7.1)$$

where $a_i \left(a_i^\dagger \right)$ represents the driver of the *nano* point break mode, $b_k \left(b_k^\dagger \right)$ represents the driver of the photodynamic modes of the photon *nano* structure, and coefficient V_{ik} represents the magnitude of the photonic mode among the *nano* break points and photon *nano* structure.

$$H = \sum \omega_{ci} a_i^\dagger a_i + \sum_K \omega_k b_k^\dagger b_k + \sum_{ik} \left(V_{ik} a^\dagger b_k + V_{ik}^* b_k^\dagger a_i \right) \quad (7.2)$$

where $a_i \left(a_i^\dagger \right)$ represents the driver of the nanoscale point break mode, $b_k \left(b_k^\dagger \right)$ represents the driver of the photodynamic modes of the nanoscale structures of the cooling photons, and coefficient V_{ik} represents the magnitude of the cool photonic mode among the nanoscale break points and photon band structures.

Thus, the induced solar photon will be formed as HcP^- where point break photon module is built by helium by utilizing photon energy (current), two diodes, and two resistors to cool the building (Fig. 7.2).

In more detail, it can be explained by the *I-V* equation of photon cells for the single-diode mode by expressing as

$$I = I_L - I_O \left\{ \exp^{\left[\frac{q(V + I_{RS})}{AkT_C} \right]} - 1 \right\} - \frac{(V + I_{RS})}{R_{Sh}} \quad (7.3)$$

I_L represents the photon generating current, *IO* represents the saturated current in the diode, R_s represents the resistance in the series, *A* represents the passive function

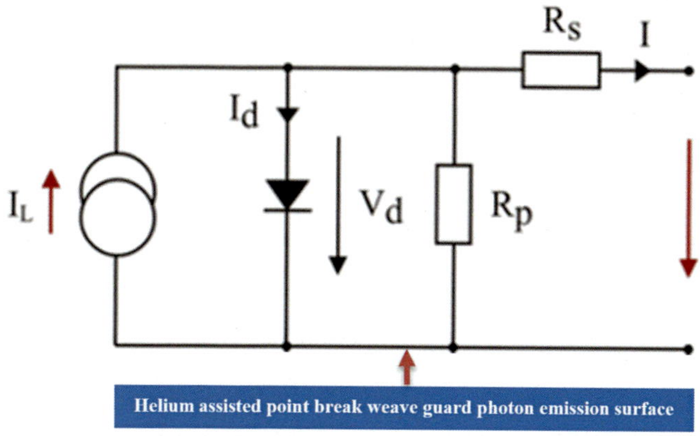

Fig. 7.2 The mechanism of two-diode model of solar irradiance receptor to cooling down electron into the curtain wall skin by photon induction assisted by helium-assisted point break

of the diode, k (=1.38 × 10^{-23} W/m^{2}K) represents Boltzmann's constant, q (=1.6 × 10^{-19} C) represents the magnitude of the charge of an electron, and T_C represents the functional cell temperature. Subsequently, the I-q relationship in the photon cells varies owing to the diode current and/or saturation current, which can be expressed as [19, 20].

$$I_O = I_{RS}\left(\frac{T_C}{T_{ref}}\right)^3 \exp\left[\frac{qEG\left(\frac{1}{T_{ref}} - \frac{1}{T_C}\right)}{KA}\right] \quad (7.4)$$

where I_{RS} represents the saturation current, considering the functional temperature and solar irradiance speed, and qEG represents the band-gap energy moving into the photon cell, respectively, per unit area, and thus considering the I-V curbs, it can be further explained in Fig. 7.3 as follows:

Considering the photon module, the I-V equation, apart from the I-V curve, is a conjunction of the I-V curves of all the cells in the photon emission panel, and thus, the equation can be rewritten as follows to determine the V-R relationship:

$$V = -IR_s + K \log\left[\frac{I_L - I + I_O}{I_O}\right] \quad (7.5)$$

Here, K is as a constant $\left(= \frac{AkT}{q}\right)$ and I_{mo} and V_{mo} are the current and voltage in the PV panel, respectively. Therefore, the relationship between I_{mo} and V_{mo} is the same as the PV cell I-V relationship:

$$V_{mo} = -I_{mo}R_{Smo} + K_{mo}\log\left(\frac{I_{Lmo} - I_{mo} + I_{Omo}}{I_{Omo}}\right) \quad (7.6)$$

where I_{Lmo} represents the photon-generated current, I_{Omo} represents the saturated current into the diode, R_{smo} represents the resistance in series, and K_{mo} represents the factorial constant. Once all non-series (NS) cells are connected in series, the series resistance is counted as the summation of the resistance of each cell in the series $R_{smo} = N_S \times R_S$, and the constant factor can be expressed as $K_{mo} = N_S \times K$. There is a certain amount of current flowing into the series of connected cells; thus, the current flow in Eq. 7.5 remains the same for each component, i.e., $I_{omo} = I_o$ and $I_{Lmo} = I_L$. Thus, the module I_{mo}-V_{mo} equation for the N_S series of connected cells is written as

$$V_{mo} = -I_{mo}N_S R_S + N_S K \log\left(\frac{I_L - I_{mo} + I_o}{I_o}\right) \quad (7.7)$$

Similarly, the current-voltage calculation can be rewritten for the parallel connections once all N_P cells are connected in parallel and can be expressed as follows [21, 22]:

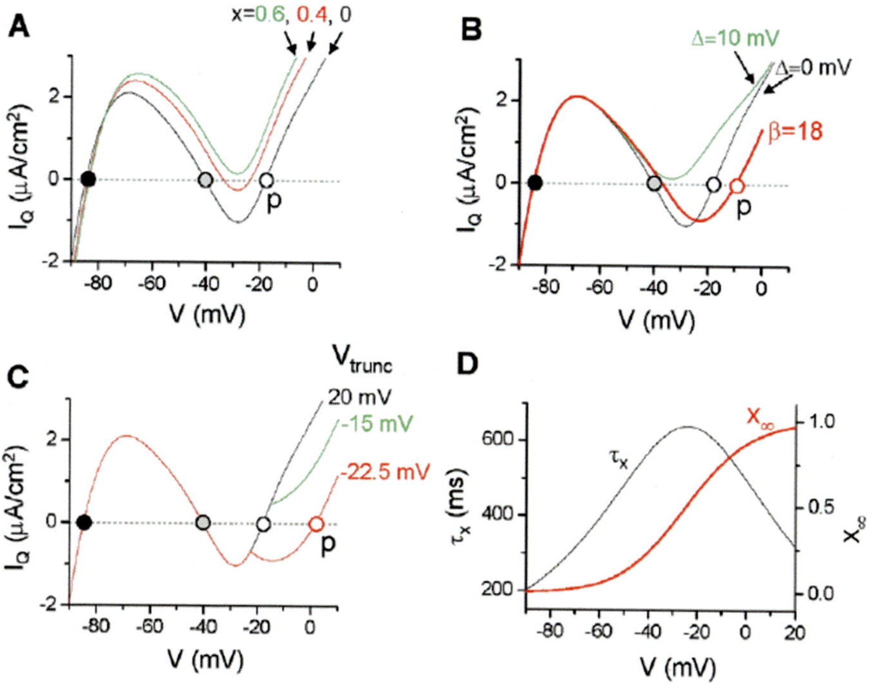

Fig. 7.3 The calculative model of *I-V* characteristics described the cooling mechanism on respectively light speed and unit of area (**a**) two-diode model for curtain wall skin cell at normal state, (**b**) double-diode model for curtain wall skin cell at normalized condition, (**c**) two-diode model for curtain wall skin cell module normal condition, and (**d**) double-diode model for curtain wall skin cell module at normalized condition

$$V_{\mathrm{mo}} = -I_{\mathrm{mo}}\frac{R_{\mathrm{s}}}{N_{\mathrm{p}}} + K\log\left(\frac{N_{\mathrm{sh}}I_{\mathrm{L}} - I_{\mathrm{mo}} + N_{\mathrm{p}}I_{\mathrm{o}}}{N_{\mathrm{p}}I_{\mathrm{o}}}\right) \tag{7.8}$$

Because the photon-generated current will depend primarily on the solar irradiance and relativistic temperature conditions of the photon emission panel, the current can be calculated using the following equation:

$$I_{\mathrm{L}} = G[I_{\mathrm{SC}} + K_{\mathrm{I}}(T_{\mathrm{cool}})] * V_{\mathrm{mo}}$$

$$T_{\mathrm{cool}} = \left(\frac{I_{\mathrm{L}}}{(G^{*}V_{\mathrm{mo}}) \times (I_{\mathrm{sc}} + K_{\mathrm{I}})}\right) \tag{7.9}$$

where I_{sc} represents the photon current at 25 °C and per unit area, K_{I} represents the relativistic photon panel coefficient, T_{cool} represents the photon cell's cooling temperature determination, and G represents the solar energy in per unit area [23–25].

Heating Mechanism

Essentially, a detailed mathematical calculation associated with the Higgs boson electromagnetic field, accuracy, and parameters of a photon-heating relationship has been performed to convert cooling photon into heating state photons [26, 27]. Thus, to create Higgs quantum field locally (into the curtain wall skin), I have implemented Abelian local symmetries using MATLAB software. Therefore, the gauge field symmetry will be broken down due to the penetration of solar irradiance, and the Goldstone scalar will become longitudinal mode of the vector [22, 28]. In the Abelian case, for each spontaneously broken particle T^{α} of the local symmetry will be the corresponding gauge field of $A_{\mu}^{\alpha}(x)$ where the Higgs quantum field started to work at a local U(1) phase symmetry [21, 29, 30]. Thus, the model can comprise a complex scalar field $\Phi(x)$ of electric charge q coupled to the EM field $A^{\mu}(x)$, which can be expressed by *Lagrangian* as:

$$\mathcal{L} = -\frac{1}{4}F_{\mu\nu}F^{\mu\nu} + D_{\mu}\Phi^{*}\, D^{\mu}\Phi - V(\Phi^{*}\Phi) \tag{7.10}$$

where

$$\begin{aligned} D_{\mu}\Phi(x) &= \partial_{\mu}\Phi(x) + iqA_{\mu}(x)\Phi(x) \\ D_{\mu}\Phi^{*}(x) &= \partial_{\mu}\Phi^{*}(x) - iqA_{\mu}(x)\Phi^{*}(x) \end{aligned} \tag{7.11}$$

and

$$V(\Phi^{*}\Phi) = \frac{\lambda}{2}(\Phi^{*}\Phi)^{2} + m^{2}(\Phi^{*}\Phi) \tag{7.12}$$

Suppose $\lambda > 0$ but $m^2 < 0$, so that $\Phi = 0$ is a local maximum of the scalar potential, while the minima form a degenerate circle $\Phi = \frac{v}{\sqrt{2}} * e^{i\theta}$,

$$v = \sqrt{\frac{-2m^{2}}{\lambda}}, \quad \text{any real } \theta \tag{7.13}$$

Consequently, the scalar field Φ develops a nonzero vacuum expectation value $\langle\Phi\rangle \neq 0$, which spontaneously create the U(1) symmetry of the magnetic field. The breakdown would lead to a massless Goldstone scalar stemming from the phase of the complex field $\Phi(x)$. The local U(1) symmetry, the phase of $\Phi(x)$ - not just the phase of the expectantion value $\langle\Phi\rangle$ but the x-dependent phase of the dynamical $\Phi(x)$ field.

To confirm this mechanism, I have used polar coordinates in the scalar field space; thus

$$\Phi(x) = \frac{1}{\sqrt{2}}\Phi_{\mathrm{r}}(x) * e^{i\Theta(x)}, \quad \text{real } \Phi_{\mathrm{r}}(x) > 0, \text{real } \Phi(\mathrm{x}) \tag{7.14}$$

This field redefinition is singular when $\Phi(x) = 0$, so I never used it for theories with $\langle\Phi\rangle \neq 0$, but it's OK for spontaneously broken theories where we expect $\Phi\langle x\rangle \neq 0$ almost everywhere. In terms of the real fields $\phi_r(x)$ and $\Theta(x)$, the scalar potential depends only on the radial field ϕ_r,

$$V(\phi) = \frac{\lambda}{8}\left(\phi_{\mathrm{r}}^2 - v^2\right)^2 + \text{const}, \tag{7.15}$$

or in terms of the radial field shifted by its VEV, $\Phi_{\mathrm{r}}(x) = v + \sigma(x)$,

$$\phi_{\mathrm{r}}^2 - v^2 = (v+\sigma)^2 - v^2 = 2v\sigma + \sigma^2 \tag{7.16}$$

$$V = \frac{\lambda}{8}\left(2v\sigma - \sigma^2\right)^2 = \frac{\lambda v^2}{2} * \sigma^2 + \frac{\lambda v}{2} * \sigma^3 + \frac{\lambda}{8} * \sigma^4 \tag{7.17}$$

At the same time, the covariant derivative $D_\mu\phi$ becomes

$$D_\mu\phi = \frac{1}{\sqrt{2}}\left(\partial_\mu\left(\phi_{\mathrm{r}}e^{i\Theta}\right) + iqA_\mu * \phi_{\mathrm{r}}e^{i\Theta}\right) = \frac{e^{i\Theta}}{\sqrt{2}}\left(\partial_\mu\phi_{\mathrm{r}} + \phi_{\mathrm{r}} * i\partial_\mu\Theta + \phi_{\mathrm{r}} * iqA_\mu\right) \tag{7.18}$$

$$\begin{aligned} \left|D_\mu\phi\right|^2 &= \frac{1}{2}\left|\partial_\mu\phi_{\mathrm{r}} + \phi_{\mathrm{r}} * i\partial_\mu\Theta + \phi_{\mathrm{r}} * iqA_\mu\right|^2 = \frac{1}{2}\left(\partial_\mu\phi_{\mathrm{r}}\right) + \frac{\phi_{\mathrm{r}}^2}{2} \\ &* \left(\partial_\mu\Theta qA_\mu\right)^2 = \frac{1}{2}\left(\partial_\mu\sigma\right)^2 + \frac{(v+\sigma)^2}{2} * \left(\partial_\mu\Theta + qA_\mu\right)^2 \end{aligned} \tag{7.19}$$

Altogether,

$$\mathcal{L} = \frac{1}{2}\left(\partial_\mu\sigma\right)^2 - v(\sigma) - \frac{1}{4}F_{\mu\nu}F^{\mu\nu} + \frac{(v+\sigma)^2}{2} * \left(\partial_\mu\Theta + qA_\mu\right)^2 \tag{7.20}$$

To determine the heating ($\mathcal{L}_{\text{heat}}$) into magnetic field properties of this *Lagrangian*, it has been expanded in powers of the fields (and their derivatives) and focus on the quadratic part describing the free particles,

$$\mathcal{L}_{\text{heat}} = \frac{1}{2}\left(\partial_\mu\sigma\right)^2 - \frac{\lambda v^2}{2} * \sigma^2 - \frac{1}{4}F_{\mu\nu}F^{\mu\nu} + \frac{v^2}{2} * \left(qA_\mu + \partial_\mu\Theta\right)^2 \tag{7.21}$$

Here this *Lagrangian* ($\mathcal{L}_{\text{free}}$) function obviously will suggest a real scalar particle of positive mass$^2 = \lambda v^2$ involving the $A_\mu(x)$ and the $\Theta(x)$ fields to initiate to create high heating within the quantum field of the curtain wall (Fig. 7.4).

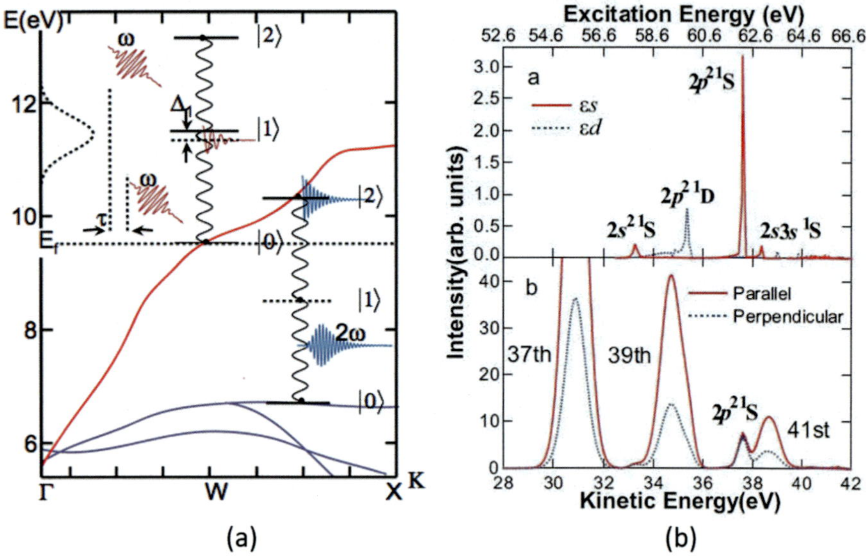

Fig. 7.4 (**a**) The mechanism of photon (2ω) transformation from (ω) at energy lever (eV) into the two-diode feed semiconductors. (**b**) The electron spectra through the photon excitation energy (eV) into the quantum field intensity (arb. units) that transform into kinetic energy (eV) by the conversion of heating state of photon

Results and Discussion

Cooling Mechanism

To mathematically determine the formation of cooling photon by the helium-assisted curtain wall skin, I have initially solved the dynamic photon proliferation by integrating Eqs. (7.16) and (7.17). It is assumed that owing to the cooling unit areal condition $J(\omega)$, the curtain wall skin produces photon proliferation dynamics [9, 21] since it has a persistent weak-coupling limit and the Weisskopf-Wigner approximation rule and/or Markovian master equation, which is equal to the photon proliferation production. Consequently, all *HcP* photon proliferation induction will have a dynamic photon state mode (1D, 2D, 3D) in the curtain wall skin, which can be expressed as described in Table 7.1 [31, 32].

Subsequently, a fine frequency cutoff Ω_C is employed to avoid bifurcation of the DOS in a 3D curtain wall skin. Similarly, a sharpened frequency cutoff at Ω_d maintains a positive DOS in 2D and 1D curtain wall skin (Fig. 7.5). Hence Li_2 (x) acts as a dilogarithm variable and e_{rfc} (x) acts as an additive variable. Subsequently, the DOS of various curtain wall skin, denoted as $\varrho_{PC}(\omega)$, is determined by calculating photon eigenfrequencies and eigenfunctions of Maxwell's rules considering the *nano* structure [15, 19, 21]. For a 1D curtain wall skin, the corresponding DOS is

Table 7.1 The photonic structures in different DOS dimensional modes in the curtain wall skin

Photon	Unit area $J(\omega)$ for different DOS	Reservoir-induced self-energy correction $\Sigma(\omega)$
1D	$\frac{C}{\pi}\frac{1}{\sqrt{\omega-\omega_e}}\Theta(\omega-\omega_e)$	$-\frac{C}{\sqrt{\omega_e-\omega}}$
2D	$-\eta\left[\ln\left\lvert\frac{\omega-\omega_0}{\omega_0}\right\rvert-1\right]\Theta(\omega-\omega_e)\Theta(\Omega_d-\omega)$	$\eta\left[\mathrm{Li}_2\left(\frac{\Omega_d-\omega_0}{\omega-\omega_0}\right)-\mathrm{Li}_2\left(\frac{\omega_0-\omega_e}{\omega_0-\omega}\right)-\ln\frac{\omega_0-\omega_e}{\Omega_d-\omega_0}\ln\frac{\omega_e-\omega}{\omega_0-\omega}\right]$
3D	$\chi\sqrt{\frac{\omega-\omega_e}{\Omega_C}}\exp\left(-\frac{\omega-\omega_e}{\Omega_C}\right)\Theta(\omega-\omega_e)$	$\chi\left[\pi\sqrt{\frac{\omega_e-\omega}{\Omega_C}}\exp\left(-\frac{\omega-\omega_e}{\Omega_C}\right)e_{\mathrm{rfc}}\sqrt{\frac{\omega_e-\omega}{\Omega_C}}-\sqrt{\pi}\right]$

They correspond with different unit area $J(\omega)$ and self-energy induction at reservoir $\Sigma(\omega)$, which is determined by the photon dynamics into the extreme relativistic curtain wall skin. The variables C, η and χ function like coupled forces between the point break and PV of 1D, 2D, and 3D into the curtain wall skin [9, 21]

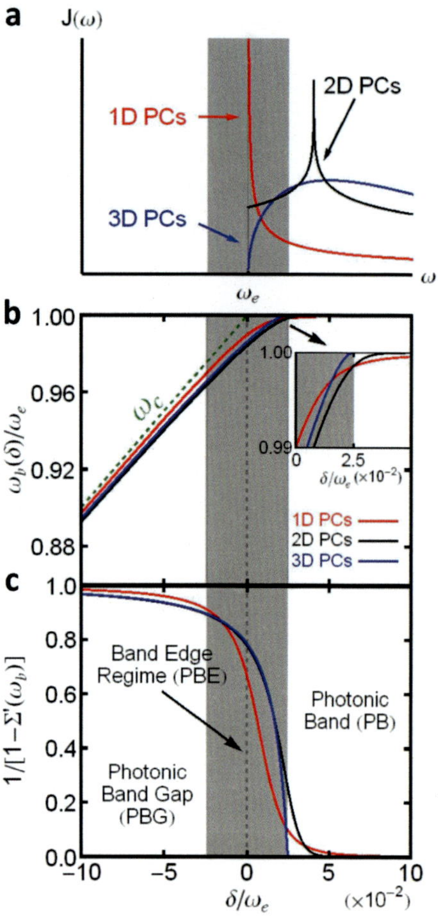

Fig. 7.5 The photonic band structure and mode for energy conversion. (**a**) Unit area at various DOS values for 1D, 2D, and 3D curtain wall skin. (**b**) Photonic mode of frequencies for functional tuning. (**c**) Photonic mode of magnitude to release energy calculated using Eq. (7.2). The photonic modes depict the crossover at 1D and 2D into curtain wall skin and a complex transitional state at 3D into the PV cell once the point break frequency *vc* transforms from a PBG area to a photonic band (PB) area [21]

denoted as $\varrho_{\mathrm{PC}}(\omega) \propto \frac{1}{\sqrt{\omega - \omega_{\mathrm{e}}}}\Theta(\omega - \omega_{\mathrm{e}})$, where $\Theta(\omega - \omega_{\mathrm{e}})$ represents the Heaviside step function and ω_{e} e represents the frequency in the PBE considering the DOS.

This DOS is therefore calculated to conduct 3D isotropic analysis in curtain wall skin to predict the error-free qualitative state of the non-Weisskopf-Wigner mode and the photon-cooling state in the photon cell considering density of states (DOS) and projected density of states (PDOS) (Fig. 7.6). Therefore, for a 3D curtain wall skin, the DOS close to the PBE is implemented by anisotropic DOS: $\varrho_{\mathrm{PC}}(\omega) \propto \frac{1}{\sqrt{\omega - \omega_{\mathrm{e}}}}\Theta(\omega - \omega_{\mathrm{e}})$, which is then clarified with respect to the electromagnetic field (EMF) vector [1, 9, 21, 30]. For 2D and 1D curtain wall skin, the cooling photon DOS exhibits a pure logarithm divergence close to the PBE, which is approximated as $\varrho_{\mathrm{PC}}(\omega) \propto -[\ln|(\omega - \omega_0)/\omega_0| - 1]\Theta(\omega - \omega_{\mathrm{e}})$, where ω_{e} represents the central point of peak logarithm.

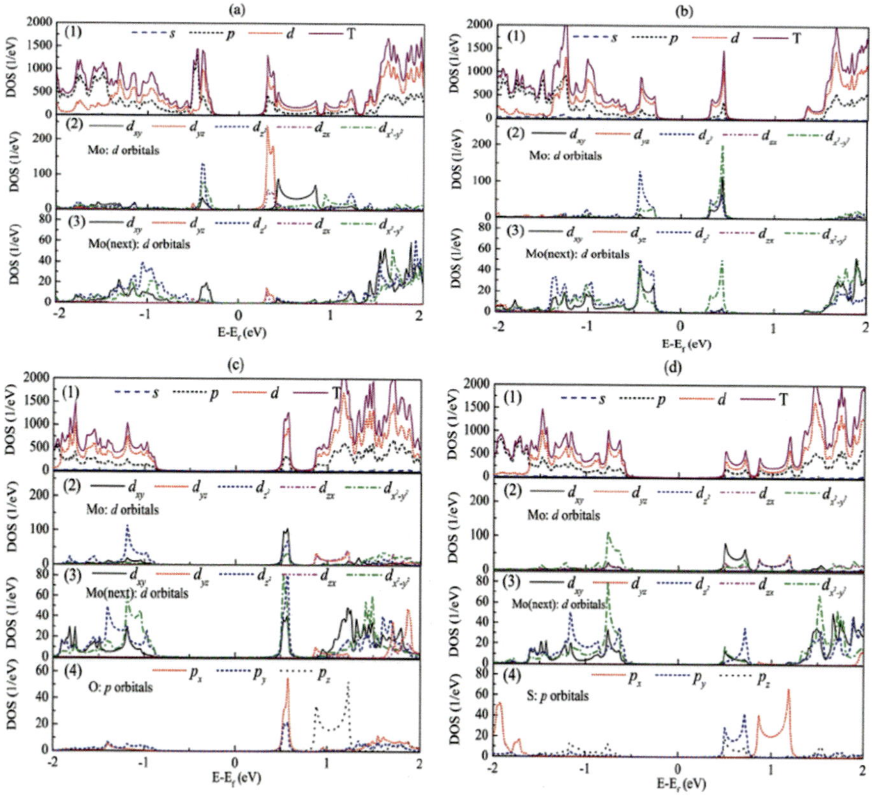

Fig. 7.6 The total density of states (DOS) and projected density of states (PDOS) of decoded photon to transform into cooling state: where (1) total DOS and DOS projected on orbitals. Notations are s for s orbital DOS, p for p orbital DOS, d for d orbital DOS, and T for total DOS, and (2) PDOS of d orbitals on the very edge Mo atoms, and (3) PDOS of d orbitals on the next edge Mo atoms. (b) The symbols are the same as in (a). (c) From (1) to (3), the symbols are same as in (a), and (4) is for PDOS of p orbitals of O atoms. (d) From (1) to (3), the symbols are the same as in (a), and (4) is for PDOS of p orbitals of external S atoms

The quantum field area $J(\omega)$ is clarified as the production field of the DOS in the PV cell by the fine cooling photonic magnitude $V(\omega)$ within the PB and PV cell [9, 20, 22],

$$J(\omega) = \varrho(\omega)|V(\omega)|^2 \tag{7.22}$$

Hereafter, I consider the PB frequency ω_c and proliferative photon dynamics using the function $u(t, t_0)$ for photon structure in the relation $\langle a(t)\rangle = u(t, t_0)\langle a(t_0)\rangle$. It is calculated using the dissipative integro-differential equation given in Eq. (7.19) and expressed as

$$u(t,t_0) = \frac{1}{1 - \Sigma'(\omega_b)} e^{-i\omega(t-t_0)} + \int_{\omega_e}^{\infty} d\omega \frac{J(\omega)e^{-i\omega(t-t_0)}}{[\omega - \omega_c - \Delta(\omega)]^2 + \pi^2 J^2(\omega)} \quad (7.23)$$

where $\Sigma'(\omega_b) = [\partial\Sigma(\omega)/\partial\omega]_{\omega=\omega_b}$ and $\Sigma(\omega)$ represent the reservoir-induced PB photon self-energy correction,

$$\Sigma(\omega) = \int_{\omega_e}^{\infty} d\omega' \frac{J(\omega')}{\omega - \omega'} \quad (7.24)$$

Here, the frequency ω_b in Eq. (7.2) represents the cooling photonic frequency mode in the PBG ($0 < \omega_b < \omega_e$) and is calculated using the pole condition: $\omega_b - \omega_c - \Delta(\omega_b) = 0$, where $\lesssim\!\Delta(\omega) = \mathcal{P}\left[\int d\omega' \frac{J(\omega')}{\omega-\omega'}\right]$ is a principal-value integral.

Furthermore, the detailed cooling photonic dynamics, considering the proliferation magnitude $|u(t,t_0)|$, have been calculated and are shown in Fig. 7.5a for 1D, 2D, and 3D photon cells with respect to various detuning δ integrated from the PBG area to the PB area [33–35]. The cooling photonic dynamic rate $\kappa(t)$ is shown in Fig. 7.5b, neglecting the function $\delta = 0.1 \ \omega_e$. The calculated result indicates that dynamic photons are produced at a high rate once ω_c crosses from the PBG to PB area. Because the range in $u(t,t_0)$ is $1 \geq |u(t,t_0)| \geq 0$, I have defined the crossover area as related to the condition $0.9 \gtrsim |u(t \rightarrow \infty, t_0)| \geq 0$. This corresponds to $-0.025\omega_e \lesssim \delta \lesssim 0.025\omega_e$, with a cooling photon induction rate $\kappa(t)$ within the PBG ($\delta < -0.025\omega_e$) and near the PBE($-0.025\omega_e \lesssim \delta \lesssim 0.025\omega_e$).

To be more specific, I have considered first the PB as the Fock cooling determination n_0, i. e., $\rho(t_0) = |n_0\rangle\langle n_0|$, which is obtained theoretically through the real-time quantum feedback control [2, 21, 36] and then by solving the Eq. (7.2), considering the state of cooling photon induction at time t:

$$\rho(t) = \sum_{n=0}^{\infty} \mathcal{P}_n^{(n_0)}(t)|n_0\rangle\langle n_0| \quad (7.25)$$

$$\mathcal{P}_n^{(n_0)}(t) = \frac{[v(t,t)]^n}{[1+v(t,t)]^{n+1}} [1 - \Omega(t)]^{n_0} \times \sum_{k=0}^{\min\{n_0,n\}} \binom{n_0}{k}$$

$$\times \binom{n}{k} \left[\frac{1}{v(t,t)} \frac{\Omega(t)}{1-\Omega(t)}\right]^k \quad (7.26)$$

where $\Omega(t) = \frac{|u(t,t_0)|^2}{1+v(t,t)}$. Therefore, the result suggests that a Fock state cooling photon will be induced into dynamic states of $|n_0\rangle$ is $\mathcal{P}_n^{(n_0)}(t)$. The proliferation of photon dissipation $\mathcal{P}_n^{(n_0)}(t)$ in the primary state $|n_0 = 5\rangle$ and steady-state limit,

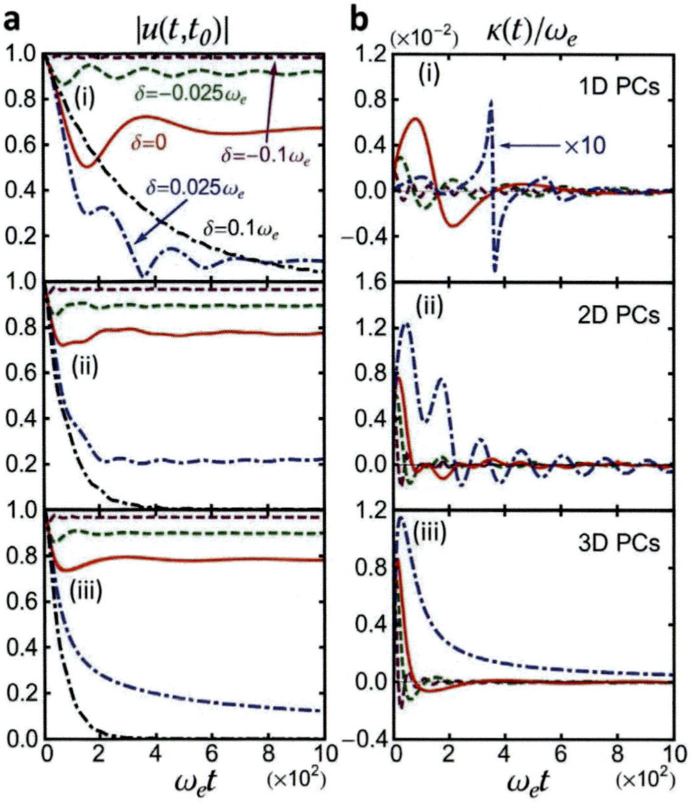

Fig. 7.7 Proliferation of dynamic photons in PV cells. (**a**) Considering the PB area, $<a(t)> = 5u(t, t_0) <a(t_0)>$ and (**b**) the dynamic photonic rate $k(t)$, plotted for (i) 1D, (ii) 2D, and (iii) 3D PV cells [22]

$\mathcal{P}_n^{(n_0)}(t \to \infty)$, is thus shown in Fig. 7.7. Therefore, the proliferation of cooling photon production will ultimately reach non-equilibrium cooling state with the photonic structure to cool the building.

Heating Mechanism

Since the Higgs boson quantum field creates electromagnetic field employed by two-diode semiconductors, the local U(1) gauge invariant (QED) did allow to add a mass term for the gauge particle under $\varnothing' \to e^{i\alpha(x)}\varnothing$ to transform cooling photons into heating photons. In detail, it can be explained by a covariant derivative with a special transformation rule for the scalar field expressed by [37–39]:

$$\partial_\mu \rightarrow D_\mu = \partial_\mu = ieA_\mu \quad \text{[covariant derivatives]}$$
$$A'_\mu = A_\mu + \frac{1}{e}\partial_\mu \alpha \quad [A_\mu \text{ derivatives}] \tag{7.27}$$

where the local U(1) gauge-invariant *Lagrangian* for a complex scalar field is given by:

$$\mathcal{L} = (D^\mu)^\dagger \left(D_\mu \varnothing\right) - \frac{1}{4}F_{\mu\nu}F^{\mu\nu} - V(\varnothing) \tag{7.28}$$

The term $\frac{1}{4}\, F_{\mu\nu}F^{\mu\nu}$ is the kinetic term for the gauge field (heating photon) and $V(\varnothing)$ is the extra term in the Lagrangian that can be seen as $V(\varnothing^*\varnothing) = \mu^2(\varnothing^*\varnothing) + \lambda(\varnothing^*\varnothing)^2$.

Therefore, the Lagrangian ($\mathcal{L}$) under perturbations into the quantum field has initiated with the massive scalar particles $\phi_1$and ϕ_2 along with a mass μ. In this situation $\mu^2 < 0$ had an infinite number of quantum; each has been satisfied by $\phi_1^2 + \phi_2^2 = -\mu^2/\lambda = v^2$ and the Lagrangian through the covariant derivatives using again the shifted fields η and ξ defined the quantum field as $\phi_0 = \frac{1}{\sqrt{2}}[(\upsilon + \eta) + i\xi]$.

$$\begin{aligned} \text{Kinetic term :} \quad \mathcal{L}_{\text{kin}}(\eta, \xi) &= (D^\mu \phi)^\dagger (D^\mu \phi) \\ &= (\partial^\mu + ieA^\mu)\phi^* \left(\partial_\mu - ieA_\mu\right)\phi \end{aligned} \tag{7.29}$$

Potential term: $V(\eta, \xi) = \lambda\, \upsilon^2\eta^2$, up to second order in the fields, and thus, the full Lagrangian can be written as:

$$\begin{aligned} \mathcal{L}_{\text{kin}}(\eta, \xi) &= \frac{1}{2}\left(\partial_\mu \eta\right)^2 - \lambda \upsilon^2\eta^2 + \frac{1}{2}\left(\partial_\mu \xi\right)^2 \\ &\quad - \frac{1}{4}F_{\mu\nu}F^{\mu\nu} + \frac{1}{2}e^2\upsilon^2A_\mu^2 - e\upsilon A_\mu(\partial^\mu \xi) + \text{int.terms} \end{aligned} \tag{7.30}$$

Here massive η, massless ξ (as before), and also a mass term for the quantum and $A_\mu$are fixed up to a term $\partial_\mu \alpha$ as can be seen from Eq. (7.28). In general, A_μ and ϕ change simultaneously and thus can be redefined to accommodate the heating photon particle spectrum within the quantum field by expressing:

$$\begin{aligned} \mathcal{L}_{\text{scalar}} &= (D^\mu \phi)^\dagger (D^\mu \phi) - V\left(\phi^\dagger \phi\right) \\ &= (\partial^\mu + ieA^\mu)\frac{1}{\sqrt{2}}(v + h)\left(\partial_\mu - ieA_\mu\right)\frac{1}{\sqrt{2}}(v + h) - V\left(\phi^\dagger \phi\right) \end{aligned} \tag{7.31}$$

$$= \frac{1}{2}\left(\partial_\mu h\right)^2 + \frac{1}{2}e^2A_\mu^2 (v + h)^2 - \lambda v^2h^2 - \lambda vh^3 - \frac{1}{4}\lambda h^4 + \frac{1}{4}\lambda h^4 \tag{7.32}$$

Thus, this expanding term in the *Lagrangian* associated to the scalar field is suggesting that Higgs boson quantum field is prepared to initiate heating photon into its quantum field.

To confirm this heating photon transformation, the isotropic distribution of movement on the differential cone has to be calculated considering the angle θ and within θ and $\theta + d\theta$ is $\frac{1}{2}$ sin $\theta d\theta$. Then, the differential photon density at energy ϵ and angle θ is

$$dn = \frac{1}{2}n(\in)\sin\theta d \in d\theta. \tag{7.33}$$

Consequently, the functional speed of the high-energy photons was calculated considering the directional form of c (1 – $cos\theta$), where the absorption per unit path length would be

$$\frac{d\tau_{\text{abs}}}{dx} = \int\int\frac{1}{2}\sigma n\ (\in)(1-\cos\theta)\sin\theta d \in d\theta. \tag{7.34}$$

Modifying the functions to an integration over s instead of θ by (3) and (5), I have calculated

$$\frac{d\tau_{\text{abs}}}{dx} = \pi r_0^2\left(\frac{m^2c^4}{E}\right)^2\int_{\frac{m^2c^4}{E}}^{\infty} \in^{-2} n(\in)\overline{\varphi}\ [s_0(\in)]de, \tag{7.35}$$

where

$$\overline{\varphi}\ [s_0(\in)] = \int_1^{s_0(\in)} s\overline{\sigma}(s)ds, \quad \overline{\sigma}(s) = \frac{2\sigma(s)}{\pi r_0^2}. \tag{7.36}$$

This result has been identified as the dimensional variable $\overline{\varphi}$ and dimensionless cross section $\overline{\sigma}$. The variable $\overline{\varphi}\ [s_0]$ is calculated based on a detailed graphical frame for $1 < s_0 < 10$. I gave a reliable functional asymptotic calculation for $\overline{\varphi}$, where $s_0 - 1 \ll 1$ and $s_0 \gg 1$, which I have expressed as

$$\overline{\varphi}[s_0] = \frac{1+\beta_0^2}{1-\beta_0^2}\ln\omega_0 - \beta_0^2\ln\omega_0 - \ln^2\omega_0 - \frac{4\beta_0}{1-\beta_0^2} + 2\beta_0 + 4\ln\omega_0\ \ \ln(\omega_0+1) - \mathrm{L}(\omega_0), \tag{7.37}$$

where

$$\beta_0^2 = \frac{1-1}{s_0}, \quad \omega_0 = \frac{(1+\beta_0)}{(1-\beta_0)},$$
$$L(\omega_0) = \int_1^{\omega_0} \omega^{-1}\ln(\omega+1)d\omega. \tag{7.38}$$

The last integral can be written as

$$(\omega+1)=\omega\left(\frac{1+1}{\omega}\right),\quad L(\omega_0)=\frac{1}{2}\ln^2\omega_0+L'(\omega_0),$$

where

$$\begin{aligned} L'(\omega_0) &= \int_1^{\omega_0} \omega^{-1}\ln\left(1+\frac{1}{\omega}\right)d\omega, \\ &= \frac{\pi^2}{12}-\sum_{n=1}^{\infty}(-1)n^{-1}n^{-2}\omega_0^{-n}. \end{aligned} \tag{7.39}$$

The accurate representation of heating photon here readily allows the calculation of $\overline{\varphi}[s_0]$ to any needed accuracy for the expected value of s_0. Thus, the corrective functional asymptotic formulas are being used as follows:

$$\overline{\varphi}[s_0]=2s_0(\ln 4s_0\text{–}2)+\ln 4s_0(\ln 4s_0\text{–}2)\text{–}\frac{(\pi^2-9)}{3}+s_0^{-1}\left(\ln 4s_0+\frac{9}{8}\right)+\cdots(s_0\gg 1); \tag{7.40}$$

$$\overline{\varphi}[s_0]=\left(\frac{2}{3}\right)(S_0-1)^{\frac{3}{2}}+\left(\frac{5}{3}\right)(S_0-1)^{\frac{5}{2}}\text{–}\left(\frac{1507}{420}\right)(S_0-1)^{\frac{7}{2}}+\cdots(s_0\text{–}1\ll 1). \tag{7.41}$$

The function $\frac{\overline{\varphi}[s_0]}{(s_0-1)}$ is shown in Fig. 7.5 for $1 < s_0 < 10$; for larger s_0, it contains natural logarithmic dependence on s_0. Here the heating photon spectrum absorption by the power law has been calculated with respect to the form $n(\in)\propto\in^m$ considering two system is in a pristine, (b) system with BN in sheet of graphene.

Thus, I have considered that the function of light absorption for the spectrum will follow the features of a high-energy cutoff with $m > 0$.

Here, the heating photon spectra with a high-energy cutoff

Consider now a spectrum of the form

$$n(\in)=D\in^{\beta},\quad \in\ <\in_m,\quad \beta_{\leq}0 \tag{7.42}$$

$$=0,\quad \in\ >\in_m. \tag{7.43}$$

For this spectrum, I have found

$$\frac{d\tau_{\text{abs}}}{dx}=\pi r_0^2 D\left(\frac{m^2c^4}{E}\right)^{1+\beta}\times\begin{cases}0, & E<E_m\\ F_\beta(\sigma_m), & E>E_m,\end{cases} \tag{7.44}$$

where

$$\sigma_m = \frac{E}{E_m} = \frac{\epsilon_m E}{m^2c^4}, \tag{7.45}$$

$$F_\beta(\sigma_m) = \int_1^{\sigma m} s_0^{\beta-2}\overline{\varphi}[s_0]ds_0. \tag{7.46}$$

Again, by Eqs. 7.42 and 7.43, we can obtain the asymptotic forms

$$\begin{aligned} &\beta=0 : F_\beta(\sigma_m) \to A_\beta + \ln^2\sigma_m - 4\ \ln\sigma_m + \cdots, \\ &\beta\neq 0 : F_\beta(\sigma_m) \to A_\beta + 2\beta^{-1}\ \sigma_m^\beta\left(\ln 4\sigma_m - \beta^{-1} - 2\right) + \cdots, \sigma_m > 10 \end{aligned} \tag{7.47}$$

$$\text{All } \beta : F_\beta(\sigma_m) \to \left(\frac{4}{15}\right)(\sigma_m - 1)^{\frac{5}{2}} + \left[\frac{2(2\beta+1)}{21}\right](\sigma_m - 1)^{\frac{7}{2}} + \cdots, \sigma_m - 1 \ll 1. \tag{7.48}$$

$\sigma_m^{-\beta}F_\beta(\sigma_m)$ is shown in Fig. 7.6 for $\beta = 0, 0.5, 1.0, 1.5, 2.0, 2.5$, and 3.0 A_β, which contribute to the integral of the region [7, 13]. The values here have been calculated as $A_\beta = 8.111, 13.53, 9.489, 15.675, 34.54, 85.29$, and 222.9 for $\beta = 0, 0.5, 1.0, 1.5, 2.0, 2.5$, and 3.0, respectively. Subsequently, I have considered heating photon terms corresponding to the spectra for both negative and positive indices:

$$\begin{aligned} n(\epsilon) &= 0, && \epsilon < \epsilon_0 \\ &= C_\epsilon^{-\alpha} \text{ or } D_\epsilon^\beta, && \epsilon_0 < \epsilon < \epsilon_m \\ &= 0, && \epsilon > \epsilon_m. \end{aligned} \tag{7.49}$$

Then, I have calculated

$$\left(\frac{d\tau_{\text{abs}}}{dx}\right)_\alpha = \pi r_0^2 C\left(\frac{m^2c^4}{E}\right)^{1-\alpha} \times \begin{cases} 0, & E < E_m \\ [\mathrm{F}_\alpha(1) - \mathrm{F}_\alpha(\sigma_m)], & E_m < E < E_0 \\ [\mathrm{F}_\alpha(\sigma_0) - \mathrm{F}_\alpha(\sigma_m)], & E > E_0; \end{cases} \tag{7.50}$$

$$\left(\frac{d\tau_{\text{abs}}}{dx}\right)_\beta = \pi r_0^2 D\left(\frac{m^2c^4}{E}\right)^{1+\beta} \times \begin{cases} 0, & E < E_m \\ [\mathrm{F}_\beta(\sigma_\mathrm{m}), & E_m < E < E_0\left[\mathrm{F}_\beta(\sigma_\mathrm{m}) - \mathrm{F}_\beta(\sigma_0)\right], \ E > E_0. \end{cases} \tag{7.51}$$

Asymptotic formulas for this case are quite valid for analysis of the heating photon spectrum, where I can further clarify Γ_{γ}^{LPM} as denoting the photon's contribution to emitting irradiance with a rate per unit volume according to bremsstrahlung processes [40, 41].

$$\Gamma_{\gamma} \equiv \frac{dn_{\gamma}}{dVdt} \tag{7.52}$$

The contribution Γ_{γ}^{LPM} was summed in order to confirm the rate of $O(\alpha_{\mathrm{EM}}\ \alpha s)$. Here, I have assumed that the result of polarized emission rate $\Gamma_{\gamma}^{\mathrm{LPM}}$ at the thermodynamically controlled equilibrium of the plasma surface at temperature T and the photo-physical reaction is μ; thus, the equation is expressed as

$$\frac{d\Gamma_{\gamma}^{\mathrm{LPM}}}{d^3k} = \frac{d_F q_s^2 \alpha_{\mathrm{EM}}}{4\pi^2 k} \int_{-\infty}^{\infty} \frac{dp_{\parallel}}{2\pi} \int \frac{d^2\mathrm{p}_{\perp}}{(2\pi)^2} A\left(p_{\parallel}, k\right) \ \mathrm{Re}\left\{2\mathrm{P}_{\perp} \cdot \mathrm{f}\left(\mathrm{p}_{\perp}; p_{\parallel}, k\right)\right\} \tag{7.53}$$

where d_{F} represents the functional strategy of a quark particle [N_c in SU(N_c)] and q_{s} represents the Abelian charge of a quark, $k \equiv |k|$, and the kinetic function $A(p_{\parallel}, k)$ of the emitted particle

$$A\left(p_{\parallel}, k\right) \equiv \begin{cases} \dfrac{n_b\left(k + p_{\parallel}\right)\left[1 + n_b\left(p_{\parallel}\right)\right]}{2p_{\parallel}\left(p_{\parallel} + k\right)}, & \text{scalars} \\ \dfrac{n_f\left(k + p_{\parallel}\right)\left[1 - n_f\left(p_{\parallel}\right)\right]}{2\left[p_{\parallel}\left(p_{\parallel} + k\right)\right]^2}\left[{p_{\parallel}}^2 + \left(p_{\parallel} + k\right)^2\right], & \text{fermions} \end{cases} \tag{7.54}$$

with

$$n_{\mathrm{b}}(p) \equiv \frac{1}{\exp[\beta(p-\mu)] - 1}, \quad n_{\mathrm{f}}(p) \equiv \frac{1}{\exp[\beta(p-\mu)] + 1} \tag{7.55}$$

where the calculation $f(p\perp; p_{\parallel}, k)$ has been integrated into Eq. (7.54) to solve the following linear integral equation to confirm that three-diode scattering occurred for multiple photon production [42–44]:

$$\begin{aligned} 2\mathrm{p}_{\perp} = {} & i\delta E\, \mathrm{f}\left(\mathrm{p}_{\perp}; p_{\parallel}, k\right) + \frac{\pi}{2} C_F g_s^2 m_D^2 \int \frac{d^2 q_{\perp}}{(2\pi)^2} \frac{dq_{\parallel}}{2\pi} \frac{dq^0}{2\pi} 2\pi\delta\left(q^0 - q_{\parallel}\right) \times \frac{\mathrm{T}}{|\mathrm{q}|} \\ & \times \left[\frac{2}{|\mathrm{q}^2 - \Pi_{\mathrm{L}}(\mathrm{Q})|^2} + \frac{\left[1 - \left(q^0/\left|\mathrm{q}_{\parallel}\right|\right)^2\right]^2}{\left|(q^0)^2 - \mathrm{q}^2 - \Pi_{\mathrm{T}}(\mathrm{Q})\right|^2}\right] \times \left[\mathrm{f}\left(\mathrm{p}_{\perp}; p_{\parallel}, k\right) - \mathrm{f}\left(\mathrm{q} + \mathrm{p}_{\perp}; p_{\parallel}, k\right)\right] \end{aligned} \tag{7.56}$$

C_F is a quadratic quark [$C_F = (N_c^2 - 1)/2N_c = 4/3$ in QCD], m_D is the leading-order Debye mass, and δE is the difference in quasiparticle energies, which has also been calculated considering the photon emission and the state of thermodynamic temperature equilibrium,

$$\delta E \equiv k^0 + E_p \mathrm{sign}\left(p_{\|}\right) - E_{p+k} \mathrm{sign}\left(p_{\|} + k\right) \tag{7.57}$$

For an SU(N) gauge theory with N_s complex scalars and N_f Dirac fermions in the fundamental representation, the Debye mass is given by [33]

$$m_D^2 = \frac{1}{6}(2N + N_s + N_f)g^2T^2 + \frac{N_f}{2\pi^2} g^2\mu^2 \tag{7.58}$$

At the last integration (7.53), to confirm the photon energy emission rate accurately in the region $p_{\|} > 0$, I have calculated the distribution of $n(k + p_{\|})[1 \pm n(p_{\|})]$ that contains $A(p_{\|}, k)$, which confirms the production of pair annihilation by using the following equation:

$$n_b(-p) = -[1 + \bar{n}_b(p)], \qquad n_f(-p) = [1 - \bar{n}_f(p)], \tag{7.59}$$

with $n(p) \equiv 1/[e\beta(p + \mu) \mp 1]$ as the appropriate antiparticle distribution function; the factor $A(p_{\|}, k)$ in this interval may be rewritten in the form:

$$A\left(p_{\|}, k\right) \equiv \begin{cases} \dfrac{n_b\left(k - \left|p_{\|}\right|\right)\bar{n}_b\left(\left|p_{\|}\right|\right)}{2\left|p_{\|}\right|\left(k - \left|p_{\|}\right|\right)}, & \text{scalars} \\[2ex] \dfrac{n_f\left(k - \left|p_{\|}\right|\right)\bar{n}_f\left(\left|p_{\|}\right|\right)}{2\left[\left|p_{\|}\right|\left(k - \left|p_{\|}\right|\right)\right]^2}\left[p_{\|}^2 + \left(k - \left|p_{\|}\right|\right)^2\right], & \text{fermions} \end{cases} \tag{7.60}$$

Thus, the explicit form of the energy E_p of a hard quark with momentum $|p|$ is given by:

$$E_p = \sqrt{p^2 + m_\infty^2} \simeq |\mathrm{p}| + \frac{m_\infty^2}{2|\mathrm{p}|} \simeq \left|p_{\|}\right| + \frac{\mathrm{p}_\perp^2 + m_\infty^2}{2\left|p_{\|}\right|} \tag{7.61}$$

where the asymptotic thermal "mass" is

$$m_\infty^2 = \frac{C_f g^2 T^2}{4} \tag{7.62}$$

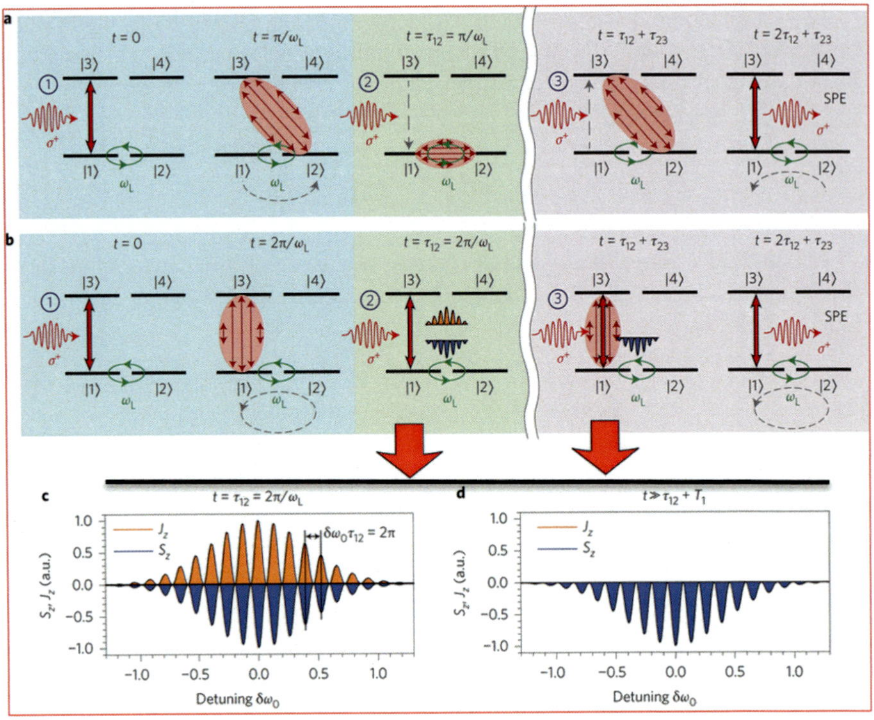

Fig. 7.8 Schematic presentation of the main mechanisms responsible for magnetic-field-induced photon (**a**) shows the heating of photons is simultaneously coupled into the fundamental mode and higher-order mode of the quantum field. (**b**) Coincidence rate of the fundamental mode output. (**c**) Coincidence rate of the higher-order mode output respectively detuning. (**d**) Classical coincidence rates are shown as a function of heating photons with respect to detuning [9]

Substituting the explicit form of E*p* into the definition (7.63) gives the following equation:

$$\delta E = \left[\frac{p_{\perp}^{2} + m_{\infty}^{2}}{2}\right]\left[\frac{k}{p_{\parallel}\left(k + p_{\parallel}\right)}\right] \tag{7.63}$$

Thus, the results here present a derivation of Eqs. (7.54) and (7.57) and are shown in Fig. 7.8 considering the leading-order heating photoemission rate with respect to power counting analysis and electron time of flight for the photon into the curtain wall skin (Fig. 7.9).

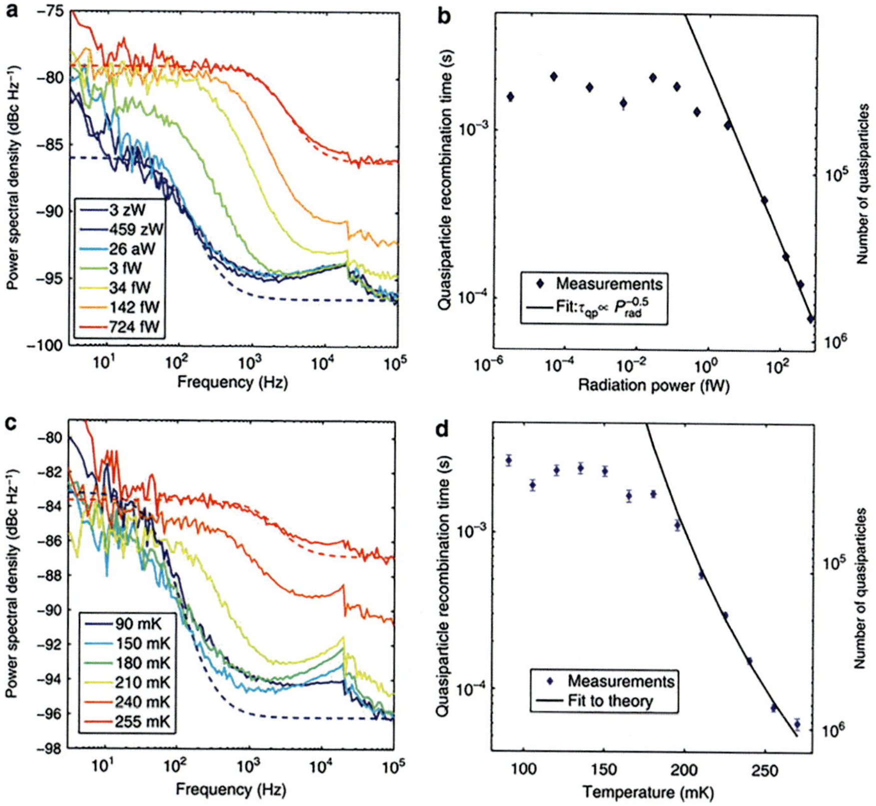

Fig. 7.9 (**a**) Thermal photon power spectral density of the resonator amplitude as a function of frequencies. (**b**) The quasiparticle recombination time as a function of thermal photon power obtained from the roll-off frequency in the measured spectra. (**c**) The thermal photon power spectral density of the resonator amplitude as a function of frequency for different bath temperatures. (**d**) The average heating photon energy measurement (DQD) of **a**, **b**, and **c**, respectively, counting photon excitation into the Higgs boson quantum B field

Conclusions

The traditional cooling and heating system for the building sectors is certainly problematic, as it is not only causing climate change but also destroying our ozone layer. To avoid these two deadly effects, a series of mathematical tests have been performed using MATLAB software to transform solar irradiation into the cooling state of photons by implementing the Bose-Einstein photon distribution mechanism on helium-assisted curtain walls to cool the building. In addition, the Higgs boson [BR ($H \rightarrow \gamma\gamma^{-}$] quantum field has been created by two thermal semiconductor diodes through the helium-assisted curtain wall to transform the cooling photon into heating state photon to naturally heat the building. All mathematical calculation suggested that formation of cooling photon from sunlight and transformation heating photon from cooling photon is quite feasible into the building exterior curtain wall skin to naturally cool and heat the building. Simply, it can be said that naturally the cooling and heating process is but a new field of science to decode the photon thermodynamics to modify the solar irradiance into cooling state photon (HcP^{-}) and transform the cooling state (HcP^{-}) photon into the heating state photon (HtP^{-}) to mitigate the global energy, environmental, and ozone layer crisis.

Acknowledgments The author, Md. Faruque Hossain, declares that any findings, predictions, and conclusions described in this article are solely performed by the author and it is confirmed that there is no conflict of interest for publishing this research paper in a suitable journal and/or publisher.

References

1. Hossain, F. (2016). Solar energy integration into advanced building design for meeting energy demand and environment problem. *Journal of Energy Research, 17*, 49–55.
2. Sivasankar, G., & Kumar, V. S. (2013). Improving low voltage ride through of wind generators using STATCOM under symmetric and asymmetric fault conditions. *International Review on Modelling & Simulations, 6*(4), 1212–1218.
3. Boukhezzar, B., & Siguerdidjane, H. (2009). Nonlinear control with wind estimation of a DFIG variable speed wind turbine for power capture optimization. *Energy Conversion and Management, 50*(4), 885–892.
4. Gopal, C., Mohanraj, M., Chandramohan, P., & Chandrasekar, P. Renewable energy source water pumping systems – A literature review. *Renewable and Sustainable Energy Reviews.* https://doi.org/10.1016/j.rser.2013.04.012
5. Kamal, E., Koutb, M., Sobaih, A. A., & Abozalam, B. (2010). An intelligent maximum power extraction algorithm for hybrid wind-diesel-storage system. *International Journal of Electrical Power & Energy Systems, 32*(3), 170–177.
6. Gupta, N., Singh, S. P., Dubey, S. P., & Palwalia, D. K. (2011). Fuzzy logic controlled three-phase three-wired shunt active power filter for power quality improvement. *International Review of Electrical Engineering (IREE), 6*(3), 1118–1129.
7. Faida, H., & Saadi, J. (2010). Modelling, control strategy of DFIG in a wind energy system and feasibility study of a wind farm in Morocco. *International Review on Modelling and Simulations (IREMOS), 3*(6), 1350–1362.
8. Ghennam, T., Berkouk, E. M., & Francois, B. (2007). A vector hysteresis current control applied on three-level inverter. Application to the active and reactive power control of doubly fed induction generator based wind turbine. *International Review of Electrical Engineering (IREE), 2*(2), 250–259.
9. Langer, L., Poltavtsev, S. V., Yugova, I. A., Salewski, M., Yakovlev, D. R., Karczewski, G., Wojtowicz, T., Akimov, I. A., & Bayer, M. (2014). Access to long-term optical memories using photon echoes retrieved from semiconductor spins. *Nature Photonics, 8*, 851–857.
10. Reinhard, A. (2012). Strongly correlated photons on a chip. *Nature Photonics, 6*, 93–96.
11. Belkacem, H., Gould, B., Feinberg, R., Bossingham, W. E., & Meyerhof. (1993). Semiclassical dynamics and relaxation. *Physical Review Letters, 71*, 1514.
12. Douglas, J. S., Habibian, H., Hung, C.-L., Gorshkov, A. V., Kimble, H. J., & Chang, D. E. (2015). Quantum many-body models with cold atoms coupled to photonic crystals. *Nature Photonics, 9*, 326–331.
13. Agger, A. K., & Sørensen, A. H. (1997). Atomic and molecular structure and dynamics. *Physical Review A, 55*, 402.

14. Baur, G., Hencken, K., Trautmann, D., Sadovsky, S., & Kharlov, Y. (2002). Dense laser-driven electron sheets as relativistic mirrors for coherent production of brilliant X-ray and γ-ray beams. *Physics Reports, 364*, 359.
15. Celik, A. N., & Acikgoz, N. (2007). Modelling and experimental verification of the operating current of mono-crystalline photovoltaic modules using four- and five-parameter models. *Applied Energy, 84*(1), 1–15.
16. Robyns, B., Francois, B., Degobert, P., & Hautier, J. P. (2012). *Vector control of induction machines*. Springer-Verlag.
17. Park, J., Kim, H.-g., Cho, Y., & Shin, C. (2014). Simple modeling and simulation of photovoltaic panels using Matlab/Simulink. *Advanced Science and Technology Letters, 73*(FGCN 2014), 147–155.
18. De Soto, W., Klein, S. A., & Beckman, W. A. (2006). Improvement and validation of a model for photovoltaic array performance. *Solar Energy, 80*(1), 78–88.
19. Li, Q., Xu, D. Z., Cai, C. Y., & Sun, C. P. (2013). *Recoil effects of a motional scatterer on single-photon scattering in one dimension*. Scientific Reports.
20. Artemyev, N., Jentschura, U. D., Serbo, V. G., & Surzhykov, A. (2012). Strong electromagnetic field effects in ultra-relativistic heavy-ion collisions. *European Physical Journal C: Particles and Fields, 72*, 1935.
21. Güçlüa, M. C., Lib, J., Umarb, A. S., Ernstb, D. J., & Strayer, M. R. (1999). Electromagnetic lepton pair production in relativistic heavy-ion collisions. *Ann. Phys., 272*, 7.
22. Najjari, A. B., Voitkiv, A., & Artemyev, A. S. (2009). Simultaneous electron capture and bound-free pair production in relativistic collisions of heavy nuclei with atoms. *Physical Review A, 80*, 012701.
23. Baur, G., Hencken, K., & Trautmann, D. (2007). Revisiting unitarity corrections for electromagnetic processes in collisions of relativistic nuclei. *Physics Reports, 453*, 1.
24. Eichler, J., & Stöhlker, T. (2007). Radiative electron capture in relativistic ion-atom collisions and the photoelectric effect in hydrogen-like high-Z systems. *Physics Reports, 439*, 1.
25. Eichler, J., & T. Stöhlker. (2007). Radiative electron capture in relativistic ion-atom collisions and the photoelectric effect in hydrogen-like high-Z systems. *Physics Reports, 439*, 1.
26. Soon, J. J., & Low, K-S. (2012). *Optimizing photovoltaic model parameters for simulation*. IEEE International Symposium on Industrial Electronics.
27. Yang, L., Wang, S., Zeng, Q., Zhang, Z., Pei, T., Li, Y., & Peng, L.-M. (2011). Efficient photovoltage multiplication in carbon nanotubes. *Nature Photonics*, 672–676.
28. Sharma, K. G., Bhargava, A., & Gajrani, K. (2013). Stability analysis of DFIG based wind turbines connected to electric grid. *International Review on Modelling & Simulations, 6*(3), 879–887.
29. Hencken, K. (2006). Transverse momentum distribution of vector mesons produced in ultraperipheral relativistic heavy ion collisions. *Physical Review Letters*.
30. Benavides, N. D., & Chapman, P. L. (2008). Modeling the effect of voltage ripple on the power output of photovoltaic modules. *IEEE Transactions on Industrial Electronics, 55*(7), 2638–2643.
31. Klein, S. A. (1975). Calculation of flat-plate collector loss coefficients. *Solar Energy, 17*, 79–80.
32. Tame, M. S., McEnery, K. R., Özdemir, Ş. K., Lee, J., Maier, S. A., & Kim, M. S. (2013). Quantum plasmonics. *Nature Physics, 9*, 329–340.
33. Valluri, S. R., Becker, U., Grün, N., & Scheid, W. (1984). Relativistic collisions of highly-charged ions. *Journal of Physics B: Atomic and Molecular Physics, 17*, 4359.
34. Gould, R. J. (1967). Pair production in photon-photon collisions. *Physical Review, 155*, 1404.
35. Zhu, Y., Xiaoyong, H., Yang, H., & Gong, Q. (2014). *On-chip plasmon-induced transparency based on plasmonic coupled nanocavities* (Vol. 4, p. 3752). Scientific Reports.
36. Soedibyo, Pamuji, F. A., & Ashari, M. (2013). Grid quality hybrid power system control of microhydro, wind turbine and fuel cell using fuzzy logic. *International Review on Modelling & Simulations, 6*(4), 1271–1278.

37. Becker, U., Grün, N., & Scheid, W. (1987). K-shell ionisation in relativistic heavy-ion collisions. *Journal of Physics B: Atomic and Molecular Physics, 20*, 2075.
38. Yan, W.-B., & Fan, H. (2014). *Single-photon quantum router with multiple output ports* (Vol. 4, p. 4820). Scientific Reports.
39. Tan, Y. T., Kirschen, D. S., & Jenkins, N. (2004). A model of PV generation suitable for stability analysis. *IEEE Transactions on Energy Conversion, 19*(4), 748–755.
40. Tu, M. W. Y., & Zhang, W. M. (2008). Non-Markovian decoherence theory for a double-dot charge qubit. *Physical Review B, 78*, 235311.
41. Xiao, W., Dunford, W. G., & Capal, A. (2004). *A novel modeling method for photovoltaic cells*. 35th Annula IEEE Power Electronics Specialists Conference, Aachen, Germany, pp. 1950–1956.
42. Pregnolato, T., Lee, E. H., Song, J. D., Stobbe, S., & Lodahl, P. (2015). Single-photon non-linear optics with a quantum dot in a waveguide. *Nature Communications, 6*, 8655.
43. Xiao, Y. F., et al. (2010). Asymmetric Fano resonance analysis in indirectly coupled microresonators. *Physical Review A, 82*, 065804.
44. Zhang, W. M., Lo, P. Y., Xiong, H. N., Tu, M. W. Y., & Nori, F. (2012). General non-Markovian dynamics of open quantum systems. *Physical Review Letters, 109*, 170402.

Chapter 8
Photon Application in the Design of Sustainable Building Technology to Console Global Energy Perplexity

Introduction

Massive development of conventional urbanization throughout the world is consuming fossil energy tremendously. Consequently, it is causing severe environmental perplexity such as acid deposition, stratospheric ozone depletion, and climate change severely where traditional building development is responsible for 40% of this environmental and climate change disaster [1–3]. Besides, conventional energy deposition is getting finite level. At present the total amount of fossil fuel reserved in the whole world is 36,600 EJ (crude oil 1.65×1011 t or unit energy 4.2×10^{10} J/t is equivalent to 6930 EJ; natural gas 1.81×10^{14} t or unit energy 3.6×107 J/m^3 is equivalent to 6500 EJ; high-quality coal 4.90×10^{11} t or unit energy 3.1×10^{10} J/t is equivalent to 15,000 EJ; low-quality coal 4.3×10^{11} t or 1.9×10^{10} J/t unit energy is equivalent to 8200 EJ) [4–6]. The annual energy consumption worldwide was 283 EJ in 1980, 347 EJ in 1990, 400 EJ in 2000, 511 EJ in 1994, and 590 EJ in 2025. This rate is expected to increase at 607 EJ in the year 2020, 702 EJ in 2030, 855 EJ in 2040, and 988 EJ in 2050 [7–9]. In the year 2017, the total fossil energy consumption was 560 EJ where building sector alone consumed 224 EJ. This means if the current level of fossil fuel consumption continues, the total fossil fuel energy source will be run out in 65 years. Since the fossil fuel causes severe environmental disaster and also gets finite level rapidly, clean and renewable energy source is an urgent demand for the sustainability of the Earth. Here, the solar radiation has been defined as a sustainable energy to implement in the building exterior skin by the process of photophysical reaction to act PV panel to produce the clean energy to satisfy its total energy demand without any outsource connection. The average solar radiation on the Earth's surface is 1366 W/m^2, commonly known as solar constant. The radius of the Earth is $(2/\pi) \times 107$ m, and thus, the total solar radiation reaching the Earth is solar $= 1366 \times (4/\pi) \times 1014 \cong 1.73 \times 10^{17}$ W [6, 10]. There are 86,400 s

M. F. Hossain, *Global Sustainability*, https://doi.org/10.1007/978-3-031-34575-3_8

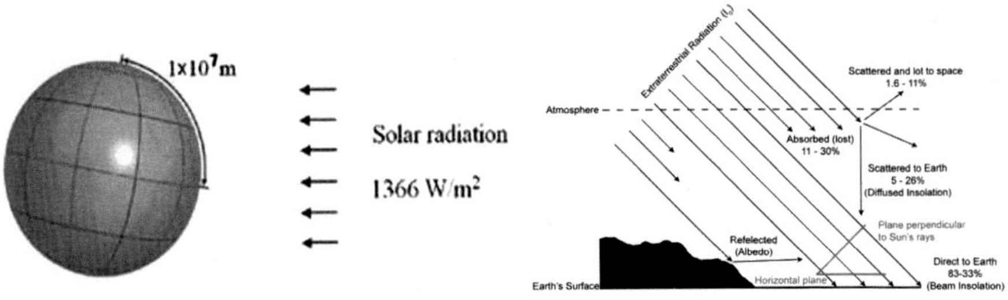

Fig. 8.1 (**a**) The yearly solar irradiance arrives on surface of Earth. The mean solar energy on the Earth is 1366 W/m^2. The length of the meridian of the Earth is 10,000,000 m. The total solar irradiance that arrives at the surface of the Earth per year is 5,460,000 EJ calculatively. (**b**) The effect of the atmosphere on the solar radiation reaching the Earth's surface

a day, with an average of 365.2422 days a year. Total annual solar radiation energy per year = 1.73 × 1017 × 86,400 × 365.2422 = 5.46 × 1024 J, equivalent to 5,460,000 EJ/year (Fig. 8.1). The world energy consumption in 2017 was 5.60 × 10^{20} J = 560 EJ, of which nearly 40% was spent by building sector. Only 0.01% of the solar energy reaching the Earth can meet the global energy demand where building sector can play a vital role to harvest solar energy by its exterior skin to fulfill the mission of clean energy technology for the sustainability of the mother Earth.

In the past several decades, huge research has been performed on solar energy and its supply technology to the national grid for commercial application as the source of alternative energy technology [11–14]. Shi el al. showed that solar energy can be harvested by installing massive solar panel in a particular places tropical or subtropical area and then supply to the national grid as a source of sustainable energy technology [15]. Gleyzes et al. suggested an advanced mechanism of solar panel by the application of graphene silicon surface can have the ability to maximum 30% efficiency to capture the solar energy to convert into clean energy [16]. Reinhard et al. show that the breakdown of photon energy and its application by quantum machines into a PV panel can produce tremendous amount of clean energy [17]. All these research findings are indeed interesting, but these technologies required additional place and technology and supply mechanism to utilize the solar energy, and none of these investigations revealed that this solar energy can be utilized directly by building itself by using its exterior curtain wall to act as the PV panel to produce energy. In this research, therefore, an innovative technology has been proposed to design all buildings to have at least 25% of the exterior is curtain walls to be used as the photovoltaic (PV) panel to capture the solar radiation and then convert it into clean energy to meet the total energy demand for a building.

Methods and Materials

The building is proposed to design of which 25% of the exterior curtain walls is to be built with solar panels. Prior to that this solar panel acting curtain wall skin must be determined the factor involved angle, latitude, longitude, and coordinate transformation in a Cartesian coordinate system to ensure that the maximum solar thermal radiation can be captured by the panel (Fig. 8.2). Considering angle, acting solar panel needs to be designed for the effect of latitude and module tilt on the solar radiation received throughout the year and the module is to be facing south in the

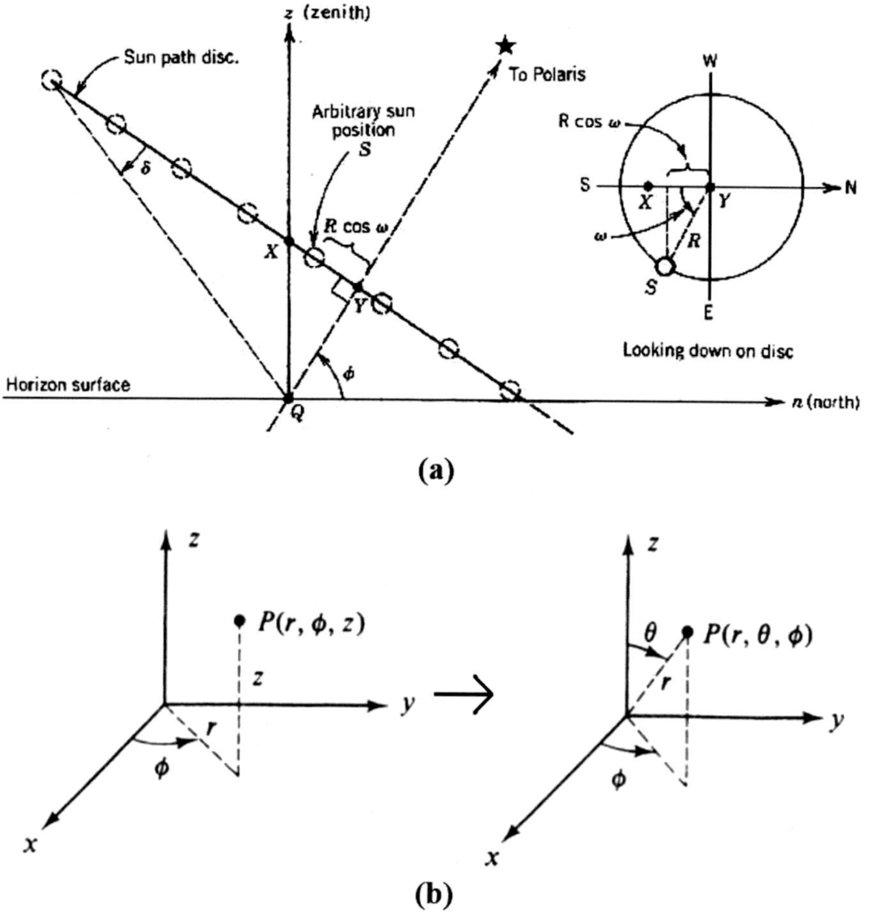

Fig. 8.2 (**a**) Photovoltaic (PV) panel angle has been calculated in consideration of the impact of latitude and mode tilt on the solar energy received within the year. (**b**) The Cartesian coordinate has been utilized for the horizon clarification considering use of the conventions where south is x, west is y, and z is the position of a celestial body, and it is determined by two angles of the equatorial systems

northern hemisphere and north in the southern hemisphere [18–20]. Cartesian coordinates for the horizon system need to be used accordingly where south should be x, west should be y, and the zenith should be z [16, 21, 22]. These positions of the celestial body are to be determined by two angles, height h and azimuth angle A; Cartesian coordinates for the equatorial system of z'-axis points to the North Pole, east-west axis y'-axis, and the x'-axis are perpendicular to both directors. Then the position of the celestial body is to be determined by declination δ and hour angle ω; refer to the figure in the preceding texts.

Since the light is an electromagnetic wave that is produced when an electric charge vibrates due to a hot object, acting PV panel installation must follow the longitude and latitude and polar coordinates and three dimensions axis (x, y, z) to get maximum sunlight obtainable on the building whole year considering the Stefan–Boltzmann laws. The clarification of Stefan–Boltzmann laws is that the electromagnetic waves follow the equal-partition radiation intensity once it emits on the plane of the PV panel [23, 24].

Once the angles and Cartesian coordinates of the solar panel are determined, the panel wiring diagram is focused on ensuring that the photovoltaic DC current can transmit into AC current at high efficiency [25, 26]. Thus, two of the Kyocera plates are mounted on aluminum bars to carry light from photovoltaic modules DC and then multiplies into the short-circuit current by 125% and uses this value for all 80% efficiency and then get AC current 125% of the continuous flow of the current by using three-parallel circuit.

Thereafter, the number of parameters of the current-voltage (I-V) characteristic should be explained considering two/single-diode equivalent circuit model of the PV cell [27–29]. Subsequently, the photovoltaic array at various parameters (*voltage proliferation, transformation rate, and PVVI curves*) and least control strategy are being used to confirm the active electricity production (I_{v+}) from solar energy to the utilization it by the building itself (Fig. 8.3).

The next step is to determine the photovoltaic current production by I_{pv} calculation from the model of one diode (Fig. 8.4a), considering I-V-R relationship (Fig. 8.4b) and using the illumination received by the photovoltaic array to convert from DC to AC and then use for domestic energy and low-voltage current demand (Fig. 8.4c).

Since the solar thermal energy is taken by the photophysical reaction of photovoltaic panel (PV), which is a continuum flow of photons into the PV panel, it is necessary to determine the excellent photophysical reaction into the curtain wall acting PV panel. For this, it is essential to clarify solar thermal conductivity and solar cell anti-reflective coatings by analyzing quantum electrodynamics, the most effective field in modern physics to capture much more solar energy [30–32]. Subsequently, classical statistical physics has been used to determine the radiation energy density of inner surface on its curtain wall considering the electromagnetic wave and Maxwell-Boltzmann constant statistics (Fig. 8.4). Therefore, a mathematical model of photovoltaic dynamic, in this research, has been developed to capture maximum solar energy by the acting PV panel of building exterior skin curtain wall where the following equation calculates the energy output of a photovoltaic (PV) cell:

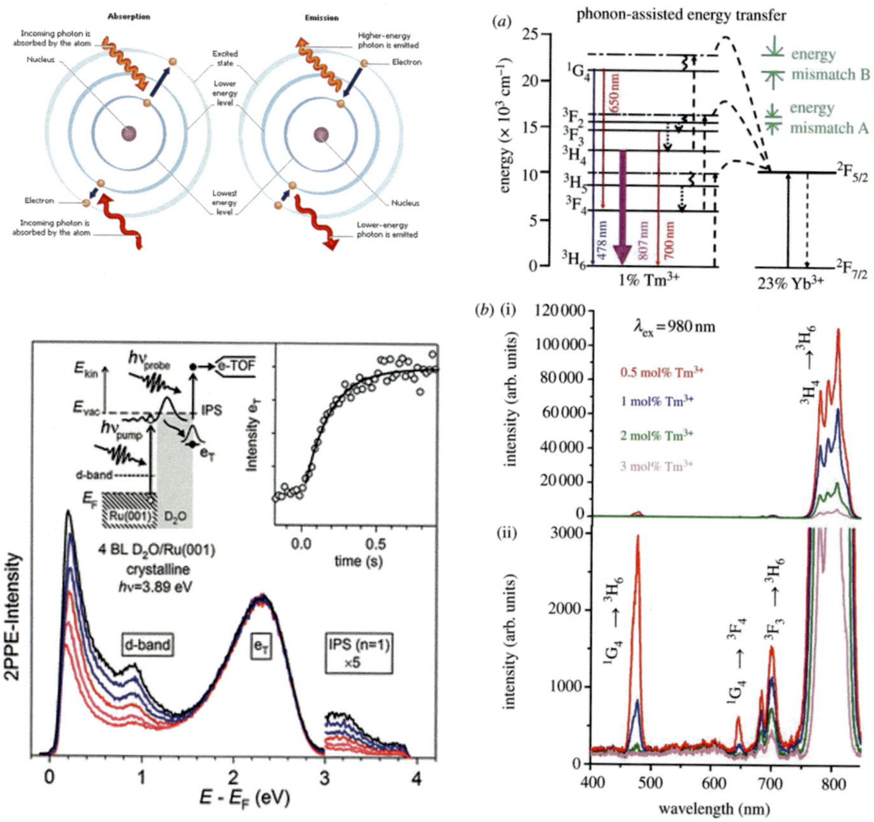

Fig. 8.3 Diagram of PV system model. (**1**) The module photon absorption and emission reaction mechanism once solar irradiance is on the photovoltaic mode, (**2**) solar intensity rate, (**3**) (*a*) the irradiation energy at 980 nm; solid arrows represent photon absorption or emission, dotted arrows represent multi-phonon relaxation, dashed arrows represent phonon-assisted energy transfer, and wavy arrows represent energy mismatches. (*b*) (i) The up-conversion spectrum of UCNPs excited at 980 nm (90 mW). (ii) Vertical axis is magnified to show the details of the visible emission spectra

$$P_{pv} = \eta_{pvg} A_{pvg} G_t \tag{8.1}$$

In this equation, η_{pvg} refers to the PV generation efficiency, A_{pvg} refers to the PV generator area (m^2), and G_t refers to the solar radiation in a titled module plane (W/m^2). η_{pvg} can be further defined as:

$$\eta_{pvg} = \eta_r \eta_{pc}[1 - \beta(T_c - T_{cref})] \tag{8.2}$$

η_{pc} refers to the power conditioning efficiency; when MPPT applied, it is equal to 1; β refers to temperature coefficient (0.004–0.006 per °C); η_r refers to the reference module efficiency; and T_{cref} refers to the reference cell temperature in °C. The reference cell temperature (T_{cref}) can be obtained from the relation below:

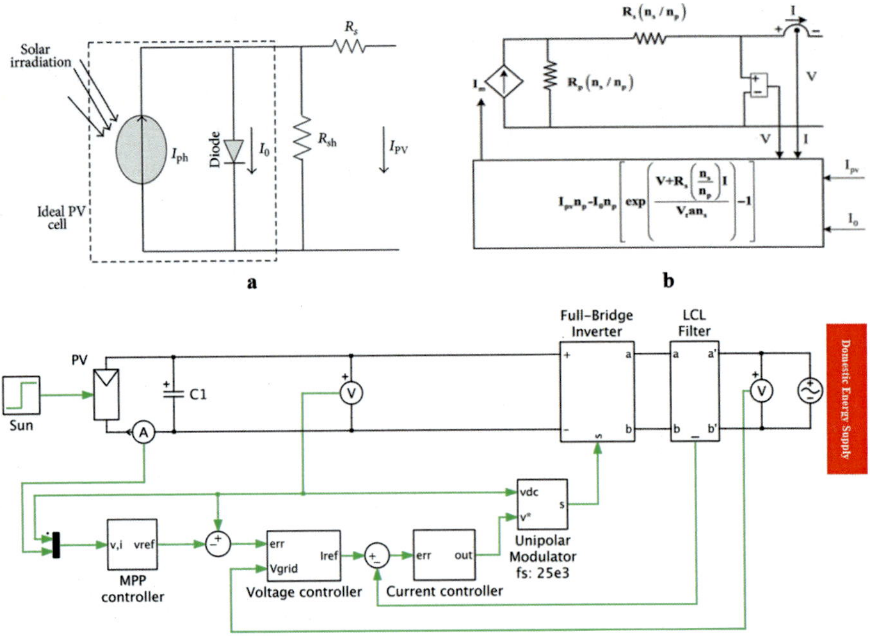

Fig. 8.4 Single-diode circuit of a photovoltaic (PV) cell modeled by MATLAB simulation, (**a**) the photovoltaic current production, (**b**) the model with a diode considering *I-V-R* relationship (**c**), the conversion process of DC to AC for the use of domestic energy and low-voltage current for the building

$$T_c = T_a + \left(\frac{\text{NOCT} - 20}{800}\right) G_t \tag{8.3}$$

T_a refers to the ambient temperature in °C, G_t refers to the solar irradiance in a tilted module plane (W/m^2), and NOCT refers to the standard operating cell temperature in Celsius (°C) degree. The total irradiance in the solar cell, considering both standard and diffuse solar irradiance, can be estimated by the following equation:

$$I_t = I_b R_b + I_d R_d + (I_b + I_d) R_r \tag{8.4}$$

The solar cells, which is essentially a P-N junction semiconductor able to produce electricity via the PV effect, which is interconnected in a series-parallel configuration to form a photovoltaic (PV) cell [33–35]. Besides, to improve the efficiency of the resulting photovoltaic (PV), graphene is integrated into the PV module [15, 36, 37] (Fig. 8.5).

Using a standard single diode, as depicted in Fig. 8.2, for a cell with N_s series-connected arrays and N_p parallel-connected arrays, the cell current must be related to the cell voltage as

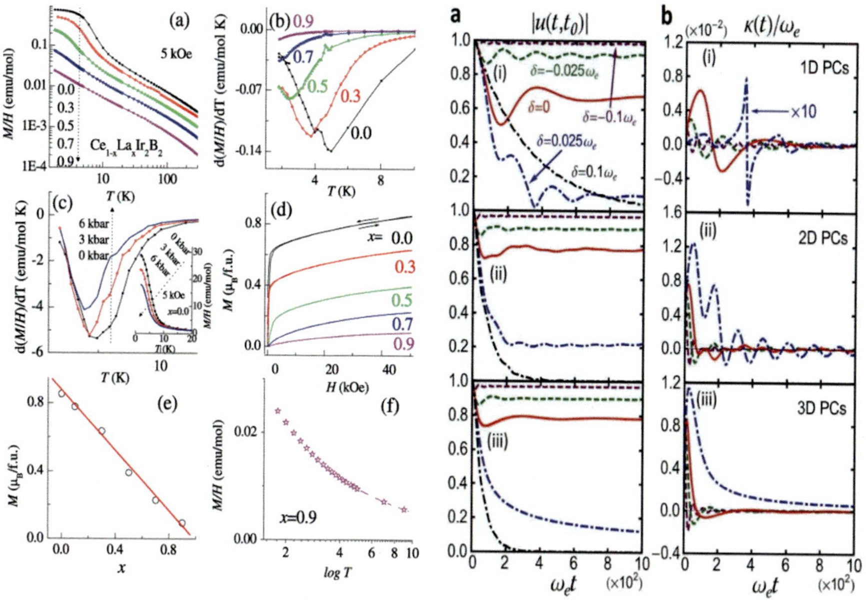

Fig. 8.5 (1) Photonic thermal activation mechanism in various PB parameters, (2) dynamic photon emission in PV cells. (a) shows the charge generation in the PB area $<a(t)> = 5\ u(t, t_0) < a(t_0)>$ and (b) demonstrates the dynamic photon emission rate $k(t)$, plotted for (i) 1D, (ii) 2D, and (iii) 3D areal field into the PV cells [38, 39]

$$I = N_{\mathrm{p}}\left[I_{\mathrm{ph}} - I_{\mathrm{rs}}\left[\exp\left(\frac{q(V + IR_{\mathrm{s}})}{AKTN_{\mathrm{s}}} - 1\right)\right]\right] \tag{8.5}$$

where

$$I_{\mathrm{rs}} = I_{\mathrm{rr}}\left(\frac{T}{T_{\mathrm{r}}}\right)^3 \exp\left[\frac{E_{\mathrm{G}}}{AK}\left(\frac{1}{T_{\mathrm{r}}} - \frac{1}{T}\right)\right] \tag{8.6}$$

In Eqs. 8.5 and 8.6, q refers to the electron charge (1.6×10^{-9} C), K refers to Boltzmann's constant, A refers to the diode idealist factor, and T refers to the cell temperature (K). I_{rs} refers to the cell reverse saturation current at T, T_{r} refers to the cell referred temperature, I_{rr} refers to the reverse saturation current at T_{r}, and E_{G} refers to the band-gap energy of the semiconductor used in the cell. The photocurrent I_{ph} varies with the cell's temperature and radiation as follows:

$$I_{\mathrm{ph}} = \left[I_{\mathrm{SCR}} + k_{\mathrm{i}}(T - T_{\mathrm{r}})\frac{S}{100}\right] \tag{8.7}$$

I_{SCR} refers to the cell short-circuit current at the reference temperature and irradiance, k_{i} refers to the short-circuit current temperature coefficient, and S refers to the solar irradiance (mW/cm^2). The *I-V* characteristics of the photovoltaic (PV) cell can be derived using a single-diode model that includes an additional shunt resistance concurrent with the optimal shunt diode model as follows:

$$I = I_{ph} - I_D \tag{8.8}$$

$$I = I_{ph} - I_0\left[\exp\left(\frac{q(V + R_s I)}{AKT} - 1\right) - \frac{V + R_s I}{R_{sh}}\right] \tag{8.9}$$

I_{ph} refers to the photocurrent (A), I_D refers to the diode current (A), I_0 refers to the inverse saturation current (A), A refers to the diode constant, q refers to the charge of the electron (1.6×10^{-9} C), K refers to Boltzmann's constant, T refers to the cell temperature (°C), R_s refers to the series resistance (ohm), R_{sh} refers to the shunt resistance (Ohm), I refers to the cell current (A), and V refers to the cell voltage (V). The output current of the PV cell using the diode model can be described as follows:

$$I = I_{PV} - I_{D1} - I_{D2} - \left(\frac{V + IR_s}{R_{SH}}\right) \tag{8.10}$$

where

$$I_{D1} = I_{01}\left[\exp\left(\frac{V + IR_s}{a_1 V_{T1}}\right) - 1\right] \tag{8.11}$$

$$I_{D2} = I_{02}\left[\exp\left(\frac{V + IR_s}{a_2 V_{T2}}\right) - 1\right] \tag{8.12}$$

I_{01} and I_{02} are the reverse saturation currents of diode 1 and diode 2, respectively, and V_{T1} and V_{T2} are the thermal voltages of the respective diodes. The diode idealist constants are represented by a_1 and a_2. The simplified model of the photovoltaic (PV) system model is presented below:

$$v_{oc} = \frac{V_{oc}}{cK\,T/q} \tag{8.13}$$

$$P_{max} = \frac{\frac{V_{oc}}{cK\,T/q} - \ln\left(\frac{V_{oc}}{cK\,T/q} + 0.72\right)}{\left(1 + \frac{V_{oc}}{nK\,T/q}\right)}\left(1 - \frac{V_{oc}}{\frac{V_{oc}}{I_{SC}}}\right)\left(\frac{V_{oc0}}{1 + \beta\ln\frac{G_0}{G}}\right) \times \left(\frac{T_o}{T}\right)^y I_{sc0}\left(\frac{G}{G_o}\right)^a \tag{8.14}$$

where ν_{oc} refers to the normalized value of the open-circuit voltage V_{oc} related to the thermal voltage $V_t = nkT/q$, K refers to Boltzmann's constant, n refers to the idealist factor ($1 < n < 2$), T refers to the temperature of the photovoltaic (PV) module in Kelvin, α refers to the factor responsible for all the nonlinear effects on which the photocurrent depends, q refers to the electron charge, γ refers to the factor representing all the nonlinear temperature-voltage effects, while β refers to a photovoltaic (PV) module technology-specific dimensionless coefficient. Equation (8.14) only represents the maximum energy output of a single photovoltaic (PV) module, while a real system consists of several photovoltaic (PV) modules connected in series and in parallel. Therefore, the equation of total power output for an array with N_s cells connected in series and N_p cells connected in parallel with power P_M for each module would be for PV panel

$$P_{array} = N_s N_p P_M \tag{8.15}$$

Naturally the movement of photon flux applied to the solar panel will be activated by the photophysical reactions to deliver energy level charges [40, 41]. Since the energy density of the solar radiation considering the photon wave frequency has been modeled by using the classical statistical physics in Fig. 8.7a, the maximum solar energy formation considering a single photon excitation at the rate of 1.4 eV with an energy value of 27.77 MW/m^2 eV has been determined in Fig. 8.6b.

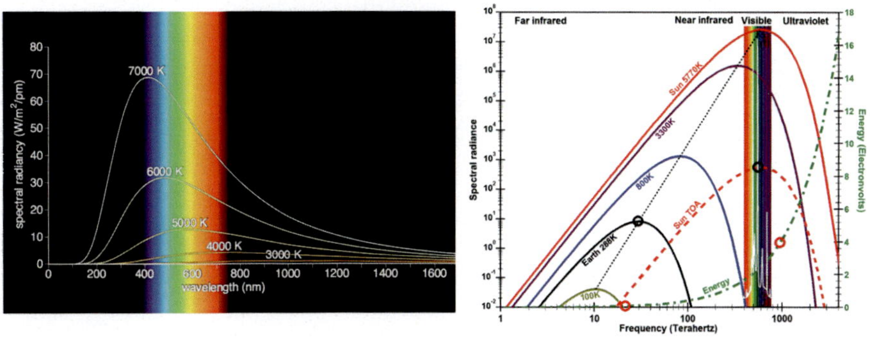

(a) (b)

Fig. 8.6 The thermal energy density of the solar radiation frequencies shown by the classical statistical physics and the figure depicted the solar radiation in various temperatures. (**a**) The spectral irradiance of the light in difference wavelength. (**b**) The radiation in different frequencies at different temperatures where the maximum irradiance by sun nearly at 5800 K (actual 5770 K) power is equivalent to $6.31 \times 10^7 (W/m^2)$; peak E is 1.410 (eV); peak λ is 0.88 (μm); peak μ is $2.81 \times 10^7 (W/m^2 \cdot eV)$

Results and Discussion

The result of the PV model is determined by the *I-V* equation of PV cells in the single-diode mode. The *I-V* relationship equation in the PV panel can be expressed as

$$I = I_L - I_O\left\{ \exp\left[\frac{q(V+I_{RS})}{AkT_C}\right] - 1\right\} - \frac{(V+I_{RS})}{R_{Sh}} \tag{8.16}$$

I_L represents the photon-generating current, *IO* represents the saturated current in the diode, R_s represents resistance in a series, *A* represents the diode passive function, k (= 1.38 × 10^{-23} W/m^{2}K) represents Boltzmann's constant, q (=1.6 × 10^{-19}C) represents the charge amplitude of an electron, and T_C represents the functional cell temperature. Subsequently, the *I-q* relationship in the PV cells varies owing to the diode current and/or saturation current, which can be expressed as [44, 45]:

$$I_O = I_{RS}\left(\frac{T_C}{T_{ref}}\right)^3 \exp\left[\frac{qEG\left(\frac{1}{T_{ref}} - \frac{1}{T_C}\right)}{KA}\right] \tag{8.17}$$

where I_{RS} represents the saturated current considering the functional temperature and solar irradiance and qEG represents the band-gap energy into the silicon and graphene PV cell considering the normal, normalized, and perfect modes.

Considering a PV module, the *I-V* equation, apart from the *I-V* curve, is a conjunction of *I-V* curves among all cells of the PV panel. Therefore, the equation can be rewritten as follows to determine the *V-R* relationship:

$$V = -IR_s + K\log\left[\frac{I_L - I + I_O}{I_O}\right] \tag{8.18}$$

Here, *K* is as a constant $\left(= \frac{AkT}{q}\right)$ and I_{mo} and V_{mo} are the current and voltage in the PV panel, respectively. Therefore, the relationship between I_{mo} and V_{mo} shall be same as the PV cell *I-V* relationship:

$$V_{mo} = -I_{mo}R_{Smo} + K_{mo}\log\left(\frac{I_{Lmo} - I_{mo} + I_{Omo}}{I_{Omo}}\right) \tag{8.19}$$

where I_{Lmo} represents the photon-generated current, I_{Omo} represents the saturated current into the diode, R_{smo} represents the resistance in series, and K_{mo} represents the factorial constant. Once all non-series (NS) cells are interconnected in series, the series resistance shall be counted as the summation of each cell series resistance

$R_{smo} = N_S \times R_S$, and the constant factor can express as $K_{mo} = N_S \times K$. There is a certain amount of current flow into the series-connected cells; thus, the current flow in Eq. 8.5 remains the same in each component, i.e., $I_{omo} = I_o$ and $I_{Lmo} = I_L$. Thus, the module I_{mo}-V_{mo} equation for the N_S series of connected cells will be written as

$$V_{mo} = -I_{mo}N_S R_S + N_S K \log\left(\frac{I_L - I_{mo} + I_o}{I_o}\right) \tag{8.20}$$

Similarly, the current-voltage calculation can be rewritten for the parallel connection once all N_P cells are connected in parallel mode and can be expressed as follows [46–48]:

$$V_{mo} = -I_{mo}\frac{R_S}{N_p} + K \log\left(\frac{N_{sh}I_L - I_{mo} + N_p I_o}{N_p I_o}\right) \tag{8.21}$$

Because the photon-generated current primarily will depend on the solar irradiance and relativistic temperature conditions of the PV panel, the current can be calculated using the following equation:

$$I_L = G[I_{SC} + K_I(T_C - T_{ref})] * V_{mo} \tag{8.22}$$

where I_{sc} represents PV current at 25 °C and KW/m^2, K_I represents the relativistic PV panel coefficient factor, T_{ref} represents the PV panel's functional temperature, and G represents the solar energy in KW/m^2 (Fig. 8.7).

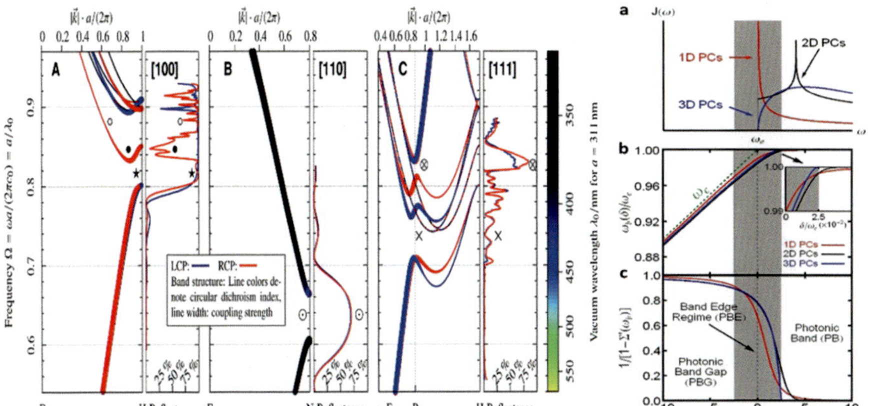

Fig. 8.7 (**1**) Photonic thermal energy conversion modes. **2**(a) Unit area versus frequency at various DOSs in 1D, 2D, and 3D glazing wall skins. **2**(b) Photonic modes of frequencies for functional tuning. **2**(c) Photonic modes of magnitudes to release the energy calculated by Eq. (8.2). The photonic modes depict the crossover into the glazing wall skin in 1D and 2D. In the complex 3D transitional state, they depict the crossover into the PV cell once the point break frequency *vc* transforms from a PBG area to a photonic band (PB) area [42, 43]

Conversion of Electricity

To convert solar thermal energy into electricity, a single-diode circuit has been used consisting of a small disk of semiconductor attached by wire to a circuit consisting of a positive and a negative film of silicon placed under a thin slice of glass and attached to graphene, using a building's exterior curtain wall skin. Necessarily, the PV panel shall have an *open-circuit point*, where the current and voltage are at the maximum, open-circuit voltage V_{oc}, and the *maximum power point* can be clarified using the maximum current and voltage calculation immediately upon capturing the non-equilibrium photons [49–51]. The power delivered by a PV panel will thus have the capability to attain a maximum value at the points (I_{mp},V_{mp}) [52–54]. It is confirmed by the PV panel's capability of such current-voltage flow to get net energy production by the PV panel by analyzing circuit models of the PV module (a) normal, (b) normalized, and (c) perfect modes where perfect modes are the best option.

To establish a connection between the number of light-quanta solar energy by steady state, the intensity of solar irradiance is considered to convert it into electricity energy by PV panel [55–57]. The number of stationary states of light quanta is a certain type of polarization whose frequency is in the range of ν_r to $\nu_r + d\nu_r$ [17, 58, 59]. From their maximum solar radiation, it can be achieved at 1.4 eV with an energy value of 27. 77 mW/m^2 eV based on an average of 5-h solar irradiance harvesting in a day peak levels, which is the equivalent of 27,770 kW/year or 7.6 kW/day energy [60, 61]. Due to physical principles, there are losses in the conversion of solar energy into DC power and conversion of direct current into alternating current (AC). This ratio of AC to DC is called "derating factor," which is typically 0.8 [62, 63]. Thus, the surface texture of selective solar metal is excellent in energy conversation [26, 27, 47], since the current net conversion by solar panels is 125% higher level with an efficiency of 80% [11, 13, 14] of solar panels, which means that (27,770 × 1.25 × 0.8) = 27,770 kW/year or 7.6 kW/day. Energy remains equal to the solar initially what was before the introduction to the solar panel. Necessarily, the maximum solar irradiance is depicted 1.4 eV with an energy value of 27.77 mW/m^2 eV in Fig. 8.8 per year in an average of 5 h a day maximum levels for 365 days referents by solar panel and black body [48, 64, 65]. A standard residential house requires average 6 kW/day [46, 66, 67]. Since the produced energy is equivalent to 27,770 kW/year or 7.6 kW/day, which in fact will meet the energy demand for a residential house required 6 kW/day by using only one solar panel 1 m^2. The average energy consumption rate monthly of commercial office or buildings is about 10,000 kW/day for a foot printing 32 m × 31 m with 30 m (ten floors), respectively [68–70]. In calculation of a building is an average of 32 m × 31 m footprint with a height of 30 m, total installed 1 m^2 PV panels requires 1195 units (945 + 250) with the capacity of 7.6 kW/unit energy production can provide total energy × 1195 = 9082 kW/day to meet the daily energy demand of about 10,000 kW/day for a commercial office or building.

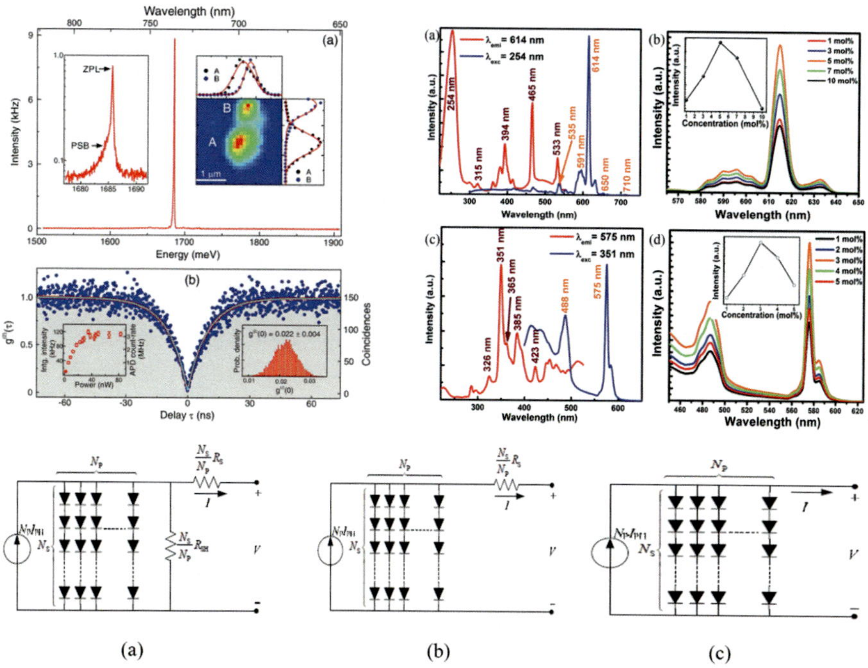

Fig. 8.8 Solar cell current-voltage characteristic features for conceptual function of (**1**) net optimal solar energy production intensity with respect to the delay of coincidental factor and (**2**) intensity of solar energy production in various wavelengths, (**3**) the current-voltage module of the current source near the short-circuit point and as a voltage source in the vicinity of the open-circuit point at the condition of the PV module is (a) normal, (b) normalized, and (c) perfect modes

Design and Construction Cost of the Curtain Wall Acting Solar Panel for Energy Production of a Ten-Storey Building (Table 8.1)

Table 8.1 The estimates (1 m^2 each solar panel and total 1195 panels) are prepared by confirming recent (June 2017) cost of material for selective manufacturer and labor rate added in accordance with international of union specified trade workers considering US location

List of component	Material cost	Labor cost	Equipment cost	GC and OH cost	Total cost
Acting solar panel	$10,000	$5000	$2500	$3500	$21,000
Instrumentation	$2000	$1000	$2000	$1000	$6000
Electrical and mechanical control	$2500	$1000	$1000	$900	$5400
Supply for 30 years cost at $0.05/kwh for monthly 4000 kwh for 100 people					$72,000
				Total cost	**$104,400**

The equipment rental cost is calculated as current rental market in conjunction with the standard practice of construction of the production rate

Savings on Energy Cost

On the other hand, the net cost for 30 years energy purchase from a traditional utility sources for a standard industry (100 people capacity) at 0.12/kwh of 4000 kwh per month is ($30 \times 12 \times 4000 \times 0.12$) \$172,800. This difference between traditional energy use and curtain wall-assisted PV panel energy production clearly indicates the cost saving of \$68,400 once curtain wall-assisted PV panel is used as the energy source.

Conclusions

In recent decades, concern about the deadly risks of greenhouse gases (GHGs) has been growing due to their level of accumulation in the atmosphere and the adverse impact on Earth. Given that conventional energy consumption is the final factor that leads to environmental perplexity and climate change, the intelligent deployment of solar energy, in effect, for a better methodological application to mitigate global environmental perplexity and climate changes is an urgent demand. In addition, the limited nature of the fossil fuel reserve poses a serious challenge for future energy supply as the consumption of fossil fuels will end in the next 65 years. In this study, it is therefore proposed to ensure an economical, reliable sustainable energy technology to meet the future energy demand, which is also climate-friendly. Simply capturing solar thermal energy using exterior curtain wall demonstrated in this research shall indeed be the innovative technology to meet the total energy demand for the building sector and contribute with the quota to mitigate the global energy needs. Therefore, application of this sustainable energy technology is very much computational that need to be encouraged globally for the maximum utilization of solar energy harvesting by building sector to secure a sustainable and environmentally friendly Earth.

Acknowledgments The author, Md. Faruque Hossain, declares that any findings, predictions, and conclusions described in this article are solely performed by the author and it is confirmed that there is no conflict of interest for publishing this research paper in a suitable journal and/or publisher.

References

1. Artemyev, N., Jentschura, U. D., Serbo, V. G., & Surzhykov, A. (2012). Strong electromagnetic field effects in ultra-relativistic heavy-ion collisions. *European Physical Journal C: Particles and Fields, 72*, 1935.
2. Birnbaum, K. M., et al. (2005). Photon blockade in an optical cavity with one trapped atom. *Nature, 436*, 87–90.
3. Xiao, Y. F., et al. (2010). Asymmetric Fano resonance analysis in indirectly coupled microresonators. *Physical Review A, 82*, 065804.

4. Busch, K., von Freymann, G., Linden, S., Mingaleev, S. F., Tkeshelashvili, L., & Wegener, M. (2007). Periodic nanostructures for photonics. *Physics Reports, 444*, 101.
5. Chang, D. E., Sørensen, A. S., Demler, E. A., & Lukin, M. D. (2007). A single-photon transistor using nanoscale surface plasmons. *Nature Physics, 3*, 807–812.
6. Cheng, M., & Song, Y. (2012). Fano resonance analysis in a pair of semiconductor quantum dots coupling to a metal nanowire. *Optics Letters, 37*, 978–980.
7. Armani, D. K., Kippenberg, T. J., Spillane, S. M., & Vahala, K. J. (2003). Ultra-high-Q toroid microcavity on a chip. *Nature, 421*, 925.
8. Chen, J., Wang, C., Zhang, R., & Xiao, J. (2012). Multiple plasmon-induced transparencies in coupled-resonator systems. *Optics Letters, 37*, 5133–5135.
9. Gupta, N., Singh, S. P., Dubey, S. P., & Palwalia, D. K. (2011). Fuzzy logic controlled three-phase three-wired shunt active power filter for power quality improvement. *International Review of Electrical Engineering (IREE), 6*(3), 1118–1129.
10. Liao, J. Q., & Law, C. K. (2010). Correlated two-photon transport in a one-dimensional waveguide side-coupled to a nonlinear cavity. *Physical Review A, 82*, 053836.
11. Dayan, B., et al. (2008). A photon turnstile dynamically regulated by one atom. *Science, 319*, 1062–1065.
12. Douglas, J. S., Habibian, H., Hung, C., Gorshkov, A., Kimble, H., & Chang, D. (2015). Quantum many-body models with cold atoms coupled to photonic crystals. *Nature Photonics, 9*, 326–331.
13. Gould, R. J. (1967). Pair production in photon-photon collisions. *Physical Review, 446*, 297–300.
14. Guerlin, C., et al. (2007). Progressive field-state collapse and quantum non-demolition photon counting. *Nature, 448*, 889.
15. Shi, T., Fan, S., & Sun, C. P. (2011). Two-photon transport in a waveguide coupled to a cavity in a two-level system. *Physical Review A, 84*, 063803.
16. Gleyzes, S., et al. (2007). Quantum jumps of light recording the birth and death of a photon in a cavity. *Nature, 446*, 297.
17. Reinhard, A. (2011). Strongly correlated photons on a chip. *Nature Photonics, 6*, 93–96.
18. Eichler, J., & Stöhlker, T. (2007). Radiative electron capture in relativistic ion-atom collisions and the photoelectric effect in hydrogen-like high-Z systems. *Physics Reports, 439*, 1.
19. Hossain, M. F. (2016). Solar energy integration into advanced building design for meeting energy demand. *International Journal of Energy Research, 40*, 1293–1300.
20. Hossain, M. F. (2017). Design and construction of ultra-relativistic collision PV panel and its application into building sector to mitigate total energy demand. *Journal of Building Engineering*. https://doi.org/10.1016/j.jobe.2016.12.005
21. Englund, D., et al. (2010). Resonant excitation of a quantum dot strongly coupled to a photonic crystal nanocavity. *Physical Review Letters, 104*, 073904.
22. Yan, W., & Fan, H. (2014). Single-photon quantum router with multiple output ports. *Scientific Reports, 4*, 4820.
23. Beyer, H. F., Gassner, T., Trassinelli, M., Heß, R., Spillmann, U., Banaś, D., Blumenhagen, K.-H., Bosch, F., Brandau, C., Chen, W., Chr Dimopoulou, E., Förster, R. E., Grisenti, A. G., Hagmann, S., Hillenbrand, P.-M., Indelicato, P., Jagodzinski, P., Kämpfer, T., Chr Kozhuharov, M., Lestinsky, D. L., Litvinov, Y. A., Loetzsch, R., Manil, B., Märtin, R., Nolden, F., Petridis, N., Sanjari, M. S., Schulze, K. S., Schwemlein, M., Simionovici, A., Steck, M., Th Stöhlker, C. I., Szabo, S. T., Uschmann, I., Weber, G., Wehrhan, O., Winckler, N., Winters, D. F. A., Winters, N., & Ziegler, E. (2015). Crystal optics for precision x-ray spectroscopy on highly charged ions—Conception and proof. *Journal of Physics B Atomic Molecular and Optical Physics, 48*, 144010.
24. Li, Q., Xu, D. Z., Cai, C. Y., & Sun, C. P. (2013). Recoil effects of a motional scatterer on single-photon scattering in one dimension. *Scientific Reports, 3*, 3144.
25. Hossain, M. F. (2018). Photonic thermal energy control to naturally cool and heat the building. *Advanced Thermal Engineering, 131*, 576–586.

26. Liao, J. Q., & Law, C. K. (2013). Correlated two-photon scattering in cavity optomechanics. *Physical Review A, 87*, 043809.
27. Li, Q., Xu, D. Z., Cai, C. Y. and Sun, C. P. Recoil effects of a motional scatterer on single-photon scattering in one dimension., Scientific Reports, 2013.
28. Lo, P., Xiong, H., and Zhang, W. Breakdown of Bose-Einstein distribution in photonic crystals., Scientific Reports, 2015.
29. Sayrin, C., et al. (2011). Real-time quantum feedback prepares and stabilizes photon number states. *Nature, 477*, 73.
30. Han, Z., & Bozhevolnyi, S. I. (2011). Plasmon-induced transparency with detuned ultracompact Fabry-Pérot resonators in integrated plasmonic devices. *Optics Express, 19*, 3251–3257.
31. Hossain, M. F. (2017). Green science: Independent building technology to mitigate energy, environment, and climate change. *Renewable and Sustainable Energy Reviews*. https://doi.org/10.1016/j.rser.2017.01.136
32. Hossain, M. F. (2018). Green science: Advanced building design technology to mitigate energy and environment. *Renewable and Sustainable Energy Reviews, 81*(2), 3051–3060.
33. Yan, W., Huang, J., & Fan, H. (2013). Tunable single-photon frequency conversion in a Sagnac interferometer. *Scientific Reports, 3*, 3555.
34. Yan, W.-B., & Fan, H. (2014). Single-photon quantum router with multiple output ports. *Scientific Reports, 4*, 4820.
35. Yu, G. (2004). A novel two-mode MPPT control algorithm based on comparative study of existing algorithms. *Solar Energy, 76*(4), 455–463.
36. Shen, J. T., & Fan, S. (2007). Strongly correlated two-photon transport in a one-dimensional waveguide coupled to a two-level system. *Physical Review Letters, 98*, 153003.
37. Yang, L., Wang, S., Zeng, Q., Zhang, Z., Pei, T., Li, Y., & Peng, L. (2011). Efficient photovoltage multiplication in carbon nanotubes. *Nature Photonics, 5*, 672–676.
38. Hossain, M. F. (2018). Transforming dark photon into sustainable energy. *International Journal of Energy and Environmental Engineering*. https://doi.org/10.1007/s40095-017-0257-1
39. Huang, J. F., Shi, T., Sun, C. P., & Nori, F. (2013). Controlling single-photon transport in waveguides with finite cross section. *Physical Review A, 88*, 013836.
40. Huang, Y., Min, C., & Veronis, G. (2011). Subwavelength slow-light waveguides based on a plasmonic analogue of electromagnetically induced transparency. *Applied Physics Letters, 99*, 143117.
41. Lang, C., et al. (2011). Observation of resonant photon blockade at microwave frequencies using correlation function measurements. *Physical Review Letters, 106*, 243601.
42. Zhang, W. M., Lo, P. Y., Xiong, H. N., Tu, M. W. Y., & Nori, F. (2012). General non-Markovian dynamics of open quantum systems. *Physical Review Letters, 109*, 170402.
43. Rauh, H. (2010). Optical transmittance of photonic structures with linearly graded dielectric constituents. *New Journal of Physics, 12*, 073033.
44. Jentschura, U., Hencken, K., & Serbo, V. (2008). Revisiting unitarity corrections for electromagnetic processes in collisions of relativistic nuclei. *The European Physical Journal C, 58*(2), 281–289.
45. Lei, C. U., & Zhang, W. M. (2012). A quantum photonic dissipative transport theory. *Ann. Phys., 327*, 1408.
46. Zhu, Y., Hu, X., Yang, H., & Gong, Q. (2014). On-chip plasmon-induced transparency based on plasmonic coupled nanocavities. *Scientific Reports, 4*, 3752.
47. Tang, J., Geng, W., & Xiulai, X. (2015). Quantum interference induced photon blockade in a coupled single quantum dot-cavity system. *Scientific Reports, 5*, 9252.
48. Saloux, E., Teyssedou, A., & Sorin, M. (2011). Explicit model of photovoltaic panels to determine voltages and currents at the maximum power point. *Solar Energy, 85*(5), 713–722.
49. Joannopoulos, J. D., Villeneuve, P. R., & Fan, S. (1997). Photonic crystals: Putting a new twist on light. *Nature, 386*, 143.

50. Sánchez Muñoz, C., Laussy, F., Valle, E., Tejedor, C., & González-Tudela, A. (2018). Filtering multiphoton emission from state-of-the-art cavity quantum electrodynamics. *Optica, 5*(1), 14–26.
51. Tame, M. S., McEnery, K. R., Özdemir, Ş. K., Lee, J., Maier, S. A., & Kim, M. S. (2013). Quantum plasmonics. *Nature Physics, 9*, 329.
52. Hencken, K. (2006). Transverse momentum distribution of vector mesons produced in ultraperipheral relativistic heavy ion collisions. *Physical Review Letters, 96*(1), 012303.
53. Klein, S. A. (1975). Calculation of flat-plate collector loss coefficients. *Solar Energy, 17*, 79–80.
54. Roy, D. (2013). Two-photon scattering of a tightly focused weak light beam from a small atomic ensemble: An optical probe to detect atomic level structures. *Physical Review A, 87*, 063819.
55. Manzoni, M. T., Chang, D. E., & Douglas, J. S. (2017). Simulating quantum light propagation through atomic ensembles using matrix product states. *Nature Communications*, 8, 1743.
56. Poshakinskiy, A. V., & Poddubny, A. N. (2016). Biexciton-mediated superradiant photon blockade. *Physical Review A, 93*, 033856.
57. Matteo Mariantoni, H., Wang, R. C., Bialczak, M. L., et al. (2011). Photon shell game in three-resonator circuit quantum electrodynamics. *Nature Physics, 7*, 287–293.
58. O'Shea, D., Junge, C., Volz, J., & Rauschenbeutel, A. (2013). Fiber-optical switch controlled by a single atom. *Physical Review Letters, 111*, 193601.
59. Ruiz, A. (2014). Partial recovery of a potential from backscattering data. In *Communications in partial differential equations* (Springer tracts in modern physics). Taylor & Francis Group.
60. Kofman, A. G., Kurizki, G., & Sherman, B. (1994). Spontaneous and induced atomic decay in photonic band structures. *Journal of Modern Optics, 41*, 353.
61. Zhou, W. (2007). A novel model for photovoltaic array performance prediction. *Applied Energy, 84*(12), 1187–1198.
62. Kolchin, P., Oulton, R. F., & Zhang, X. (2011). Nonlinear quantum optics in a waveguide: Distinct single photons strongly interacting at the single atom level. *Physical Review Letters, 106*, 113601.
63. Reiserer, A., Kalb, N., Rempe, G., & Ritter, S. (2014). A quantum gate between a flying optical photon and a single trapped atom. (research: letter) (report). *Nature, 508*, 237–240.
64. Longo, P., Schmitteckert, P., & Busch, K. (2011). Few-photon transport in low-dimensional systems. *Physical Review A, 83*, 063828.
65. Wang, Y., Zhang, Y., Zhang, Q., Zou, B., & Schwingenschlogl, U. (2016). Dynamics of single photon transport in a one-dimensional waveguide twopoint coupled with a Jaynes-Cummings system. *Scientific Reports, 6*, 33867.
66. Valluri, S. R., Becker, U., Grün, N., & Scheid, W. (1984). Relativistic collisions of highly-charged ions. *Journal of Physics B: Atomic and Molecular Physics, 17*, 4359.
67. Yan, W.-B., Huang, J.-F., & Fan, H. (2013). Tunable single-photon frequency conversion in a Sagnac interferometer. *Scientific Reports, 3*, 3555.
68. Lü, X., Zhang, W., Ashhab, S., Wu, Y., & Nori, F. (2013). Quantum-criticality-induced strong Kerr nonlinearities in optomechanical systems. *Scientific Reports, 3*, 2943.
69. Najjari, B., Voitkiv, A., Artemyev, A., & Surzhykov, A. (2009). Simultaneous electron capture and bound-free pair production in relativistic collisions of heavy nuclei with atoms. *Physical Review A, 80*, 012701.
70. Tu, M. W. Y., & Zhang, W. M. (2008). Non-Markovian decoherence theory for a double-dot charge qubit. *Physical Review B, 78*, 235311.

Chapter 9
Modeling of Time and Space for Designing Massless House and Building in the Air

Introduction

There is a lack of knowledge about the perception of the nature of the time and space in the air that a massless house and building design in the air is impossible to live mankind in the air comfortably. In fact, this knowledge gap is nothing but mystery since no one had done any research about the requirement to design a massless house and building in the time and space, and thus, the problem remains unresolved; therefore, the myth remains a controversy [4, 11, 20]. In this research, thus, a novel mathematical modeling is being proposed considering primarily the activation of the light energy (L_e), dark energy (D_e), and antimatter (A_m) in the time and then, secondarily, how energies of the time and space of the air can be decoded to design a comfortable housing and building in the air. Simply, the energy momentum of the time and space is being modeled to absorb its energy band edges (PBE) by its nano-point break waveguides using its QED to model massless houses and buildings in the time and space in the air [16, 27, 48].

Subsequently, these massless houses and buildings are being comforted by the implementation of Higgs-Bosons ($H \rightarrow \gamma\gamma^{-}$) electromagnetic fields of these activated energies to control the air and temperature of its surroundings' environment to confirm to design a massless house and building in the air.

Methods and Simulation

Computation of Light Energy (L_e) *in the Time and Space*

The computation of light energy considering the sphericity and the orbital parameters of the time and space in the air, a mathematical modeling is being conducted to activate the light energy in the time and space by expressing law of cosines as:

M. F. Hossain, *Global Sustainability*, https://doi.org/10.1007/978-3-031-34575-3_9

$$\rho\cos(c) = \rho\cos(a)\cos(b) + \rho\sin(a)\sin(b)\cos(C) \tag{9.1}$$

Here *a, b,* and *c* represent the arc lengths in radians corresponding to the sides of a spherical triangle where *C* represents the angle of the vertex corresponding to the other side of *arc* with the length *c* [8, 21, 43]. Naturally, considering the implementation of zenith angle Θ, the following terms can be written as the laws of cosines, $C = h = r$; $c = \Theta$; $a = \frac{1}{2}\pi - \phi$; $b = \frac{1}{2}\pi - \delta$, and thus, the equation can be expressed as:

$$\rho\cos(\Theta) = \sin(\phi)\sin(\delta) + \cos(\phi)\cos(\delta)\cos(r) \tag{9.2}$$

To modify this calculation, a further clarification has been conducted considering a general derivative as below:

$$\begin{aligned}\rho\cos(\theta) = {} & \rho\sin(\phi)\sin(\delta)\cos(\beta) + \sin(\delta)\cos(\phi)\sin(\beta)\cos(\gamma) \\ & + \cos(\phi)\cos(\delta)\cos(\beta)\cos(h) - \cos(\delta)\sin(\phi)\sin(\beta)\cos(\gamma)\cos(h) \\ & - \cos(\delta)\sin(\beta)\sin(\gamma)\sin(h)\end{aligned} \tag{9.3}$$

Here, β represents the angle corresponding to the horizon and γ corresponding to the azimuth angle and thus the sphere of any time and space in the multiverse, which is originated from its energy dynamics, denoted here R_E and the mean distance is denoted as R_0 represents here as the average distant corresponding of astronomical unit (*a.u.*) and the energy constant is named here S_0 [7, 12, 33]. Thus, the energy flux onto the plane tangent of the sphere of the time and space is calculated as:

$$Q = \begin{cases} S_0\dfrac{R_0^2}{R_E^2}\cos(\theta) & \cos(\theta) > 0 \\ 0 & \cos(\theta) \leq 0 \end{cases} \tag{9.4}$$

Here, the mean Q over a time is the mean value of Q in relation to its rotation within the time angle of $h = \pi$ to $h = -\pi$, and thus, calculation can be expressed as:

$$Q^{-\text{time}} = -\frac{1}{2\pi}\int_{\pi}^{-\pi} Q\,dh \tag{9.5}$$

Since h_0 is the time-related angle once Q is the positive, it will be feasible during the light energy presence duration in the time and space when $\Theta = 1/2\pi$ or for h_0, and thus the equation can be solved by:

$$\rho \sin(\phi)\sin(\delta) + \cos(\phi)\cos(\delta)\cos\left(h_{0)}\right) = 0 \tag{9.6}$$

or

$$\rho \cos(h_0) = -\tan(\phi)\tan(\delta) \tag{9.7}$$

Since $\tan(\varphi)\tan(\delta) > 1$, the light energy will confirm its existence at $h = \pi$, so $h_0 = \pi$. Subsequently, in a $Q^{\text{-time}} = 0$; $\frac{R_0^2}{R_E^2}$ would remain mostly constant during the period of light energy presence in a specific space in the multiverse, which can be expressed by considering the following integral:

$$\begin{aligned}\int_{\pi}^{-\pi} Q dh &= \int_{h_0}^{-h_0} Q dh \\ &= S_0 \frac{R_0^2}{R_E^2} \int_{h_0}^{-h_0} \cos(\theta) dh = S_0 \frac{R_0^2}{R_E^2} \left[h \sin(\phi)\sin(\delta) + \cos(\phi)\cos(\delta)\sin(h)\right]_{h=h_0}^{h=-h_0} \\ &= -2S_0 \frac{R_0^2}{R_E^2} \left[h_0 \sin(\phi)\sin(\delta) + \cos(\phi)\cos(\delta)\sin\left(h_{0)}\right)\right]\end{aligned} \tag{9.8}$$

Therefore:

$$Q^{-\text{time}} = \frac{S_0}{\pi} \frac{R_0^2}{R_E^2} \left[h_0 \sin(\phi)\sin(\delta) + \cos(\phi)\cos(\delta)\sin\left(h_{0)}\right)\right] \tag{9.9}$$

Since, the θ is represented here as mainstream angle that describe the orbit of the time and space; thus, $\theta = 0$ corresponding to its vernal equinox of δ at its orbital position that can be written as:

$$\delta = \varepsilon \sin(\theta) \tag{9.10}$$

Here ε represents the mainstream longitude of sphere ϖ that is related to the vernal equinox, and thus, the equation can be written as:

$$R_E = \frac{R_0}{1 + e\cos(\theta - \omega)} \tag{9.11}$$

or

$$\frac{R_0}{R_E} = 1 + e\cos(\theta - \omega) \tag{9.12}$$

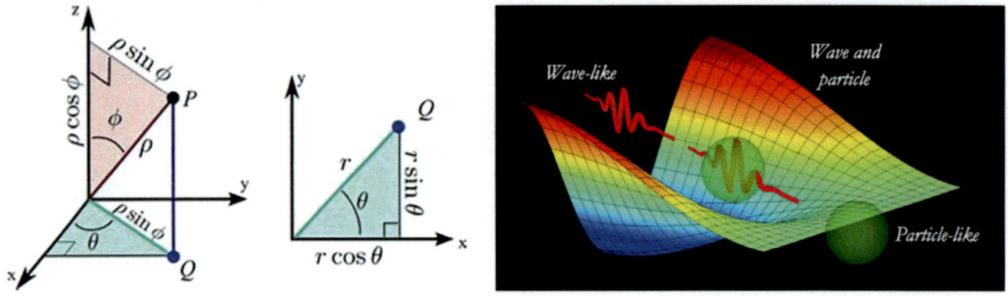

(a) (b)

Fig. 9.1 The distribution of light energy in the time and space, (**a**) clarify the presence of light photon with respect law of cosines, (**b**) schematic diagram of the light energy dynamic on the time and space's sphericity and orbital parameter

Given the above clarification of ϖ, ε, s_o, and e from astrodynamical computation, light energy L_e can be computed for any latitude φ and θ of the time and space (Fig. 9.1). Here, $\theta = 0°$ is representing the precise time of the vernal equinox, $\theta = \infty$ is representing the precise space of the equinox, and thus, the equation can be simplified for light energy irradiance presence on a given time in the space that can be expressed as:

$$L_p = S_0\left(1 + \cos\left(2\pi\frac{n}{x}\right)\right) \tag{9.13}$$

Here S_o is the energy constant, n is the unit of time, x is the duration of time, and π is the sphere of the space that optimized the activation of light energy (L_e) in the time and space, respectively.

Computation of Dark Energy (**D**$_e$) *in the Time and Space*

Since the dark energy is a hidden particle, which is a force resemble to the light energy, the particle of the dark energy is being clarified considering their quantum electrodynamics (QED) in relation to the time and space by analyzing its minimal way of new U(1) gauge field considering its vectorial (V) dynamics of energy of its tangential (T) field of unknown Z dimension of the time and space in the air [1, 9, 10]. Here, the new U(1) gauge symmetry and its kinetic energy proliferation corresponding to the dark energy field, the *standard model hypercharge field*, are being computed to confirm the determination of striking electron flow from the dark energy that will initial to activate of this energy (Fig. 9.2). After striking the

$$V = \sqrt{V_R^2 + (V_L - V_C)^2} \qquad Z = \sqrt{R^2 + (X_L - X_C)^2}$$

$$\phi = \tan^{-1}\frac{V_L - V_C}{V_R} \qquad \phi = \tan^{-1}\frac{X_L - X_C}{R}$$

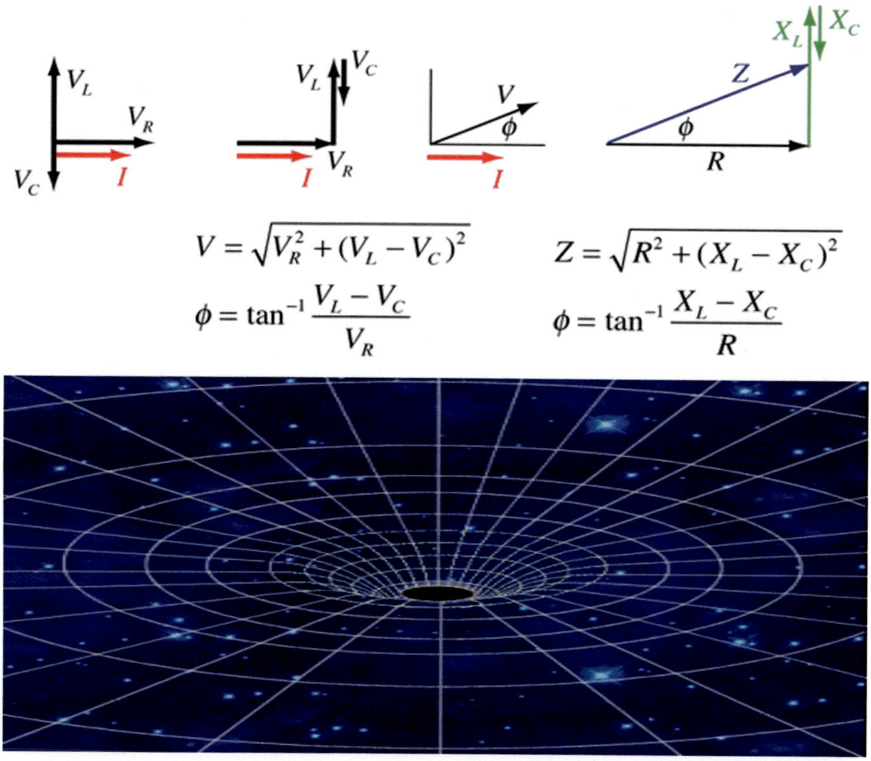

Fig. 9.2 The basic model of dark energy U(1) gauge symmetry in the time and space corresponding to its vectorial dynamics of tangential striking electron within the kinetic energy field parameters

electron, the dark energy ($\overrightarrow{De}$) will be converted to a regular light energy and scatters at an angle θ with speed of c, which is being expressed as the real fermion of the energy momentum of four vectors of each constituent particle of the time and space:

$$\overrightarrow{De} = \left[\frac{Edp}{c}, \overrightarrow{Pd}\right] + \overrightarrow{Xef} \cdots \left[\frac{Eef}{c}, \overrightarrow{Pef}\right] \tag{9.14}$$

Since energy momentum is conserved and thus, $\Delta P = 0$, therefore, the sum of the initial momenta must be equal to the sum of the final momentum [41, 45, 46]. Thus, it can be rewritten the sum of the four vectors as:

$$\overrightarrow{De} + \overrightarrow{Xei} = \overrightarrow{Xp} + \overrightarrow{Xef} \tag{9.15}$$

Since the isolated of the four momentums of the outgoing electron are being calculated and thus the square of both sides of the equation can be rewritten as:

$$\left(\overrightarrow{De} + \overrightarrow{Xei} - \overrightarrow{Xp}\right)^2 = \overrightarrow{Xef}^2 \tag{9.16}$$

Multiplying the energy momentum of four vectors together the cross term here it can be rewritten as:

$$2\overrightarrow{Xd}\cdot\overrightarrow{Xei} = 2 * \left[\frac{Ed}{c}, \overrightarrow{Pd}\right]\cdot\left[m_e c, \vec{0}\right] = 2Edm_e \tag{9.17}$$

Subsequently, multiplying together the four momentums of the initially stationary electron considering the generation of dark energy, the equation can be rewritten as follows:

$$m_d^2c^2 - m_e^2c^2 + -2m_eE_p - 2\left(\frac{E_dE_p}{c^2} - \frac{P_dE_p\cos(\theta)}{c}\right) + 2E_dm_e = m_e^2c^2 \tag{9.18}$$

which can be simplified as:

$$m_d^2c^2 - 2\left(m_eE_p + \left(\frac{E_dE_p}{c^2} - \frac{P_dE_p}{c}\cos(\theta)\right) - E_dm_e\right) = 0 \tag{9.19}$$

Here, to confirm the energy generation from dark energy of the momentums, the vector energy momentum has been calculated by conducting the following mathematical probe as:

$$2\left(m_eE_p + \left(\frac{E_dE_p}{c^2}\right) - \frac{P_dE_p}{c}\cos(\theta)\;)\right) = m_d^2c^2 - 2E_dm_e \tag{9.20}$$

Here, the expression for energy generation is related to the vectorial momentum of dark energy; thus, the limit that $m_d \rightarrow 0$ that can be expressed the case of scattering. Since the $m_d \rightarrow 0$, the $P_d \rightarrow E_c$ and thus, the energy generation can be expressed by adding the differential scattering cross section of the configuration of the dark energy that will confirm the activation of dark energy at any time and space in the air by representing the following equation:

$$D_e = \frac{\partial \sigma_{kn}}{\partial \Omega} = \frac{r_0^2}{2} \frac{\left(P_d^2 c^2 + m_d^2 c^4\right)}{\left(\frac{-\left(\frac{m_d^2 c^2}{2m_e} - E_d\right)}{\left(1+\left(\frac{E_d}{m_e c^2} - \frac{P_d}{m_e^c}\cos(\theta)\right)\right)}\right)^2} \left[\left(\frac{\left(P_d^2 c^2 + m_d^2 c^4\right)^{1/2}}{\frac{-\left(\frac{m_d^2 c^2}{2m_e} - E_d\right)}{\left(1+\left(\frac{E_d}{m_e c^2} - \frac{P_d}{m_e c}\cos(\theta)\right)\right)}}\right) + \frac{\frac{-\left(\frac{m_d^2}{2m_e} - E_d\right)}{\left(1+\left(\frac{E_d}{m_e c^2} - \frac{P_d}{m_e c}\cos(\theta)\right)\right)}}{\left(P^2 C^2 + m_d^2 C^4\right)^{1/2}} - \sin(\theta)^2\right] \quad (9.21)$$

Computation of Antimatter (A_m) in the Time and Space

Since the antimatter is the matter consisting of the anti-particles of the time and space that represents particles in standard regular matter, nanoscopic numbers of anti-particles are being calculated here in relation to its particle's accelerators [2, 6, 47]. Since a standard particle and its anti-particle (proton-antiproton) are the same, its electric charge will have opposite quantum numbers in order to confirm that proton has a positive charge and anti-proton has a negative charge [13, 31, 44]. Simply, the energy content within the antimatter will be absorbed from its surrounding time and space and then will be released as an energy of $E = mc^2$. Therefore, it is confirmed evidence that any time and space in the air consists entirely of standard matter, which is equal to antimatter (H^*) once its free quantum fields (n) are in both activated and excited states (Fig. 9.3).

Here, the quantum field of the antimatter is the components of all free quantum momentum of time and space; thus, its components (t, x) are being represented into a vector $x = (ct, x)$ considering superposition of the energy momentum in a common metric signature convention of $\eta_{\mu\nu} =$ diag $(\pm1, \mp1, \mp1, \mp1)$ of Klein-Gordon equations that can be expressed by the following Table 9.1:

Here, $\pm\eta^{\mu\nu}\partial_\mu\partial_\nu$ is representing the d'Alembert function and ∇^2 is representing the Laplace variable, while c is representing the light speed and ħ is the Planck constant, while $c = ħ = 1$ (Table 9.2).

Here, the Klein-Gordon equation permits two variables where one variable is positive and another is negative, which can be obtained by describing a relativistic wavefunction of the antimatter as:

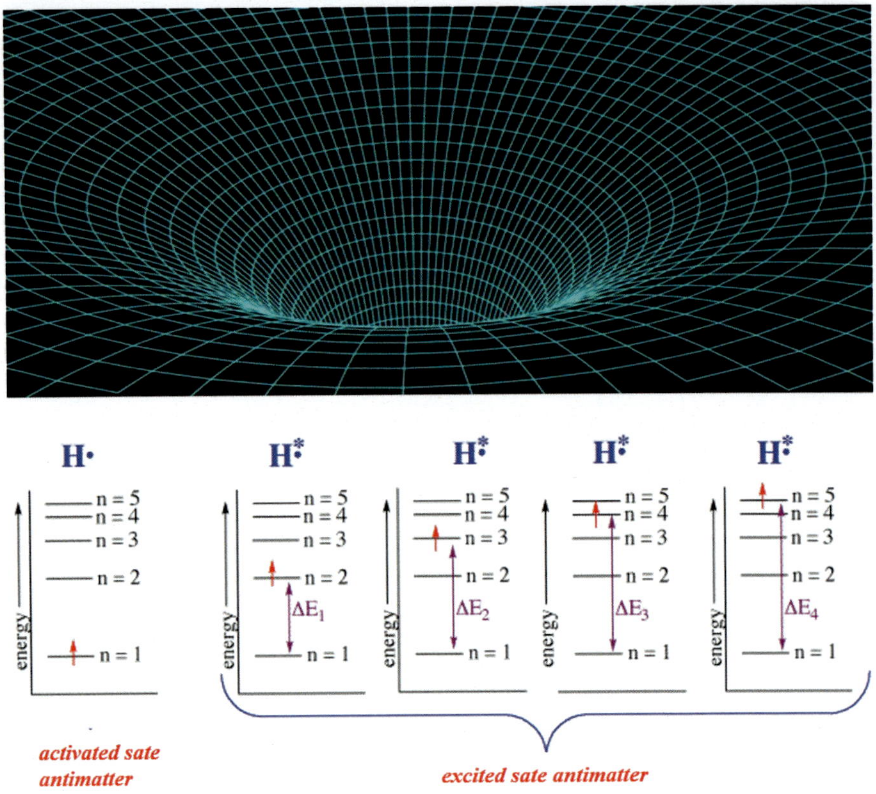

Fig. 9.3 The concept of antimatter distribution energy (H^*) in time and space due to the quantum mechanical functions of its free quantum fields (n) corresponding to its vectorial superposition of activated state and excited state

$$\left[\nabla^2 - \frac{m^2c^2}{\hbar^2}\right]\psi(r) = 0 \tag{9.22}$$

Since the Klein-Gordon function is a standard unit, $(c + m^2)\psi(x) = 0$,with the metric signature of $\eta_{\mu\nu} =$ diag $(\pm 1, -1, -1, -1)$, it is solved by Fourier transformation as:

$$\psi(x) = \int \frac{d^4p}{(2\pi)^4} e^{-ip.x}\psi(p) \tag{9.23}$$

and using spherical dimension of time and space, the complex exponentials are described here as:

Table 9.1 Klein-Gordon equation in normal units with metric of $\eta_{\mu\nu} =$ diag $(\pm 1, \mp 1, \mp 1, \mp 1)$

	Position space $x = (ct, x)$	Fourier transformation $\omega = \frac{E}{\hbar}$, $k = p/\hbar$	Momentum space $p = \left(, \frac{E}{c}, p \right)$
Time and space in the air	$\left(\frac{1}{c^2} \frac{\partial^2}{\partial t^2} - \nabla^2 + \frac{m^2 c^2}{\hbar^2} \right) \psi(t, x) = 0$	$\psi(t, x) = \int \frac{d\omega}{2\pi\hbar} \int \frac{d^3 k}{(2\pi\hbar)^3} e^{\mp i(\omega t - k.x)} \psi(\omega, k)$	$E^2 = p^2 c^2 + m^2 c^4$
Four-vector form	$(c + \mu^2)\psi(x) = 0, \ \mu = mc/\hbar$	$\psi(x) = \int \frac{d^4 p}{(2\pi\hbar)^4} e^{-ip.\frac{x}{\hbar}} \psi(p)$	$p^2 = \pm m^2 c^2$

Table 9.2 Klein-Gordon equation in natural unit metric of $\eta_{\mu\nu} =$ diag $(\pm 1, \mp 1, \mp 1, \mp 1)$

	Position space	Fourier transformation $\omega = E, k = p$	Momentum space $p = (E, p)$
Time and space in the air	$\left(\partial_t^2 - \nabla^2 + m^2\right)\psi(t,x) = 0$	$\psi(t,x) = \int \frac{dw}{2\pi} \int \frac{d^3k}{(2\pi)^3} e^{\mp i(wt - k.x)} \psi(w,k)$	$E^2 = p^2 + m^2$
Four-vector form	$(c + m^2)\psi(x) = 0$	$\psi(x) = \int \frac{d^4p}{(2\pi)^4} e^{-ip.x} \psi(p)$	$p^2 = \pm m^2$

$$p^2 = (p^0)^2 - p^2 = m^2 \tag{9.24}$$

This result will confirm the momenta of the antimatter's positive and negative energy dynamics as:

$$p^0 = \pm E(p) \text{ where } E(p) = \sqrt{p^2 + m^2}. \tag{9.25}$$

Using a new set of constant $C_{(p)}$, the solution will become

$$\psi(x) = \int \frac{d^4p}{(2\pi)^4} e^{ip.x} C(P) \delta(P^0)^2 - E(P)^2) \tag{9.26}$$

It is a standard practice to segregate the positive and negative energy dynamics while working with the positive energy only as:

$$\psi(x) = \int \frac{d^4p}{(2\pi)^4} \delta\left((p^0)^2 - E(p)^2\right)\left(A(p)e^{-ip^0x^0 + ip^ix^i} + B(p)e^{+ip^0x^0 + ip^ix^i}\right)\theta(p^0) \tag{9.27}$$

$$= \int \frac{d^4p}{(2\pi)^4} \delta\left((p^0)^2 - E(p)^2\right)\left(A(P)e^{-ip^0x^0 + ip^ix^i} + B(-P)e^{+ip^0x^0 - ip^ix^i}\right)\theta(p^0)$$

$$\to \int \frac{d^4p}{(2\pi)^4} \delta\left((p^0)^2 - E(p)^2\right)(A(p)e^{-ip.x} + B(P)e^{+ip.x})\theta(p^0) \tag{9.28}$$

Here, $B(p) \to B(-p)$ can be considered p^0- to pick up the positive frequency from its delta as:

$$\psi(x) = \int \frac{d^4p}{(2\pi)^4} \frac{\delta(p^0 - E(p))}{2E(p)} (A(p)e^{-ip.x} + B(p)e^{+ip.x})\theta(p^0) \tag{9.29}$$

$$= \int \frac{d^3p}{(2\pi)^3} \frac{1}{2E(p)} (A(p)e^{-ip.x} + B(p)e^{+ip.x})\big|_{p^0 = +E(p)} \tag{9.30}$$

Since it is a general solution to the Klein-Gordon equation, Lorentz invariant quantities of $p.\ x = p_\mu x^\mu$ are the final variable to solve the Klein-Gordon equation [21, 34, 36]. Subsequently, the Lorentz invariance will absorb the ½$E(p)$ factor into the coefficients $A(p)$and $B(p)$, and thus, the antimatter energy of a free particle can be written as:

$$\frac{p^2}{2m} = E \tag{9.31}$$

By quantizing this, it can confirm the non-relativistic Schrödinger equation for a free particle energy as:

$$\frac{\widehat{p}^2}{2m}\psi = \widehat{E}\psi \tag{9.32}$$

where $\widehat{p} = -i\hbar\nabla$ is representing the momentum function (∇ is the delta variable) and $\widehat{E} = i\hbar\frac{\partial}{\partial t}$ is the energy dynamic.

Since the Schrödinger equation is a non-relativistic invariant, it is standard to use its free particle as the momentum energy as:

$$\sqrt{p^2c^2 + m^2c^4} = E \tag{9.33}$$

Here, implementing the quantum momentum of the free particle, the energy yields can be written as:

$$\sqrt{(-i\hbar\nabla)^2c^2 + m^2c^4\psi} = i\hbar\frac{\partial}{\partial t}\psi \tag{9.34}$$

Subsequently, the electromagnetic fields of the energy yield can be further clarified by the Klein and Gordon equation of e identity, i.e., $p^2c^2 + m^2c^4 = E^2$, which will give:

$$\left((-i\hbar\nabla)^2c^2 + m^2c^4\right)\psi = \left(i\hbar\frac{\partial}{\partial t}\right)^2\psi \tag{9.35}$$

which simplifies to

$$-\hbar^2c^2\nabla^2\psi + m^2c^4\psi = \hbar^2\frac{\partial^2}{\partial t^2}\psi \tag{9.36}$$

Rearranging terms yields

$$\frac{1}{c^2}\frac{\partial^2}{\partial t^2}\psi - \nabla^2\psi + \frac{m^2c^2}{\hbar^2}\psi = 0 \tag{9.37}$$

Rewriting these functions considering the inverse metric $(-c^2, 1, 1, 1)$ in relation to Einstein's summational variable, it can be written as:

$$-\eta^{\mu\nu}\partial_\mu\partial_\nu\psi = \sum_{\mu=0}^{\mu=3}\sum_{\nu=0}^{\nu=3} -\eta^{\mu\nu}\partial_\mu\partial_\nu\psi = \frac{1}{c^2}\partial_0^2\psi - \sum_{\nu=1}^{\nu=3}\partial_\nu\partial_\nu\psi = \frac{1}{c^2}\frac{\partial^2}{\partial t^2}\psi - \nabla^2\psi \tag{9.38}$$

Therefore, the Klein-Gordon equation can be expressed as a covariant function, which is the abbreviation of the form of $(c + \mu^2)\psi = 0$, where $\mu = \frac{mc}{\mathrm{h}}$ and $c = \frac{1}{c^2}\frac{\partial^2}{\partial t^2} - \nabla^2$, and thus, the Klein-Gordon equation can be described as a field of energy dynamics $V(\psi)$ as:

$$c\psi + \frac{\partial V}{\partial \psi} = 0 \tag{9.39}$$

Here, the gauge $\cup(1)$ symmetry is a complex field $\varphi(x)\epsilon\mathbb{C}$ that can satisfy the Klein-Gordon as:

$$\partial_\mu j^\mu(x) = 0, \quad j^\mu(x) = \frac{e}{2m}(\varphi^*(x)\partial^\mu\varphi(x) - \varphi(x)\partial^\mu\varphi^*(x)) \tag{9.40}$$

Subsequently, proof Klein-Gordon equation using algebraic manipulations from to form a complex field $\varphi(x)$ of mass m can be expressed as a covariant notation as follows:

$$(c + \mu^2)\varphi(x) = 0 \tag{9.41}$$

And its complex conjugate

$$(c + \mu^2)\varphi^*(x) = 0 \tag{9.42}$$

Then, multiplying by the vectorial variables of $\varphi^*(x)$ and $\varphi(x)$, it can be expressed as:

$$\varphi^*(c + \mu^2)\varphi = 0 \tag{9.43}$$

$$\varphi(c + \mu^2)\varphi^* = 0 \tag{9.44}$$

Subtracting the former from the latter, it can be obtained

$$\varphi^* c\varphi - \varphi c\varphi^* = 0 \tag{9.45}$$

$$\varphi^*\partial_\mu\partial^\mu\varphi - \varphi\partial_\mu\partial^\mu\varphi^* = 0 \tag{9.46}$$

Then it will confirm

$$\partial_\mu(\varphi^*\partial^\mu\varphi) = \partial_\mu\varphi^*\partial^\mu\varphi + \varphi^*\partial_\mu\partial^\mu\varphi \quad (9.47)$$

From which it can be obtained the conservation law for the Klein-Gordon field:

$$\partial_\mu j^\mu(x) = 0, \quad j^\mu(x) \equiv \varphi^*(x)\partial^\mu\varphi(x) - \varphi(x)\partial^\mu\varphi^*(x) \quad (9.48)$$

Here, implementing of the vectorial variable to the Klein-Gordon field, the scalar field-derived stress-energy tensor can be calculated as:

$$T^{\mu\nu} = \frac{\hbar^2}{m}\left(\eta^{\mu\alpha}\eta^{\nu\beta} + \eta^{\mu\beta}\eta^{\nu\alpha} - \eta^{\mu\nu}\eta^{\alpha\beta}\right)\partial_\alpha\overline{\psi}\partial_\beta\psi - \eta^{\mu\nu}mc^2\overline{\psi}\psi \quad (9.49)$$

Here, the time-time component T^{00} will confirm the positive frequency related to the particles with free energy of antimatter as $\psi(x,t) = \phi v; (x,t)e^{-\frac{i}{\hbar}mc^2t}$, where $\varnothing(x,t) = u_E(x)e^{-\frac{i}{\hbar}Et}$. Thus, defining the kinetic energy $E' = E - mc^2 = \sqrt{m^2c^4 + c^2p^2} - mc^2 \approx \frac{p^2}{2m}$, $E' \ll mc^2$ in the non-relativistic limit $v \sim p \ll c$, and hence $\left|i\hbar\frac{\partial\varnothing}{\partial t}\right| = E'\phi \ll mc^2\phi$; thus, this yield of the derivative ψ can be written as:

$$\frac{\partial\psi}{\partial t} = \left(-i\frac{mc^2}{\hbar}\varnothing + \frac{\partial\varnothing}{\partial t}\right)e^{-\frac{i}{\hbar}mc^2t} \approx -i\frac{mc^2}{\hbar}\varnothing e^{-\frac{i}{\hbar}mc^2t}$$

$$\frac{\partial^2\psi}{\partial t^2} \approx -\left(i\frac{2mc^2}{\hbar}\frac{\partial\varnothing}{\partial t} + \left(\frac{mc^2}{\hbar}\right)^2\varnothing\right)e^{-\frac{i}{\hbar}mc^2t} \quad (9.50)$$

Substituting it with the free Klein-Gordon equation, $c^{-2}\partial_t^2\psi = \nabla^2\psi - m^2\psi$, yield $-\frac{1}{c^2}\left(i\frac{2mc^2}{\hbar}\frac{\partial\varnothing}{\partial t} + \left(\frac{mc^2}{\hbar}\right)^2\varnothing\right)e^{-\frac{i}{\hbar}mc^2t} \approx \left(\nabla^2 - \left(\frac{mc^2}{\hbar}\right)^2\right)\varnothing e^{-\frac{i}{\hbar}mc^2t}$ can be simplified as:

$$-i\hbar\frac{\partial\varnothing}{\partial t} = -\frac{\hbar^2}{2m}\nabla^2\varnothing \quad (9.51)$$

Since free Klein-Gordon equation is a *classical* Schrödinger field, it is further analyzed by quantum field of Klein-Gordon of the time and space [3, 40]. Subsequently, the quantum field of Klein-Gordon equation is being analyzed to confirm the symmetry of the antimatter energy wavefunction φ is in the local U(1) gauge of the sphere of $\varphi \to \varphi' = \exp(i\theta)\varphi$ of the time and space where $\theta(t,x)$ is a local vectorial angle, which transforms the wavefunction into a complex phase as $\exp(i\theta) = \cos\theta + i\sin\theta$, where derivative ∂_μ can be replaced by gauge-covariant variable $D_\mu = \partial_\mu - ieA_\mu$, while the gauge fields transform as $eA_\mu \to eA'_\mu + \partial_\mu\theta$, which confirm that presence of (−,+,+,+) metric signature of the antimatter in the time and space as:

$$D_\mu D^\mu \varphi = -(\partial_t - ieA_0)^2 \varphi + (\partial_i - ieA_i)^2 \varphi = m^2 \varphi \tag{9.52}$$

where A is the scalar potential which can be simplified as functional term of $D_\mu D^\mu \varphi + AF^{\mu\nu} D_\mu \varphi D_\nu (D_\alpha D^\alpha \varphi) = 0$, and thus, the scalar field can be multiplied by i to confirm the for a field of energy dynamics of A_m is sufficiently presence in the time and space by expressing as:

$$A_m = \int_x \eta^{\mu\nu} (\partial_\mu \varphi^* + ieA_\mu \varphi^*)(\partial_\nu \varphi - ieA_\nu \varphi) = \int_x |D\varphi|^2 \tag{9.53}$$

Since the massless light energy L_e, dark energy D_e, and antimatter A_p in the time and space are being activated by the above mathematical modeling, these energies are being further remodeled computationally to pave the time and space to control its surroundings' air and temperature by controlling its energy dynamics to design and a massless comfortable housing and building there.

Mechanism of Massless Housing and Building Design in the Time and Space

Since light energy (L_p), dark energy ($\bar{}(D_p)$), and antimatter (A_p) are the massless elements, it has been activated mathematically to deform as an activated energy state by analyzing its nano-point break waveguide energy momentum [5, 14, 37]. Then, the cluster of its energy momentum is being calculated considering its nano-point deflect waveguide to decode the quantum dynamic into massless activated state energy by expressing the following *Hossain* equation as:

$$H = C + \sum \omega_{ci} a_i^\dagger a_i + \sum_K \omega_k b_k^\dagger b_k + \sum_{ik} \left(V_{ik} a^\dagger b_k + V_{ik}^* b_k^\dagger a_i \right), \tag{9.54}$$

where $a_i \left(a_i^\dagger \right)$ and $b_k \left(b_k^\dagger \right)$ present the function of the nano-point break mode and its energy dynamic modules of its cofactor, V_{ik}, denote the magnitude of the energy mode within its nanostructures.

Simply, this massless energy state module is being formed from its energy dynamics in the time and space considering its the internuclear distances of 1D-H ($n = 1$) + H($n = 2,\ldots,\infty$), 2D-H($n = 2$) + H($n = 2,\ldots,\infty$), and 3D-H($n = \infty$) + H (n = unlimited,$\ldots,\infty$), where n and l are the primary and vectorial quanta of the massless energy. Simply, this massless energy state quanta considering time-dependent energy intensity (k) and space-dependent energy wavelength (λ) confirms the mode of energy dynamics within the time and space to activate this energy into massless energy state (Fig. 9.4).

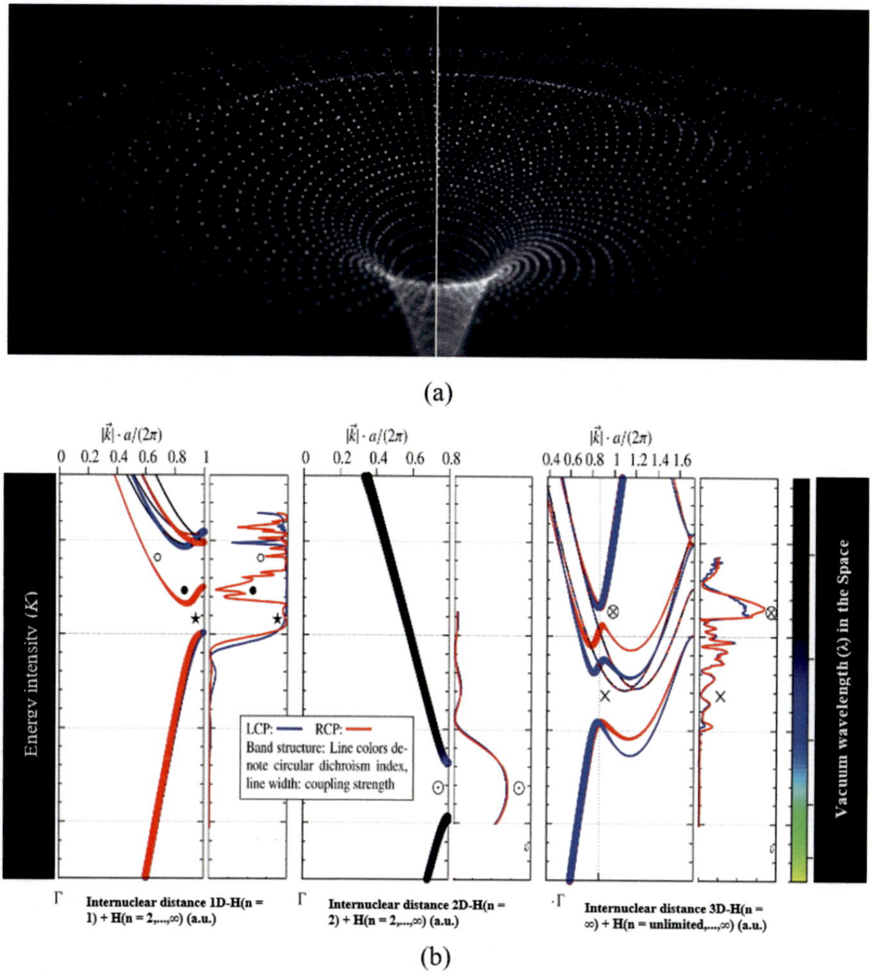

Fig. 9.4 (**a**) The conceptual array of the activated massless energy distributions in the time and space, (**b**) semiclassical pathways of energy band structure (LCP and RCP) that generate activated massless energy within the internuclear distance (a.u.) where red and blue represent the excited and activated states of energy corresponding to its wavelength in the space and energy intensity of (k) to the time

Simply, this massless energy state can thus be defined by *activated energy* transformation characteristics of the massless energy state quanta in a specific area that can be expressed by:

$$E = E_L - E_O\left\{\exp^{\left[\frac{q(V+I_{RS})}{AkT_C}\right]} - 1\right\} - \frac{(V + I_{RS})}{R_S}, \tag{9.55}$$

where E_L presents the massless energy state generation, E_O is the energy dynamics, and R_s is the variable function for the adverse condition. Here A, representation of the activated module energy in related to Boltzmann constant, k ($=1.38 \times 10^{-23}$ W/m^{2}K) where q ($=1.6 \times 10^{-19}$ C) is the electrons of activated energy and T_C is the dynamic proliferation of the energy, and thus, the energy dynamics can be expressed by:

$$E_O = E_{RS}\left(\frac{T_C}{T_{ref}}\right)^3 \exp\left[\frac{qEG\left(\frac{1}{T_{ref}} - \frac{1}{T_C}\right)}{KA}\right]. \tag{9.56}$$

Here, E_{RS} denotes the functional energy that rely on the energy speed and qEG denotes the energy flow per unit area of the massless energy state [15, 35, 38].

Simply, in this massless energy mode, the energy dynamics is related to T–R module of the space that can be expressed by:

$$\mathrm{T} = -\mathrm{IR_s} + \mathrm{K}\ \log\left[\frac{I_L - I + I_O}{I_O}\right], \tag{9.57}$$

where K represents the constant $\left(= \frac{AkT}{q}\right)$ and I_{mo} and V_{mo} are the energy dynamics. Therefore, the correlation between I_{mo} and V_{mo} is similar as the T–R module of the space, which can be expressed precisely as:

$$T_{mo} = -I_{mo}R_{Smo} + K_{mo}\log\left(\frac{I_{Lmo} - I_{mo} + I_{Omo}}{I_{Omo}}\right) \tag{9.58}$$

where I_{Lmo} denotes the net energy generation, I_{Omo} denotes the activated energy dynamics, R_{smo} is the inactivated energy dynamics, and K_{mo} is the constant cofactor related to $K_{mo} = N_S \times K$, and thus, net massless energy state dynamic can be written as:

$$T_{mo} = -I_{mo}N_S R_S + N_S K \log\left(\frac{I_L - I_{mo} + I_o}{I_o}\right). \tag{9.59}$$

Here, massless energy generation relies primarily on energy flux and relativistic thermal state of the space; thus, the thermal conductivity could be expressed as follows:

$$I_L = G[I_{SC} + K_I(T_a)]V_{mo} \tag{9.60}$$

$$T_a = T\left(\frac{I_L}{(G * V_{mo}) \times (\mathrm{I_{sc}} + K_I)}\right) \tag{9.61}$$

where I_{sc} is the energy dynamics per unit area, K_{I} is the relativistic energy co-efficient, T_{cool} is the massless energy thermal of the energy, and G is the massless energy thermal conductivity of the space of its certain area, which is referred as the massless space of the time; thus, it is a massless space for living in the air.

Mechanism of Controlling Air and Temperature of the Massless Houses and Building in the Time and Space

To control air and temperature of the massless space in the air, the local symmetry of Higgs-Boson quantum field of the time and space has been computed [4, 20]. Simply, the time and space has been simulated considering Abelian local symmetries by applying Higgs-Boson electromagnetic field of the time and space to control air and temperature of its surroundings [24, 25, 39]. Here, the momentum of energy dynamic will interrupt the gauge-field symmetries of the time and space, and thus, the Goldstone scalar field will act as the longitude modes of the Higgs-Boson field. Therefore, the local symmetries of quantum field of the time and space in the air will be broken down as particle of T^{α} related to the gauge field of $A_{\mu}^{\alpha}(x)$, and thus, the Higgs-Boson quantum fields will then start to generate local U(1) phase symmetries to control air and temperature [28, 29, 42]. So, this transformation process can be consisting of the precise scalar fields $\Phi(x)$ of energetically charged q paired with the electromagnetic field $A^{\mu}(x)$ of the time and space that could be written as:

$$\mathcal{L} = -\frac{1}{4}F_{\mu\nu}F^{\mu\nu} + D_{\mu}\Phi^{*}\, D^{\mu}\Phi - V(\Phi^{*}\Phi), \tag{9.62}$$

where

$$\begin{aligned} D_{\mu}\Phi(x) &= \partial_{\mu}\Phi(x) + iqA_{\mu}(x)\Phi(x) \\ D_{\mu}\Phi^{*}(x) &= \partial_{\mu}\Phi^{*}(x) - iqA_{\mu}(x)\Phi^{*}(x), \end{aligned} \tag{9.63}$$

and

$$V(\Phi^{*}\Phi) = \frac{\lambda}{2}\,(\Phi^{*}\Phi)^{2} + m^{2}(\Phi^{*}\Phi). \tag{9.64}$$

Here $\lambda > 0$ but $m^2 < 0$; therefore, $\Phi = 0$ is a local peak factor of the scalar field, and the precise form of quantum dynamics is $\Phi = \frac{v}{\sqrt{2}} * e^{i\theta}$ that can be expressed as:

$$v = \sqrt{\frac{-2m^{2}}{\lambda}} \text{ for any real } \theta. \tag{9.65}$$

Consequently, here, the scalar quantum dynamics $\Phi(x)$ will form a nonzero value of $\langle\Phi\rangle \neq 0$, which will form the local U(1) symmetries into electromagnetic field of time and space. Therefore, in this local U(1) symmetries, the quantum dynamics of $\Phi(x)$ will be an activated scalar field of the time and space that can be expressed as:

$$\Phi(x) = \frac{1}{\sqrt{2}}\,\Phi_{\mathrm{r}}(x) * e^{i\Theta(x)},\ \ \text{real}\ \Phi_{\mathrm{r}}(x) > 0,\ \ \text{real}\ \Phi(x). \tag{9.66}$$

Since the scalar field of time and space is a dynamic momentum of the quantum $\Phi(x) = 0$, it will meet the principle of $\langle\Phi\rangle \neq 0$ to be considered as the instinctual function of the time and space in the air. Thus, considering the momentum field of $\phi_{\mathrm{r}}(x)$ and $\Theta(x)$, the scalar field related to the radial field of ϕ_{r} of the time and space can be simplified:

$$V(\phi) = \frac{\lambda}{8}\left(\phi_{\mathrm{r}}^2 - v^2\right)^2 + \text{const.} \tag{9.67}$$

Subsequently, the momentum field can be shifted by applying the variable scalars $\Phi_{\mathrm{r}}(x) = v + \sigma(x)$ and expressed as:

$$\phi_{\mathrm{r}}^2 - v^2 = (v + \sigma)^2 - v^2 = 2v\sigma + \sigma^2 \tag{9.68}$$

$$V = \frac{\lambda}{8}\left(2v\sigma - \sigma^2\right)^2 = \frac{\lambda v^2}{2} * \sigma^2 + \frac{\lambda v}{2} * \sigma^3 + \frac{\lambda}{8} * \sigma^4. \tag{9.69}$$

Thus, the functional derivative $D_\mu\phi$ will become

$$\begin{aligned} D_\mu\phi &= \frac{1}{\sqrt{2}}\left(\partial_\mu\left(\phi_{\mathrm{r}}e^{i\Theta}\right) + iqA_\mu * \phi_{\mathrm{r}}e^{i\Theta}\right) \\ &= \frac{e^{i\Theta}}{\sqrt{2}}\left(\partial_\mu\phi_{\mathrm{r}} + \phi_{\mathrm{r}} * i\partial_\mu\Theta + \phi_{\mathrm{r}} * iqA_\mu\right) \end{aligned} \tag{9.70}$$

$$\begin{aligned} \left|D_\mu\phi\right|^2 &= \frac{1}{2}\left|\partial_\mu\phi_{\mathrm{r}} + \phi_{\mathrm{r}} * i\partial_\mu\Theta + \phi_{\mathrm{r}} * iqA_\mu\right|^2 \\ &= \frac{1}{2}\left(\partial_\mu\phi_{\mathrm{r}}\right) + \frac{\phi_{\mathrm{r}}^2}{2} * \left(\partial_\mu\Theta qA_\mu\right)^2 \\ &= \frac{1}{2}\left(\partial_\mu\sigma\right)^2 + \frac{(v+\sigma)^2}{2} * \left(\partial_\mu\Theta + qA_\mu\right)^2. \end{aligned} \tag{9.71}$$

The Lagrangian is then given by

$$\mathcal{L} = \frac{1}{2}\left(\partial_\mu\sigma\right)^2 - v(\sigma) - \frac{1}{4}F_{\mu\nu}F^{\mu\nu} + \frac{(v+\sigma)^2}{2} * \left(\partial_\mu\Theta + qA_\mu\right)^2. \tag{9.72}$$

Simply, to conduct the heat ($\mathcal{L}_{\text{heat}}$) generation through electromagnetic field of the time and space considering this *Lagrangian*, it has been further expanded $\mathcal{L}_{\text{heat}}$ as the energy dynamics of the electromagnetic field of the time and space that can be expressed as free particles of thermal conductivity:

$$\mathcal{L}_{\text{t}} = \frac{1}{2}\left(\partial_{\mu}\sigma\right)^{2} - \frac{\lambda v^{2}}{2} * \sigma^{2} - \frac{1}{4}F_{\mu\nu}F^{\mu\nu} + \frac{v^{2}}{2} * \left(qA_{\mu} + \partial_{\mu}\Theta\right)^{2}. \tag{9.73}$$

Subsequently, it will initiate to control air and temperature into the quantum field of the time and space considering the scalar particles of λv^2 corresponding to its energy spectrum of singlet and triplet system of the time and space.

Results and Discussion

Mechanism of Massless Housing and Building Design in the Time and Space

To computationally confirm to design a massless house and building design utilizing the activated energy (*HaE*¯) in the time and space, the vectorial momentum of its steady weak coupling energy field has been clarified by determining its absorbed energy rate per unit area of $J(\omega)$. Here, $J(\omega)$ is considered as the quantum field of the density of state (DOS) energy proliferation, which, in fact, is the deformed massless space of energy magnitudes $V(\omega)$ that is formed from its nano-point break waveguide energy momentum [17, 19, 23]. Thus, massless space formation follows the Weisskopf-Wigner theory is therefore all the generated *HaE*¯ shall pass through all dimensional modes (1D, 2D, and 3D) within the time and space in the air (Table 9.3).

The unit areas $J(\omega)$ and the generation of self-energy inductions in the time and space reservoir $\Sigma(\omega)$, demonstrated by the activated energy corresponding to its variables of C, η, and χ as coupled forces among three-dimensional (1D, 2D, and 3D) point break of energy band.

Subsequently, all dimensions of 1D, 2D, and 3D of the time and space are being further clarified by implanting frequency cutoff to avoid the bifurcation of the DOS to confirm the accuracy of massless space formation [30, 32]. Thus, the DOS analyzed at various dimensions of the time and space has been calculated, here named as $\varrho_{\text{PC}}(\omega)$, and is being confirmed by clarifying activated energy frequency formation of Maxwell's theory of nanostructure corresponding to the time and space (Fig. 9.5). The DOS is thus here expressed by $\varrho_{\text{PC}}(\omega) \propto \frac{1}{\sqrt{\omega - \omega_{\text{e}}}}\Theta(\omega - \omega_{\text{e}})$, where $\Theta(\omega - \omega_{\text{e}})$, where ω_{e} is denoted as the frequency of the PBE at the used DOS.

Then, the density of state (DOS) has been implemented to confirm the qualitative form of the non-Weisskopf-Wigner module that represents the precise massless space of energy, calculated by the dimensional clarification of the time and space [17, 19, 26]. This DOS is thereafter determined considering the electromagnetic field

Table 9.3 Massless space energy dynamics considering its densities of states (DOS) in time and space

Energy particle	Unit area $J(\omega)$ for different DOS*	Reservoir-induced self-energy induction $\Sigma(\omega)$*
1D	$\frac{C-1}{\pi}\frac{1}{\sqrt{\omega-\omega_e}}\Theta(\omega-\omega_e)$	$-\frac{C-1}{\sqrt{\omega_e-\omega}}$
2D	$C-\eta\left[\ln\left\|\frac{\omega-\omega_0}{\omega_0}\right\|-1\right]\Theta(\omega-\omega_e)\Theta(\Omega_d-\omega)$	$C-\eta\left[Li_2\left(\frac{\Omega_d-\omega_0}{\omega-\omega_0}\right)-Li_2\left(\frac{\omega_0-\omega_e}{\omega_0-\omega}\right)-\ln\frac{\omega_0-\omega_e}{\Omega_d-\omega_0}\ln\frac{\omega_e-\omega}{\omega_0-\omega}\right]$
3D	$\chi\sqrt{\frac{\omega-\omega_e}{\Omega_C}}\exp\left(-\frac{\omega-\omega_e}{\Omega_C}\right)\Theta(\omega-\omega_e)$	$\chi\left[\pi\sqrt{\frac{\omega_e-\omega}{\Omega_C}}\exp\left(-\frac{\omega-\omega_e}{\Omega_C}\right)erfc\sqrt{\frac{\omega_e-\omega}{\Omega_C}}-\sqrt{\pi}\right]$

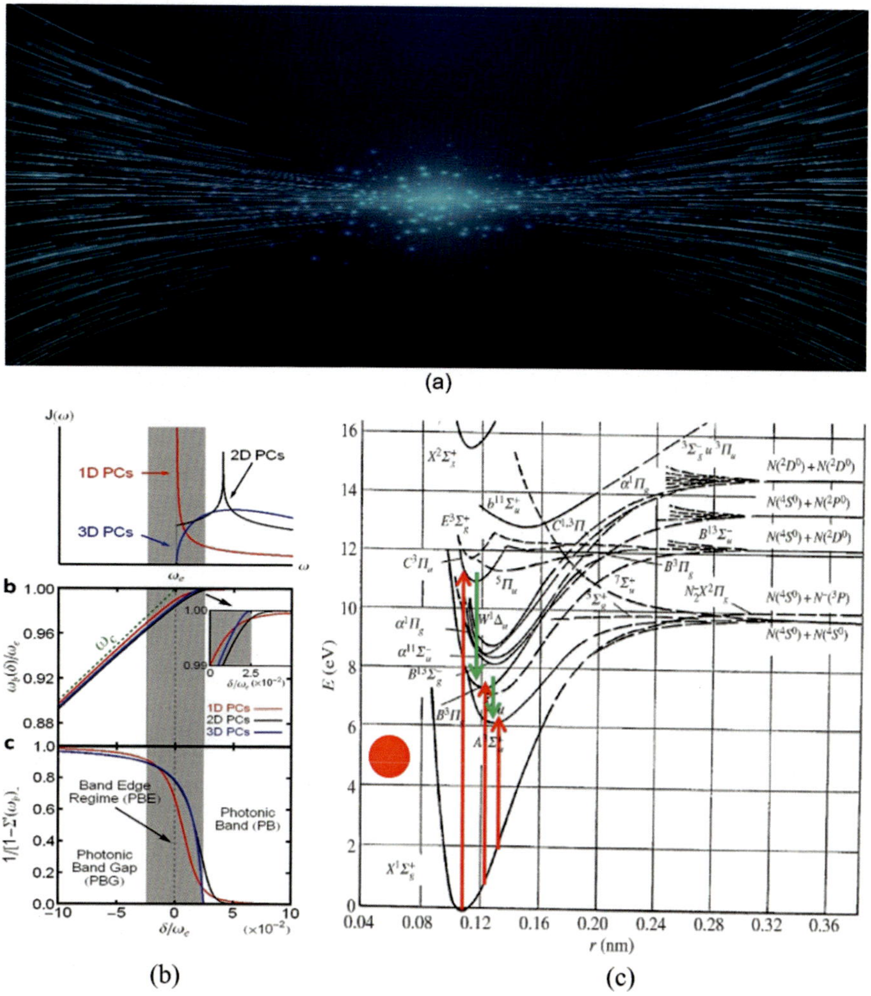

Fig. 9.5 (**a**) The concept of massless space energy dynamics in time and space, (**b**) the activated massless energy band structures and energy transformation mode at various DOSs of 1D, 2D, and 3D in the time and space (**c**) the massless activated energy formation magnitude (eV) in relation to the radial (r) dimension of time and space (Starikovskiy et al. 2015; P. T. Royal Society A: Math. Phys. and Eng. Sci.)

of time and space, which is determined by $\varrho_{\mathrm{PC}}(\omega) \propto \frac{1}{\sqrt{\omega - \omega_{\mathrm{e}}}}\Theta(\omega - \omega_{\mathrm{e}})$. Simply, here, the massless energy of DOS reveals a perfect algorithmic function close to the PBE, which is $\varrho_{\mathrm{PC}}(\omega) \propto -[\ln|(\omega - \omega_0)/\omega_0| - 1]\Theta(\omega - \omega_{\mathrm{e}})$, where ω_{e} defines the prime tip point of the DOS distribution in the time and space.

Subsequently, this distributed DOS energy will transform it into the massless space, which will confirm the total deliberation of energy dynamics of the time and space corresponding to its quantum fields, and, thus, defined above, $J(\omega)$ will confirm the quantum field of activated energy magnitudes $V(\omega)$ corresponding to its photonic band (PB), photonic band edge (PBE), and photonic band gap (PBG) in the time and space as:

$$J(\omega) = \varrho(\omega)|V(\omega)|^2. \tag{9.74}$$

Therefore, considering here the activated energy frequency of ω_c and the formation of massless space of $\langle a(t)\rangle = u(t, t_0)\langle a(t_0)\rangle$, the function $u(t, t_0)$ has been described as the photon structure as $u(t, t_0)$ by clarifying the integral-differential equation as:

$$u(t, t_0) = \frac{1}{1 - \Sigma'(\omega_b)} e^{-i\omega(t - t_0)} + \int_{\omega_e}^{\infty} d\omega \frac{J(\omega)e^{-i\omega(t - t_0)}}{[\omega - \omega_c - \Delta(\omega)]^2 + \pi^2 J^2(\omega)}, \tag{9.75}$$

where $\Sigma'(\omega_b) = [\partial\Sigma(\omega)/\partial\omega]_{\omega = \omega_b}$ and $\Sigma(\omega)$ confirm the self-energy induction into the reservoir,

$$\Sigma(\omega) = \int_{\omega_e}^{\infty} d\omega' \frac{J(\omega')}{\omega - \omega'} . \tag{9.76}$$

Here, the frequency ω_b represents the activated energy frequency mode ($0 < \omega_b < \omega_e$), calculated as $\omega_b - \omega_c - \Delta(\omega_b) = 0$, where $\lesssim\Delta(\omega) = \mathcal{P}\left[\int d\omega' \frac{J(\omega')}{\omega - \omega'}\right]$ is a primary-valued integral of the time and space. In details, it can be explained that activated energy dynamics (eV) is in the time and space is considered as the massless space n_0, i. e., $\rho(t_0) = |n_0\rangle\langle n_0|$, which is obtained computationally from the real-time quantum field of its radial (r) dimension of the space at a time t and expressed as:

$$\rho(t) = \sum_{n=0}^{\infty} \mathcal{P}_n^{(n_0)}(t)|n_0\rangle\langle n_0| \tag{9.77}$$

$$\mathcal{P}_n^{(n_0)}(t) = \frac{[v(t, t)]^n}{[1 + v(t, t)]^{n+1}}[1 - \Omega(\mathrm{t})]^{n_0} \times \sum_{k=0}^{\min\{n_0, n\}} \binom{n_0}{k}\binom{n}{k}\left[\frac{1}{v(t, t)} \frac{\Omega(\mathrm{t})}{1 - \Omega(\mathrm{t})}\right]^k \tag{9.78}$$

where $\Omega(t) = \frac{|u(t, t_0)|^2}{1 + v(t, t)}$. Thus, this result suggested that this massless space is indeed generated through its activated energy dynamic states of the time and space $\mathcal{P}_n^{(n_0)}(t)$ of $|n_0\rangle$ where the proliferation of energy deliberation $\mathcal{P}_n^{(n_0)}(t)$ is in the prime state $|n_0 = 5\rangle$ and in the steady-state limits $\mathcal{P}_n^{(n_0)}(t \to \infty)$ is being eventually confirm to design and massless house and building the time and space, respectively.

Mechanism of Controlling Air and Temperature of the Massless Houses and Building in the Time and Space

Since the electromagnetic field of the time and space is being modeled by Higgs-Boson quantum dynamics, the local symmetry of U(1) here will allow to form gauge-variable QED, in terms of gauge particles $\varnothing' \rightarrow e^{i\alpha(x)}\,\varnothing$, which is the activated state energy (HaE^{-}) that can transformed into thermal state energy in order to control the temperature of the time and space at a comfort level. This mechanism is being confirmed by the following variable derivatives considering the specific transformational laws of the scalar field, written as:

$$\partial_\mu \rightarrow D_\mu = \partial_\mu = ieA_\mu \quad \text{[covariant derivatives]}$$
$$A'_\mu = A_\mu + \frac{1}{e}\partial_\mu \alpha \quad [A_\mu \text{ derivatives}] \tag{9.79}$$

Here, the local U(1) gauge-invariant Lagrangian is being considered as the perplex scalar field that is expressed by

$$\mathcal{L} = (D^\mu)^\dagger (D_\mu \varnothing) - \frac{1}{4}F_{\mu\nu}F^{\mu\nu} - V(\varnothing). \tag{9.80}$$

The term $\frac{1}{4}F_{\mu\nu}F^{\mu\nu}$ is the kinetic motion of the gauge field for considering thermal energy and $V(\phi)$ is denoted the kinetic term written as $V(\varnothing^*\varnothing) = \mu\,(\varnothing^*\varnothing) + \lambda\,(\varnothing^*\varnothing)^2$.

Thus, the equation of the Lagrangian $\mathcal{L}$ is being considered as the quantum field of scalar particle ϕ_1 and ϕ_2 and a mass μ; thus, here, $\mu^2 < 0$ will confirm an infinite number of quanta to satisfy the equation $\phi_1^2 + \phi_2^2 = -\mu^2/\lambda = v^2$. So, the quantum field can be clarified as $\phi_0 = \frac{1}{\sqrt{2}}[(v + \eta) + i\xi]$, and then the derivative of the Lagrangian can be expressed as kinetic mode of:

$$\mathcal{L}_{kin}(\eta, \xi) = (D^\mu \phi)^\dagger (D^\mu \phi) = (\partial^\mu + ieA^\mu)\phi^*\,(\partial_\mu - ieA_\mu)\,\phi \tag{9.81}$$

Here, $V(\eta, \xi) = \lambda\, v^2\eta^2$ is the final term to the second order, and thus, the full Lagrangian can be expressed as:

$$\mathcal{L}_{\text{kin}}(\eta, \xi) = \frac{1}{2}(\partial_\mu \eta)^2 - \lambda v^2\eta^2 + \frac{1}{2}(\partial_\mu \xi)^2 - \frac{1}{4}F_{\mu\nu}F^{\mu\nu} + \frac{1}{2}e^2v^2A_\mu^2 - evA_\mu(\partial^\mu \xi) \tag{9.82}$$

Here, η represents the mass, ξ represents the massless, μ represents the mass for the quanta, and thus A_μ is defined as the term $\partial_\mu \alpha$, as is the function of the quantum field. Naturally, A_μ and ϕ can be changed spontaneously, so the equation can be rewritten as a Lagrangian scalar to confirm the formation of the thermal energy particles within the quantum field of the time and space in the air:

$$\mathcal{L}_{\text{scalar}} = (D^{\mu}\phi)^{\dagger}(D^{\mu}\phi) - V(\phi^{\dagger}\phi)$$
$$= (\partial^{\mu} + ieA^{\mu})\frac{1}{\sqrt{2}}(v+h)\left(\partial_{\mu} - ieA_{\mu}\right)\frac{1}{\sqrt{2}}(v+h) - V\left(\phi^{\dagger}\phi\right)$$

$$= \frac{1}{2}\left(\partial_{\mu}h\right)^{2} + \frac{1}{2}e^{2}A_{\mu}^{2}(v+h)^{2} - \lambda v^{2}h^{2} - \lambda v h^{3} - \frac{1}{4}\lambda h^{4} + \frac{1}{4}\lambda h^{4} \tag{9.83}$$

Here, the Lagrangian scalar thus revealed that the Higgs-Boson quantum field can certainly be initiated to from thermal energy in time and space. To confirm to determine this heat-energy generation in the time and space in the air, a further calculation considering its isotropic distributed kinetic energy has been conducted with respect to the angle θ from the dimensional axis of the time and space and the differential density of energy $\in$ and written as:

$$dn = \frac{1}{2}n(\in)\sin\theta d \in d\theta. \tag{9.84}$$

Here, the speed of high-energy frequency is being implemented as c (1–cosθ) considering the absorption of energy per unit area by expressing the following equation:

$$\frac{d\tau_{\text{abs}}}{dx} = \iint \frac{1}{2}\sigma n\,(\in)(1-\cos\theta)\,\sin\theta d \in d\theta. \tag{9.85}$$

Rewriting these variables as integral over s instead of θ, by (9.84) and (9.85), it has been determined as:

$$\frac{d\tau_{\text{abs}}}{dx} = \pi r_0^2\left(\frac{m^2c^4}{E}\right)^2\int_{\frac{m^2c^4}{E}}^{\infty} \in^{-2} n(\in)\,\overline{\phi}[s_0(\in)]de, \tag{9.86}$$

where

$$\overline{\phi}[s_0(\in)] = \int_1^{s_0(\in)} s\overline{\sigma}\,(s)ds, \overline{\sigma}(s) = \frac{2\sigma(s)}{\pi r_0^2}. \tag{9.87}$$

This result confirms the representation of the thermal energy generation in that is readily allow the calculation of $\overline{\phi}\,[s_0]$ to the determine the net thermal energy formation of s_0 [18, 22, 26] in the time and space. Therefore, the net functional asymptotic formula is expressed as follows to confirm the thermal energy formation:

$$\overline{\phi}[s_0] = 2s_0(\ln 4s_0 - 2) + \ln 4s_0(\ln 4s_0 - 2) - \frac{(\pi^2 - 9)}{3} s_0^{-1}\left(\ln 4s_0 + \frac{9}{8}\right) + \cdots (s_0 \gg 1); \tag{9.88}$$

$$\overline{\phi}[s_0] = \left(\frac{2}{3}\right)(S_0 - 1)^{\frac{3}{2}} + \left(\frac{5}{3}\right)(S_0 - 1)^{\frac{5}{2}} - \left(\frac{1507}{420}\right)(S_0 - 1)^{\frac{7}{2}} + \cdots (s_0 - 1 \gg 1). \tag{9.89}$$

The function $\frac{\overline{\phi}[s_0]}{(s_0 - 1)}$ is revealed as $1 < s_0 < 10$; at larger s_0, it confirms a standard algorithm function of s_0. Thus, the energy spectra of the thermal energy can be written as $n(\in) \propto \in^m$, and thus, the calculation of the thermal energy in terms of energy spectra in the time and space can be expressed as:

$$\begin{aligned} n(\in) &= 0, \quad \in < \in_0 \\ &= C_{\in}^{-\alpha} \text{ or } D_{\in}^{\beta}, \quad \in_0 < \in < \in_m \\ &= 0, \quad \in > \in_m \end{aligned} \tag{9.90}$$

Then, it can be transformed as:

$$\left(\frac{d\tau_{\text{abs}}}{dx}\right)_\alpha = \pi r_0^2 C \left(\frac{m^2c^4}{E}\right)^{1-\alpha} \times \begin{cases} 0 & , \quad E < E_m \\ [F_\alpha(1) - F_\alpha(\sigma_m)], & E_m < E < E_0 \\ [F_\alpha(\sigma_0) - F_\alpha(\sigma_m)], & E > E_0; \end{cases} \tag{9.91}$$

$$\left(\frac{d\tau_{\text{abs}}}{dx}\right)_\beta = \pi r_0^2 D \left(\frac{m^2c^4}{E}\right)^{1+\beta} \times \begin{cases} 0 & , \quad E < E_m \\ [F_\beta(\sigma_m) & , \quad E_m < E < E_0 \\ [F_\beta(\sigma_m) - F_\beta(\sigma_0)], & E > E_0. \end{cases} \tag{9.92}$$

In these functional variables, the thermal energy spectra on the time and space can be properly defined by asymptotic formula. Thus, the term Γ_γ^{LPM} defines the energy generation that can be rewritten as per unit area as:

$$\Gamma_\gamma \equiv \frac{dn_\gamma}{dVdt}. \tag{9.93}$$

Here, the contributions $\Gamma_\gamma^{\text{LPM}}$ and the rate of thermal energy generation are being confirmed as $O(\alpha_{\text{EM}}\ \alpha_s)$. Thus, it has been confirmed by implementing the contributed thermal energy rate $\Gamma_{\gamma}^{\mathrm{LPM}}$ to thermodynamically control the temperature T and energy-physical reaction μ of the time and space by expressing the following equation:

$$\frac{d\Gamma_{\gamma}^{\mathrm{LPM}}}{d^3k} = \frac{d_F q_s^2 \alpha_{\mathrm{EM}}}{4\pi^2 k} \int_{-\infty}^{\infty} \frac{dp_{\|}}{2\pi} \int \frac{d^2\mathrm{p}_{\perp}}{(2\pi)^2} A\left(p_{\|}, k\right) \ \mathrm{Re}\left\{2\mathrm{P}_{\perp} \cdot f\left(\mathrm{p}_{\perp}; p_{\|}, k\right)\right\} \tag{9.94}$$

where d_{F} is the variable state of the energy particles [N_{c} in SU(N_{c})] and q_s is the Abelian charge of the energy quark, $k \equiv |k|$, and thus, the kinetic functional mode $A(p\|, k)$ is being expressed by:

$$A\left(p_{\|}, k\right) \equiv \begin{cases} \dfrac{n_b\left(k + p_{\|}\right)\left[1 + n_b\left(p_{\|}\right)\right]}{2p_{\|}\left(p_{\|} + k\right)} & \text{scalars} \\ \dfrac{n_f\left(k + p_{\|}\right)\left[1 - n_f\left(p_{\|}\right)\right]}{2\left[p_{\|}\left(p_{\|} + k\right)\right]^2}\left[{p_{\|}}^2 + \left(p_{\|} + k\right)^2\right], & \text{fermions} \end{cases} \tag{9.95}$$

with

$$n_b(p) \equiv \frac{1}{\exp[\beta(p-\mu)] - 1}, \quad n_f(p) \equiv \frac{1}{\exp[\beta(p-\mu)] + 1} \tag{9.96}$$

The function $f(p\perp; p\|, k)$ is then integrated to resolve the below equation, which suggested that the thermal energy proliferation in the time and space is very much possible that can be expressed as the following derivative of energy generation:

$$\begin{aligned} 2\mathrm{p}_{\perp} &= i\delta E\, \mathrm{f}\left(\mathrm{p}_{\perp}; p_{\|}, k\right) + \frac{\pi}{2} C_F g_s^2 m_D^2 \int \frac{d^2 q_{\perp}}{(2\pi)^2} \frac{dq_{\|}}{2\pi} \frac{dq^0}{2\pi} 2\pi\delta\left(q^0 - q_{\|}\right) \\ &\times \frac{\mathrm{T}}{|\mathrm{q}|}\left[\frac{2}{\left|\mathrm{q}^2 - \Pi_{\mathrm{L}}(\mathrm{Q})\right|^2} + \frac{\left[1 - \left(q^0/\left|\mathrm{q}_{\|}\right|\right)^2\right]^2}{\left|(q^0)^2 - \mathrm{q}^2 - \Pi_{\mathrm{T}}(\mathrm{Q})\right|^2}\right] \\ &\times \left[f\left(\mathrm{p}_{\perp}; p_{\|}, k\right) - f\left(q + \mathrm{p}_{\perp}; p_{\|}, k\right)\right] \end{aligned} \tag{9.97}$$

Simply, this heating energy generation is derived from the explicit forms of heating energy that maximize the heating energy generation corresponding to its given energy function of $f(p\perp; p\|, k)$ to transform it into thermal state energy to control the temperature of the *time and space* naturally.

Conclusions

To design massless houses and buildings in the air, the activated massless light energy (L_e), dark energy (D_e), and antimatter (A_m) are being modeled independently considering their quantum electrodynamics (QED) in relation to the time and space. The results of this computational modeling suggested that this massless activated energy considering their quantum electrodynamics (QED), here named as *Hossain activated energy (HaE^-)*, can play a vital role to model a massless space in the air by the induction of its nano-point defect in order to design massless houses and buildings in the time and space in the air. Subsequently, the transformation of this HaE^- energy into *Hossain air-thermal energy ($HatE^-$)* by the implementation of its Higgs-Bosons ($H \rightarrow \gamma\gamma^-$) electromagnetic fields can control its surroundings' air and temperature. Simply, the activated energy of light, dark, and antimatter in a combined mode of HaE^- and its transformed mode energy state $HatE^-$ suggested that it has the extreme power to develop a massless house and building in the air to live mankind in the atmosphere comfortably in the near future.

Acknowledgments This research work is solely performed by the author, and it does not have any conflict of interest to publish any suitable journal and/or publisher. Special thanks to Faria Hossain for the valuable comments about this project.

References

1. Alexey, G., et al. (2020). Two-photon frequency comb spectroscopy of atomic hydrogen. *Science, 370*, 1061–1066.
2. Armani, D. K., Kippenberg, T. J., Spillane, S. M., & Vahala, K. J. (2003). Ultra-high-Q toroid microcavity on a chip. *Nature, 421*, 925–928.
3. Artemyev, N., Jentschura, U., Serbo, V., & Surzhykov, A. (2012). Strong electromagnetic field effects in ultra-relativistic heavy-ion collisions. *European Physical Journal C: Particles and Fields, 72*, 1935.
4. Baur, G., Hencken, K., & Trautmann, D. (2007). Revisiting unitarity corrections for electromagnetic processes in collisions of relativistic nuclei. *Physics Reports, 453*, 1–27.
5. Baur, G., Hencken, K., Trautmann, D., Sadovsky, S., & Kharlov, Y. (2002). Dense laser-driven electron sheets as relativistic mirrors for coherent production of brilliant X-ray and γ-ray beams. *Physics Reports, 364*, 359–450.
6. Birnbaum, K. M., et al. (2005). Photon blockade in an optical cavity with one trapped atom. *Nature, 436*, 87–90.
7. Boettcher, I., Pawlowski, J. M., & Diehl, S. (2012). Ultracold atoms and the functional renormalization group. *Nucl. Phys. B-Proc. Suppl., 228*, 63–135.
8. Broz, M., et al. (2020). A generator of forward neutrons for ultra-peripheral collisions. *Computer Physics Communications, 253*, 107181.
9. Busch, K., et al. (2007). Periodic nanostructures for photonics. *Physics Reports, 444*, 101–202.
10. Cardoso, V., Lemos, J. P., & Yoshida, S. (2004). Quasinormal modes of Schwarzschild black holes in four and higher dimensions. *Phys. Rev. D, 69*, 044004.
11. Chang, D. E., Sørensen, A. S., Demler, E. A., & Lukin, M. D. (2007). A single-photon transistor using nanoscale surface plasmons. *Nature Physics, 3*, 807–812.

12. Chen, J., Wang, C., Zhang, R., & Xiao, J. (2012). Multiple plasmon-induced transparencies in coupled-resonator systems. *Optics Letters, 37*, 5133–5135.
13. Dayan, B., et al. (2008). A photon turnstile dynamically regulated by one atom. *Science, 319*, 1062–1065.
14. Del'Haye, P. et al. (2008). *Full stabilization of a frequency comb generated in a monolithic microcavity*. 2008 Conference on Lasers and Electro-Optics.
15. Dobrynina, A., Kartavtsev, A., & Raffelt, G. (2015). Photon-photon dispersion of TeV gamma rays and its role for photon-ALP conversion. *Phys. Rev. D, 91*, 083003.
16. Douglas, J. S., et al. (2015). Quantum many-body models with cold atoms coupled to photonic crystals. *Nat. Photon., 9*, 326–331.
17. Dupuis, N. L., et al. (2021). The nonperturbative functional renormalization group and its applications. *Physics Reports, 910*, 1–114.
18. Eichler, J., & Stöhlker, T. (2007). Radiative electron capture in relativistic ion–atom collisions and the photoelectric effect in hydrogen-like high-Z systems. *Physics Reports, 439*, 1–99.
19. Englund, D., et al. (2010). Resonant excitation of a quantum dot strongly coupled to a photonic crystal nanocavity. *Physical Review Letters, 104*, 073904.
20. Fernandez, J., & Martín, F. (2009). Electron and ion angular distributions in resonant dissociative photoionization of H2 and D2 using linearly polarized light. *New Journal of Physics, 11*, 043020.
21. Gabor, M., et al. (2009). Extremely efficient multiple electron-hole pair generation in carbon nanotube photodiodes. *Science, 325*, 1367–1371.
22. Gleyzes, S., et al. (2007). Quantum jumps of light recording the birth and death of a photon in a cavity. *Nature, 446*, 297–300.
23. Gould, R. J. (1967). Pair production in photon-photon collisions. *Physics Review, 155*, 1404–1406.
24. Guerlin, C., et al. (2007). Progressive field-state collapse and quantum non-demolition photon counting. *Nature, 448*, 889–893.
25. Guo, Y., Al-Jubainawi, A., & Ma, Z. (2019). Performance investigation and optimisation of electrodialysis regeneration for LiCl liquid desiccant cooling systems. *Applied Thermal Engineering, 149*, 1023–1034.
26. Hencken, K., Baur, G., & Trautmann, D. (2006). Transverse momentum distribution of vector mesons produced in ultraperipheral relativistic heavy ion collisions. *Physical Review Letters, 96*, 012303.
27. Hübel, H., et al. (2010). Direct generation of photon triplets using cascaded photon-pair sources. *Nature, 466*, 601–603.
28. Igor, B., et al. (2012). Ultracold atoms and the functional renormalization group. *Nuclear Physics B – Proceedings Supplements, 228*, 63–135.
29. Javadi, A., et al. (2015). Single-photon non-linear optics with a quantum dot in a waveguide. *Nature Communications, 6*, 8655.
30. Johnson, B. R., et al. (2010). Quantum non-demolition detection of single microwave photons in a circuit. *Nature Physics, 6*(9). https://doi.org/10.1038/nphys1710
31. Kaneda, F., et al. (2019). High-efficiency single-photon generation via large-scale active time multiplexing. *Science Advances, 5*(10), eaaw8586.
32. Kevin, A., et al. (2017). Signatures of two-photon pulses from a quantum two-level system. *Nature Physics, 3*, 649–654.
33. Langford, K., et al. (2021). Efficient quantum computing using coherent photon conversion. *Nature, 478*, 360–363.
34. Langer, L., et al. (2014). Access to long-term optical memories using photon echoes retrieved from semiconductor spins. *Nature Photonics, 8*, 851–857.
35. Naghiloo, M., et al. (2016). Mapping quantum state dynamics in spontaneous emission. *Nature Communications, 7*, 11527.

36. Najjari, B., Voitkiv, A. B., Artemyev, A., & Surzhykov, A. (2009). Simultaneous electron capture and bound-free pair production in relativistic collisions of heavy nuclei with atoms. *Physical Review A, 80*, 012701.
37. Reinhard, A., et al. (2012). Strongly correlated photons on a chip. *Nature Photonics, 6*, 93–96.
38. Reinhard, P., et al. (2021). Nuclear charge densities in spherical and deformed nuclei: Toward precise calculations of charge radii. *Physical Review C, 103*, 054310.
39. Shalm, L. K., et al. (2012). Three-photon energy–time entanglement. *Nature Physics, 9*(1). https://doi.org/10.1038/nphys2492
40. Szafron, R., & Czarnecki, A. (2016). High-energy electrons from the muon decay in orbit: Radiative corrections. *Physics Letters B, 753*, 61–64.
41. Tame, M. S., et al. (2013). Quantum plasmonics. *Nature Physics, 9*, 329–340.
42. Ting, T. C. (2004). The polarization vector and secular equation for surface waves in an anisotropic elastic half-space. *International Journal of Solids and Structures, 41*(8), 2065–2083.
43. Tu, M. W., & Zhang, W. M. (2008). Non-Markovian decoherence theory for a double-dot charge qubit. *Physical Review B, 78*, 235311.
44. Viktor, J., et al. (2014). More on thermal probes of a strongly coupled anisotropic plasma. *Journal of High Energy Physics, 2014*, 149.
45. Wang, C. et al. (2017). *Single-satellite positioning algorithm based on direction-finding*. 2017 Progress In Electromagnetics Research Symposium – Spring (PIERS).
46. Waseem, S. B. (2011). Orbital excitation blockade and algorithmic cooling in quantum gases. *Nature, 480*, 500–503.
47. Xiao, Y. F., et al. (2010). Asymmetric Fano resonance analysis in indirectly coupled microresonators. *Physical Review A, 82*, 065804.
48. Yan, W. B., & Fan, H. (2014). Single-photon quantum router with multiple output ports. *Scientific Reports, 4*, 4820.

Part IV
Sustainable Water, Infrastructure, and Transportation Technology

Chapter 10
Sustainable Water Technology: Diversion of Transpiration Mechanism to Meet Global Water Supply Naturally

Introduction

Plants give O_2 and take CO_2 by the process of photosynthesis to keep the global environment in balance. Plants are simply the hero for the environment; unfortunately, hero plants are also the villain for the environment who play the significant role in causing global warming. In fact the body of plants needs water for the reaction of biochemical metabolism for its growth [2, 4]. This water is taken up by the cohesion-tension mechanism of the soil (groundwater) through the roots, transported by osmosis through the xylem to the leaves of the plants [11, 14]. Interestingly only a mere 0.5–3% of water is used by plants for their metabolism, and the rest of water releases into the air through stomatal cells by transpiration process [7]. This process of transpiration is not only causing the largest loss of groundwater that is also causing global warming, since this water vapor is a notable cause for global warming. Recent studies on transpiration and groundwater relationship have been discussed respectively terrestrial water fluxes especially in rural area where their water models revealed that streamflow getting lower due to the plant transpiration [10, 16]. These are very interesting findings, but no mechanism has been studied yet to trap this transpiration water for meeting global water demand. In this research, therefore, a technology has been proposed to eliminate this water loss by diverting this transpiration mechanism by collecting this water vapor instead of allowing it to enter the air and transform it into water potable and clean energy. Simply static electricity creator plastic tank near the plants has been proposed to install of each home of rural area to trap all the water vapor as the water vapor is attracted by the force of static electricity. Just because water vapor has positive and negative charges and the electrons that ended up on static electrical force have a positive charge, while water molecules have a negative charge on one side, the positive charge of static electric force and negative charges of water vapor pull each other closer together, and the positive side tug the direction and force the water to

M. F. Hossain, *Global Sustainability*, https://doi.org/10.1007/978-3-031-34575-3_10

come down to collect the water in a tank and be treated *in site* to meet the daily water demand.

Calculations revealed only four standard oak trees can meet the total water for a small family throughout the year. Since the groundwater strata are getting lower fast to finite level, and global water and global warming are getting dangerous seriously to putting earth on vulnerable condition, these two vital needs must be resolved immediately. Interestingly this new finding has the total solution to solve the global water and environmental crisis for the survival of this planet that will indeed open a new door in science.

Material, Methods, and Simulation

Static Electric Force Generation

To capture the water vapor from air that is released by stomatal cells of the plants during the day $\mathfrak{H}$time, a model has been proposed to create *Hossain static electric force (HSEF* $= \mathfrak{H}$*)* by implementing the friction of insulator into the plastic tank to pulling down the water vapor into the plastic tank [1, 5]. To create *HSEF* into the plastic tank, I have implemented Abelian local symmetries calculation by using MATLAB software considering gauge-field symmetry and the Goldstone scalar with respect to longitudinal mode of the vector [2, 4]. Thus, for each spontaneously broken particle T^{α} of the local symmetry will be corresponding gauge field of $A^{\alpha}_{\mu}(x)$ where *HSEF* will start to work at a local U (1) phase symmetry [3, 5]. Therefore, the model will be comprised as a complex scalar field $\Phi(x)$ of static electric charge q coupled to the EM field $A^{\mu}(x)$, which is expressed by $\mathfrak{H}$:

$$\mathfrak{H} = -\frac{1}{4}F_{\mu\nu}F^{\mu\nu} + D_{\mu}\Phi^{*}D^{\mu}\Phi - V(\Phi^{*}\Phi) \tag{10.1}$$

where

$$D_{\mu}\Phi(x) = \partial_{\mu}\Phi(x) + iqA_{\mu}(x)\Phi(x)$$
$$D_{\mu}\Phi^{*}(x) = \partial_{\mu}\Phi^{*}(x) - iqA_{\mu}(x)\Phi^{*}(x) \tag{10.2}$$

and

$$V(\Phi^{*}\Phi) = \frac{\lambda}{2}(\Phi^{*}\Phi)^{2} + m^{2}(\Phi^{*}\Phi) \tag{10.3}$$

Suppose $\lambda > 0$ but $m^2 < 0$, so that $\Phi = 0$ is a local maximum of the scalar potential, while the minima form a degenerate circle $\Phi = \frac{v}{\sqrt{2}} * e^{i\theta}$,

$$v = \sqrt{\frac{-2m^2}{\lambda}}, \quad \text{any real } \theta \tag{10.4}$$

Consequently, the scalar field Φ develops a nonzero vacuum expectation value $\langle \Phi \rangle \neq 0$, which spontaneously creates the U (1) symmetry of the static electric field. The breakdown would lead to a massless Goldstone scalar stemming from the phase of the complex field $\Phi(x)$. Here local U (1) symmetry, the phase of $\Phi(x)$ – not just the phase of the expectation value $\langle \Phi \rangle$ but the x-dependent phase of the dynamical $\Phi(x)$ field. To analyze this static electricity force mechanism, I have used polar coordinates in the scalar field space, thus

$$\Phi(x) = \frac{1}{\sqrt{2}} \Phi_r(x) * e^{i\Theta(x)}, \quad \text{real } \Phi_r(x) > 0, \quad \text{real } \Phi(x) \tag{10.5}$$

This field redefinition is singular when $\Phi(\text{x}) = 0$, so I never used it for theories with $\langle \Phi \rangle \neq 0$, but it's alright for spontaneously broken theories where I can expect $\Phi\langle x \rangle \neq 0$ almost everywhere. In terms of the real fields $\phi_r(x)$ and $\Theta(\text{x})$, the scalar potential depends only on the radial field ϕ_r,

$$V(\phi) = \frac{\lambda}{8} \left(\phi_r^2 - v^2\right)^2 + \text{const}, \tag{10.6}$$

or in terms of the radial field shifted by its VEV, $\Phi_r(x) = v + \sigma(x)$,

$$\phi_r^2 - v^2 = (v + \sigma)^2 - v^2 = 2v\sigma + \sigma^2 \tag{10.7}$$

$$V = \frac{\lambda}{8} \left(2v\sigma - \sigma^2\right)^2 = \frac{\lambda v^2}{2} * \sigma^2 + \frac{\lambda v}{2} * \sigma^3 + \frac{\lambda}{8} * \sigma^4 \tag{10.8}$$

At the same time,the covariant derivative $D_\mu$becomes

$$\begin{aligned} D_\mu \phi &= \frac{1}{\sqrt{2}} \left(\partial_\mu \left(\phi_r e^{i\Theta}\right) + iqA_\mu * \phi_r e^{i\Theta}\right) \\ &= \frac{e^{i\Theta}}{\sqrt{2}} \left(\partial_\mu \phi_r + \phi_r * i\partial_\mu \Theta + \phi_r * iqA_\mu\right) \end{aligned} \tag{10.9}$$

$$\begin{aligned} \left|D_\mu \phi\right|^2 &= \frac{1}{2} \left|\partial_\mu \phi_r + \phi_r * i\partial_\mu \Theta + \phi_r * iqA_\mu\right|^2 \\ &= \frac{1}{2} \left(\partial_\mu \phi_r\right) + \frac{\phi_r^2}{2} * \left(\partial_\mu \Theta q A_\mu\right)^2 \\ &= \frac{1}{2} \left(\partial_\mu \sigma\right)^2 + \frac{(v + \sigma)^2}{2} * \left(\partial_\mu \Theta + qA_\mu\right)^2 \end{aligned} \tag{10.10}$$

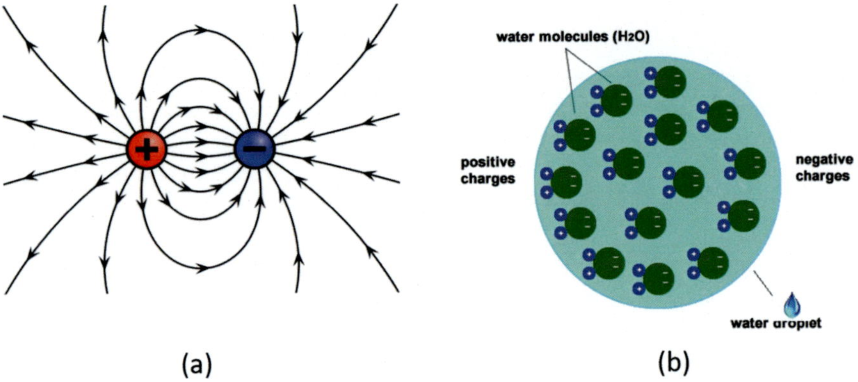

Fig. 10.1 (**a**) The creating of static electricity force and (**b**) its mechanism of conversion of static energy into an electromotive force of positive and negative charges that mobilizes the "static" electricity to tug down the water molecules

Altogether,

$$\mathfrak{h} = \frac{1}{2}(\partial_\mu \sigma)^2 - v(\sigma) - \frac{1}{4}F_{\mu\nu}F^{\mu\nu} + \frac{(v+\sigma)^2}{2} * (\partial_\mu \Theta + qA_\mu)^2 \tag{10.11}$$

To confirm the creating of this static electric force ($\mathfrak{h}_{\text{sef}}$) into the static electric field properties of this *HSEF*, it has been expanded in powers of the fields (and their derivatives), and focus on the quadratic part describing the free particles,

$$\mathfrak{h}_{sef} = \frac{1}{2}(\partial_\mu \sigma)^2 - \frac{\lambda v^2}{2} * \sigma^2 - \frac{1}{4}F_{\mu\nu}F^{\mu\nu} + \frac{v^2}{2} * (qA_\mu + \partial_\mu \Theta)^2 \tag{10.12}$$

Here this *HSEF* ($\mathfrak{h}_{\text{free}}$) function obviously will suggest a real scalar particle of positive mass^2 = λv^2 involving the $A_\mu(x)$ and the $\Theta(x)$ fields to initiate to create tremendous static electricity force within the electric field of the plastic tank (Fig. 10.1).

In Site Water Treatment

Since the collected water into the plastic tank is just nothing but the liquid form of vapor, it will not require any sedimentation, coagulation, and chlorination to clean the water. Only mixing physics (UV application) and filtration will be required to treat the water to meet the US National Primary Drinking Water Standard code [9]. It is the simplest way to treat water by using *SODIS* (*SOlar DISinfection*) system, where a transparent container is filled with water and exposed to full sunlight for

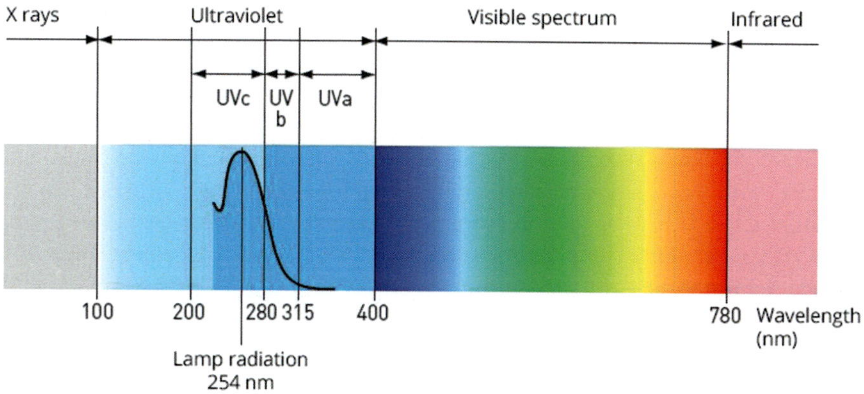

Fig. 10.2 The photophysics radiation application for the purification of water shows that once UV radiation of 320 nm applied into the water, it starts to disinfect all microorganisms immediately once temperature reaches at 50 °C

several hours. As soon as the water temperature reaches 50 °C with a UV radiation of 320 nm, the inactivation process will be accelerated to lead to complete microbiological disinfection immediately, and the treated water shall be used to meet the total domestic water demand (Fig. 10.2).

Results and Discussion

Electrostatic Force Analysis

To mathematically determine the electric static force proliferation around the plastic tank to confirm to tug down the water, I have initially solved the dynamic photon proliferation by integrating *HSEF* electric field create; thus, the local U (1) gauge invariant did allow to add a mass term for the gauge particle under $\varnothing' \rightarrow e^{i\alpha(x)}\varnothing$ to. In detail it can be explained by a covariant derivative with a special transformation rule for the scalar field expressed by [11, 12]:

$$\partial_\mu \rightarrow D_\mu = \partial_\mu = ieA_\mu \quad [\text{covariant derivatives}]$$

$$A'_\mu = A_\mu + \frac{1}{e}\partial_\mu \alpha \quad [A_\mu \text{ derivatives}] \tag{10.13}$$

where the local U (1) gauge-invariant *HSEF* for a complex scalar field is given by:

$$\mathfrak{h} = (D^{\mu})^{\dagger}\left(D_{\mu}\varnothing\right) - \frac{1}{4}F_{\mu\nu}F^{\mu\nu} - V(\varnothing) \tag{10.14}$$

The term $\frac{1}{4}F_{\mu\nu}F^{\mu\nu}$ is the kinetic term for the gauge field (heating photon) and $V(\varnothing)$ is the extra term in the HSEF that be $V(\varnothing^*\varnothing) = \mu^2(\varnothing^*\varnothing) + \lambda\,(\varnothing^*\varnothing)^2$.

Therefore, the HSEF ($\mathfrak{h}$) under perturbations into the quantum field has been initiated with the massive scalar particles $\phi_1$and ϕ_2 along with a mass μ. In this situation $\mu^2 < 0$ had an infinite number of quantum; each has been satisfied by $\phi_1^2 + \phi_2^2 = -\mu^2/\lambda = v^2$ and the $\mathfrak{h}$ through the covariant derivatives using again the shifted fields η and ξ defined the quantum field as $\phi_0 = \frac{1}{\sqrt{2}}\,[(v+\eta) + i\xi]$.

$$\begin{aligned}\text{Kinetic term}: \quad \mathfrak{h}(\eta,\xi) &= (D^{\mu}\phi)^{\dagger}(D^{\mu}\phi)\\ &= (\partial^{\mu} + ieA^{\mu})\phi^{*}\left(\partial_{\mu} - ieA_{\mu}\right)\phi\end{aligned} \tag{10.15}$$

Thus, this expanding term in the $\mathfrak{h}$ associated to the scalar field is suggesting that *HSEF* electric field is prepared to initiate the proliferation of static electricity force into its quantum field to tug down the water [7, 16].

To confirm this tug downing water by static electricity force, hereby, with readily I have implemented the calculation of $\overline{\varphi}[s_0]$ for the confirmation of the expected value of s_0 for capturing water vapor [6, 8]. Thus, the corrective functional asymptotic formulas are being used as follows:

$$\overline{\varphi}[s_0] = 2s_0(\ln 4s_0 - 2) + \ln 4s_0(\ln 4s_0 - 2) - \frac{(\pi^2 - 9)}{3} + s_0^{-1}\left(\ln 4s_0 + \frac{9}{8}\right) + \cdots (s_0 \gg 1); \tag{10.16}$$

$$\overline{\varphi}[s_0] = \left(\frac{2}{3}\right)(S_0 - 1)^{\frac{3}{2}} + \left(\frac{5}{3}\right)(S_0 - 1)^{\frac{5}{2}} - \left(\frac{1507}{420}\right)(S_0 - 1)^{\frac{7}{2}} + \cdots (s_0 - 1 \ll 1). \tag{10.17}$$

The function $\dfrac{\overline{\varphi}[s_0]}{(s_0 - 1)}$ is thus described as $1 < s_0 < 10$; for larger s_0, and it contains natural logarithmic, which is s_0 to confirm the tug down of 100% water vapor by the *HSEF* into the plastic tank.

In average 100 gallons of water is required per day per person in a standard daily life [14, 15]. Thus, it will require total (100_{gallons}/Day/Person × 4_{persons} × 365_{days}) 146,000 gallons water per year for a small family of four persons. Since a standard oak tree can transpire 40,000 gallons (151,000 L) per year, tug down of 100% water vapor by HSEF described above will require only four standard oak trees to satisfy the total water demand for a small family.

Conclusions

Water and environmental vulnerability are the top two problems globally where trees play a significant role in creating these problems by the process of transpiration. To mitigate these problems, transpiration mechanism has been proposed to transform and convert it into clean water to meet the global potable water demand and reduce the global warming by the utilization of electrostatic force to capture this transpiration water vapor and treat in site by UV application would indeed be a novel, integrated, and innovative field in science to console the rural water and global warming crisis.

Acknowledgments The author, Md. Faruque Hossain, declares that any findings, predictions, and conclusions described in this article are solely performed by the author and it is confirmed that there is no conflict of interest for publishing this research paper in a suitable journal and/or publisher.

References

1. Andreas, R. (2012). Strongly correlated photons on a chip. *Nature Photonics, 6*, 93–96.
2. Douglas, S., Habibian, H., et al. (2015). Quantum many-body models with cold atoms coupled to photonic crystals. *Nature Photonics, 9*, 326–331.
3. Langer, L., Poltavtsev, S., & Bayer, M. (2014). Access to long-term optical memories using photon echoes retrieved from semiconductor spins. *Nature Photonics, 8*, 851–857.
4. Leijing, Y., Sheng, W., Qingsheng, Z., Zhiyong, Z., Tian, P., & Yan, L. (2011). Efficient photovoltage multiplication in carbon nanotubes. *Nature Photonics, 8*, 672–676.
5. Pregnolato, T., Lee, E., Song, J., Stobbe, D., & Lodahl, P. (2015). Single-photon non-linear optics with a quantum dot in a waveguide. *Nature Communications, 6*, 8655.
6. Soto, W., Klein, S., et al. (2006). Improvement and validation of a model for photovoltaic array performance. *Solar Energy, 80*, 78–88.
7. Yan, W., & Heng, F. (2014). Single-photon quantum router with multiple output ports. *Scientific Reports, 4*, 4820.
8. Zhu, Y., Xiaoyong, H., Hong, Y., & Qihuang, G. (2014). On-chip plasmon-induced transparency based on plasmonic coupled nanocavities. *Scientific Reports, 4*, 3752.
9. Hossain, M. (2016). Solar energy integration into advanced building design for meeting energy demand and environment problem. *International Journal of Energy Research, 40*, 1293–1300.
10. Hossain, M. F. (2017). Green science: Independent building technology to mitigate energy, environment, and climate change. *Renewable and Sustainable Energy Reviews, 73*, 695–705.
11. Yuwen, W., Yongyou, Z., Qingyun, Z., Bingsuo, Z., & Udo, S. (2016). Dynamics of single photon transport in a one-dimensional waveguide two-point coupled with a Jaynes-Cummings system. *Scientific Reports, 6*, 33867.
12. Li, Q., & Xu, D. (2013). Recoil effects of a motional scatterer on single-photon scattering in one dimension. *Scientific Reports, 8*, 3144.
13. Scott, J., & Zachary, D. (2013). Terrestrial water fluxes dominated by transpiration. *Nature, 496*, 347–350.
14. Josette, M., & Scott, R. (2005). The ERECTA gene regulates plant transpiration efficiency in Arabidopsis. *Nature, 436*, 866–870.
15. Wheeler, T. D., & Stroock, A. D. (2008). The transpiration of water at negative pressures in a synthetic tree. *Nature, 455*, 208–212.
16. Hossain, F. (2016). Theory of global cooling. *Energy, Sustainability and Society, 7*, 6–24.

Chapter 11
Sustainable Sanitation Technology: Transformation of Sanitation Waste into Useful Element

Introduction

Environmental vulnerability is created much on building sector since 40% of global fossil energy is consumed by the building sector throughout the world [7, 28]. In 2018, the net energy consumption globally accounted for 5.59×10^{20} joules = 559 EJ, where 2.236×10^{20} EJ energy is alone engulfed by the building sector [14, 20]. Consequently, the building sector is triggered to release nearly 8.01×10^{11} ton of CO_2 (218 gtC by building sector of worldwide total carbon production of 545 gtC; 1 gtC = 10^9 ton C = 3.67 gt CO_2) into the atmosphere in the year 2018. The quickening of fossil fuel consumption by the building sector is getting higher and higher globally, and the situation shall remain unchanged until an innovative technology is used to power the building sector globally. At present, the atmospheric CO_2 level is 400 ppm where the building sector is the major player for creating this high level of CO_2 concentration into the atmosphere, and it is accelerating by 2.11% per year, which is a clear and present danger to the survival of all living beings in this planet in the near future [1, 3, 59]. Necessarily, the atmospheric CO_2 level must be lowered to a clean breathable level of 300 ppm CO_2. Therefore, a sustainable energy mechanism in the building sector is an urgent demand to confirm a clean and green environment on earth.

There are some recent interesting studies that show that a person can produce feces at an average of 0.4 kg/day that can form 0.4 m^3 biogas/day, and this amount of biogas (0.4 m^3/day) production is good enough to transform it into enough electricity for a family of four persons in a day [2, 26, 31]. However, no one has shown that the mechanism of using the cellar of a building as an acting bioreactor to transform biowaste into electricity energy satisfies the total energy demand of a building. Therefore, in this research, a Sustainable Sanitation Technology implementation by a building has been proposed by producing bioenergy by the building itself and transforming it into electricity energy to meet its net energy need. Simply, the domestic biowaste including human stool and wastewater of the building are being

M. F. Hossain, *Global Sustainability*, https://doi.org/10.1007/978-3-031-34575-3_11

chosen to be collected into the sealed separation chamber into the basement. Thereafter, this biowaste is isolated into (i) wastewater and (ii) sludge and transferred into two separation tanks into the cellar. Then the wastewater is being conducted for treatment process in situ by integrating all the required chemical and physical processes to use for landscaping. Consequently, the solid biowaste has been permitted to undergo for *methanogenesis* process into the bioreactor to form bioenergy and then convert it into electricity energy. Implementation of this innovative mechanism shall indeed be a promising technology to fulfill the net energy need of a building, which is delivered by the building itself.

Materials and Methods

For the conversion of domestic biowaste into bioenergy, structurally sound long-lasting bioreactor (BR) needs to be designed. Thus, load-resistant factor design (LRFD) bioreactor must be constructed for a structurally sound bioreactor to operate regularly under high water velocity pressure considering the mathematical calculation of water velocity (379 mile/h), water density (1.2 kg/m^3), and the friction loss cofactor 1.00/m^2, respectively [25, 26, 44]. As the water dynamic force is 0.5 of half of the density of the water, thus, the equation for water force into the bioreactor can be expressed as $p_w = 0.5\rho C_p v_r^2$, where p_w represents the water force (Pa), ρ represents water density (kg/m^3), C_p is denoted as water force gradient which is 1, and v_r^2 is the water velocity (m/s) into the bioreactor. Thus, the net resultant force of $P_w = 0.5 \times 1.2$ kg/m^3 $\times$ 379^2 m/s is 86,185 Pa of the water pressure resistance capacity of the bioreactor. It can be simplify as force of $F =$ area $\times$ drag coefficient (constant = 1.00) $\times$ water dynamic force by following equation $F = 1$ m^2 $\times$ 1.0 $\times$ 86,185 = 86,185 N(8788 kgf) = 19,375 ibf to confirm that the bioreactor is structurally sound which flows the water velocity must be less than 19,000 ibf to operate bioreactor normally throughout the year.

Once the sophisticated water force resistance of the two-chamber bioreactor has been constructed, then the bioreactor is to be connected into biowaste chamber in order to collect the biowaste into the shut separation chamber into the basement. The other chamber is to be connected with the separated wastewater for the process of treatment of primary, secondary, and tertiary mechanism and then implemented into UV application to disinfect the wastewater. The UV application and filtration constitutes the simplest way of treating wastewater involving *disinfection* (DIS) system in which one fills a detention chamber with water and exposes it to full UV light for a few hours. Once the wastewater temperature hits 50 °C due to the subject of UV light of approximately 320 nm, it functions immediately to kill all bacteria, viruses, and molds and disinfects the water completely through bacteriological disinfection process (Fig. 11.1).

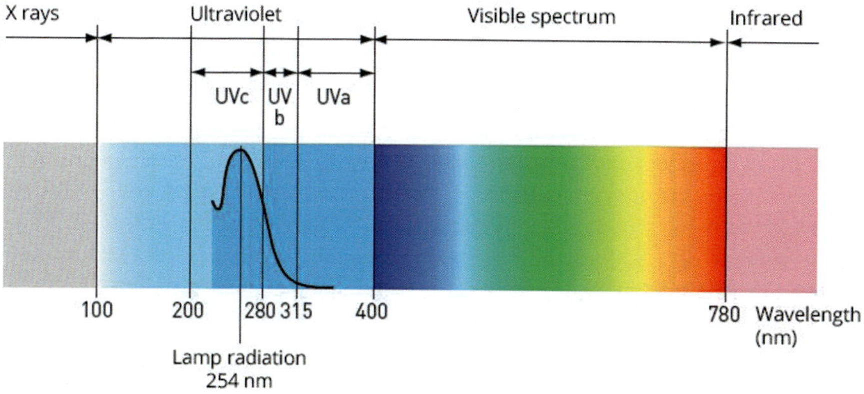

Fig. 11.1 The application of photo-physics radiation in purifying water that illustrates that once one applies UV light of 320 nm into the wastewater, it begins to kill the microorganisms once the temperature momentum hits at 50 °C

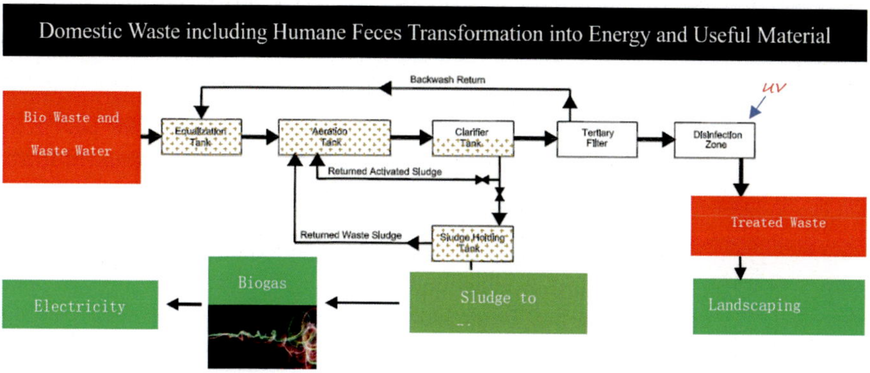

Fig. 11.2 The schematic diagram of wastewater treatment mechanism where treated water could be utilized for landscaping and the sludge is to be used for transforming process to produce the energy

This treatment mechanism removes nearly 100% microorganism and other contaminants from the wastewater effluent which could be used for local gardening (Fig. 11.2).

Then, the other product – sludge (human feces including domestic waste) – in another chamber of the bioreactor is being conducted for disinfection process in situ into an anaerobic chamber. This is the conversion mechanism performed by electrochemical filters of activated carbon nanotubes (CNT), which has the capability to electrolyze and oxidize pollutant in the anode actively from the sludge [21, 39, 43]. It is an advanced mechanism of biowaste disinfection mechanism in combining both the electrolysis and oxidation process into the anode of carbon nanotubes and catalyzed by the process of oxidation by H_2O_2 into the cathode of carbon nanotubes.

The function here is to accelerate the rate of sludge treatment, and its active oxidation process into the tank is being calculated and demonstrated a pathway that H_2O_2 flow is very much effective to disinfect the biowaste by the electrode and the cathode potential in order to achieve the content of biowaste pH, flow rate, and oxygen dissolved into a normal clean biomaterial form [35, 37, 48]. Hence, the maximum flow of H_2O_2 is being accounted for 1.38 mol L^{-1} m^{-2} C by achieving CNT L^{-1} m^{-2} with the implementation of cathode potential V −0.4 (vs. Ag/AgCl), with a pH of 6.46, with the flowing rate of 1.5 mL min^{-1} and the dissolved oxygen (DO) content of 1.95 mol L^{-1} m^{-2}. Additionally, phenol (C_6H_5OH) is being induced as an aromatic element for addressing the removal efficiency by clarifying the oxidation rate directed to the H_2O_2 flow [4, 29, 30]. Consequently, the electrochemical carbon nanotube filters activated the H_2O_2 generation tremendously in order for the carbon nanotubes to work most effectively to remove the organic contaminants from the biowaste at nearly 100%.

Once the sludge is disinfected, the product is being placed into the closed bioreactor tank to allow for anaerobic co-digestion process [5, 58]. Thereafter, the product is being heated for 95 °F for 15 days, which will stimulate the growth of anaerobic bacteria of *Desulfovibrio* and *Methanococcus*, which engulf the organic material of the sludge and produce biogas through biosynthesis process (Fig. 11.3).

Then the biogas is to be conducted for transforming process to generate electricity energy through the semiconductor diodes of the circuit panel. Hence, the electricity production from biogas into the circuit panel is being examined by detailed mathematical computations [20, 22, 38] (Fig. 11.4).

Hence, to achieve a successful conversion of biogas into electricity energy, the first-order perturbation theory has been implemented considering the production of biogas [35–37]. The first-order mechanism of the transformation of the biogas into the electricity energy needs the adequate surface into the bioreactor to separate the electrons into the semiconductor to produce the electric charge by the given term below [11, 12, 57]:

$$I = I_{\mathrm{ph}} - I_{\mathrm{ph}}\left[\exp\left(\frac{V + R_{\mathrm{s}}I}{V_{\mathrm{r}}}\right) - 1\right] - \frac{V + R_{\mathrm{s}}I}{R_{\mathrm{p}}} \tag{11.1}$$

Here I is the current and V is the voltage into the circuit panel. I_{ph} ($=N_{\mathrm{p}}I_{\mathrm{ph},\ \mathrm{cell}}$) is the electricity energy-created current running inside the circuit module which consists of N_{p} cells that are connected in parallel. I_0 ($=N_{\mathrm{pI0,\ cell}}$) is called the reverse current passing through N_{p} cells that are connected in parallel, wherein the reverse saturation current $I_{0,\ \mathrm{cell}}$ passes through each cell. Subsequently, V_{T} ($=aN_{\mathrm{s}}{\cdot}kT/q$) is represented as a matrix of thermal stress of N_{S} cells that are connected in series where (~1.5 = 1.0) keeping in mind the diode ideality factor, k (=1.38e^{-23} J/K) is a constant, q (=1.602e^{-19} C) is the charge on an electron, and T is the temperature in Kelvin. Here R_{P} is the equivalent resistance in parallel, while R_{S} is the equivalent resistance in series for circuit generator. Depending on the operational point, the

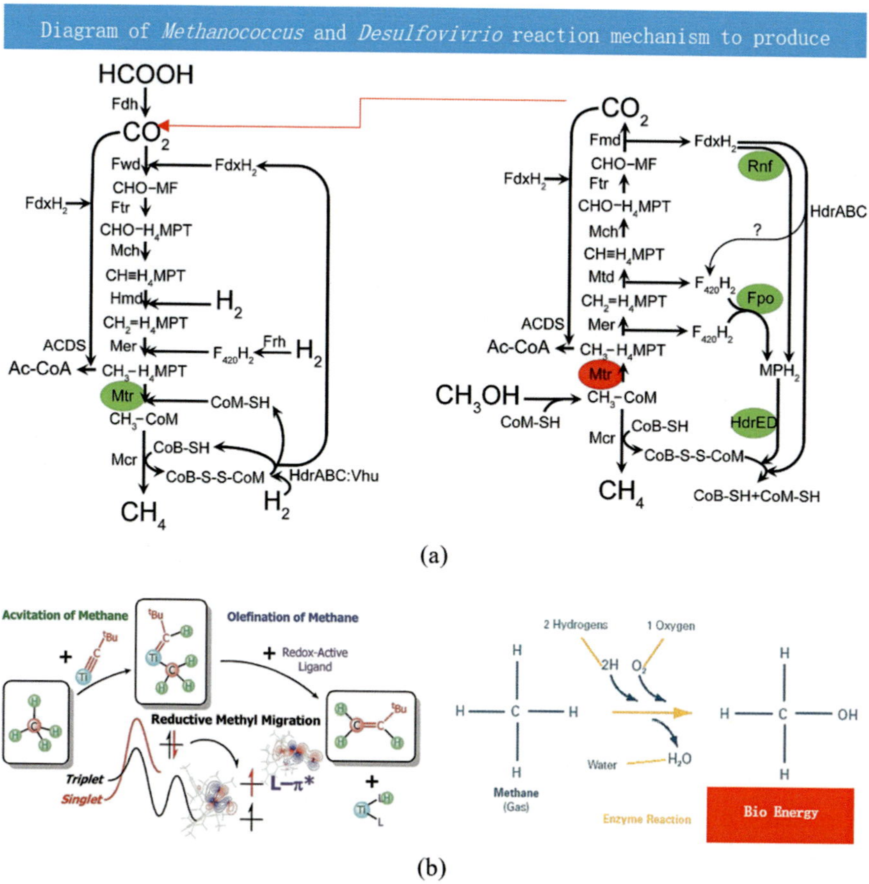

Fig. 11.3 (**a**) Biosynthesis mechanism of methanogenesis showing the conduction of chain reaction to form methane from sludge where two bacteria of Methanococcus and Desulfovibrio are the primary inhibitor to conduct this reaction. (**b**) The process showing the transformation of methane into bioenergy

circuit device, in practice, operates as a mixed performance of the current source or the voltage source [47, 51]. Practically, for the circuit panel, the effect of R_P parallel resistance will be greater in the operating area having a current source, while the R_S series resistance has a bigger effect on the functioning of the photovoltaic modules when the device works in the area having a voltage source [7, 18, 56]. Based on studies of various researchers, it can be concluded that for simplifying the model, the value of R_P can be ignored as it is very high [13, 23, 46]. Likewise, the value of R_S being very low can be neglected too [8, 9, 42]; thus, the temperature of the circuit panel can be shown as follows [17, 41]:

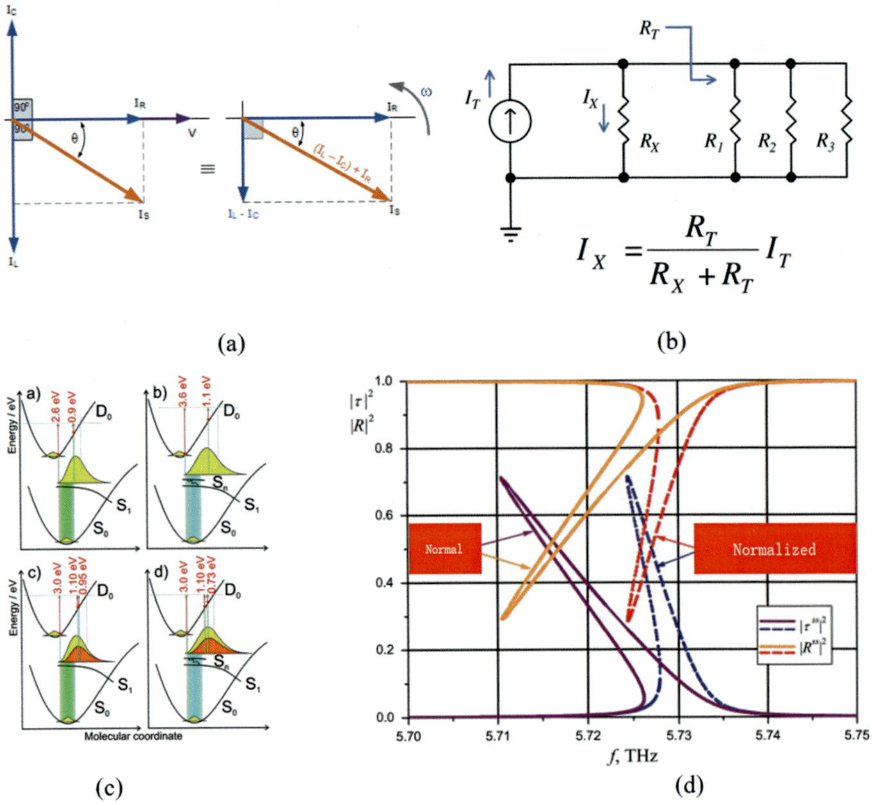

Fig. 11.4 The conversion mechanism of bioenergy into electricity energy: (**a**) mode of electricity production dynamics with respect to power factor (pf), (**b**) flow of electricity current generation, (**c**) net electricity energy production (eV) rate at molecular rate, and (**d**) the rate of electricity energy generation at normal and normalized circuit parameters, respectively

$$T = 3.12 + 0.25\frac{S}{S_{\mathrm{n}}} + 0.899T_{\mathrm{a}} - 1.3v_{\mathrm{a}} + 273 \tag{11.2}$$

Here S and S_{n} (=1000 W/m^2) are the electricity energy available in working condition, respectively, and T_{a} is the surrounding temperature, and v_{a} is the surrounding energy flow. The I–V features of photovoltaic panel are based on the internal qualities of the device, i.e., R_{S} and R_{P}; consequently, the electricity energy and surrounding temperature affect outer features. The electricity energy that is responsible for producing the electric current is linked linearly to the electricity energy and temperature and can be stated as follows [6, 10, 54]:

$$I_{\mathrm{ph}} = \left(I_{\mathrm{ph},n} + \alpha_{\mathrm{I}}\Delta T\right)\frac{S}{S_{\mathrm{n}}} \tag{11.3}$$

Here I_{ph} is the current that is produced because of biogas at STC and $\Delta T = T - T_n$, T is the temperature of the circuit panel because of the electricity energy, and T_n is the supposed temperature. For preventing any problems faced by the electricity energy current in deciding the series resistance (very low) as well as the parallel resistance (very high), it has been presumed that $I_{sc} \approx I_{ph}$ so that an explanation can be given for the complex circuit modeling and the open-circuit voltage that is dependent on the temperature can be confirmed [16, 18, 49]. This can be shown by (Fig. 11.1)

$$V_{oc} = V_{oc,n}(1 + \alpha_v \Delta T) + V_T \ln\left(\frac{S}{S_n}\right) \tag{11.4}$$

Here $V_{oc,n}$ is the open-circuit voltage that is calculated at the given conditions, and α_v is the voltage-temperature coefficient. The electrical and thermal features of the electricity energy panels can be achieved from these characteristics, which are integrated to achieve the I–V curve to produce much electricity energy (Eq. 11.1). The characteristics of the suggested electricity energy panel should consist of the following, the short-circuit current/temperature coefficient (α_I), the open-circuit voltage/temperature coefficient (α_v), the experimental peak power (P_{max}), the insignificant short-circuit current ($I_{sc,n}$), the maximum power point (MPP) voltage (V_{mp}), the MPP current (I_{mpp}), and the insignificant open-circuit voltage ($V_{oc,n}$), to calculate at the supposed conditions or standard test conditions (STC) of temperature $T = 298$ K and electricity energy of $S = 1000$ W [51, 52]. The simple equation at STC can be expressed as follows:

$$I = I_{ph,n} - I_{0,n}\left[\exp\left(\frac{V + R_s I}{V_{T,n}}\right) - 1\right] - \frac{V + R_s I}{R_p} \tag{11.5}$$

Here "n" is evaluated at STC, and the values are expected to show that the resistance in series and the resistance in parallel are not dependent on each other. Hence, the modeling in Eq. (11.5) can be simplified as below:

$$I = I_{ph,n} - I_{0,n}\left[\exp\left(\frac{V + R_s I}{V_{T,n}}\right) - 1\right] \tag{11.6}$$

There are three significant points on the I–V curve of electricity energy, maximum power point (V_{mp}, I_{mpp}), open circuit (V_{oc}, 0), and short circuit (0, I_{sc}), which can be shown as

$$I_{sc,n} = I_{ph,n} - I_{0,n}\left[\exp\left(\frac{R_s I_{sc,n}}{V_{T,n}}\right) - 1\right] \tag{11.7}$$

$$0 = I_{\text{ph},n} - I_{0,n}\left[\exp\left(\frac{V_{\text{oc},n}}{V_{\text{T},n}}\right) - 1\right] \tag{11.8}$$

$$I_{\text{mpp},n} = I_{\text{ph},n} - I_{0,n}\left[\exp\left(\frac{V_{\text{mpp},n} + R_{\text{s}}I_{\text{mpp},n}}{V_{\text{T},n}}\right) - 1\right] \tag{11.9}$$

The diode saturation current can thus be shown by its dependence on the temperature of the bioreactor [11]

$$I_0 = I_{0,n}\left(\frac{T_n}{T}\right)^3 \exp\left[\frac{qE_{\text{G}}}{ak}\left(\frac{1}{T_{\text{n}}} - \frac{1}{T}\right)\right] \tag{11.10}$$

Here E_{G} represents the band-gap energy of the electricity energy. Equation (11.8) shows that the diode saturation current at the STC and the photo-current at STC are linked:

$$I_{0,n} = \frac{I_{\text{ph},n}}{\left[\exp\left(\frac{V_{\text{oc},n}}{V_{\text{T},n}}\right) - 1\right]} \tag{11.11}$$

The electricity generation model can be further enhanced if Eq. (11.8) is substituted by

$$I_0 = \frac{I_{\text{sc},n} + \alpha_{\text{I}}\Delta T}{\exp\left(\frac{V_{\text{oc},n+\alpha_{\text{V}}\Delta T}}{V_{\text{T}}}\right) - 1} \tag{11.12}$$

By assuming $V_{\text{oc},n}/V_{\text{T},n} \gg 1$, $I_{0,n}$ can be shown as

$$I_{0,n} = I_{\text{ph},n} \exp\left(-\frac{V_{\text{oc},n}}{V_{\text{T},n}}\right) \tag{11.13}$$

Using Eqs. (11.13) and (11.6), it can be shown that

$$V = V_{\text{oc},n} + V_{\text{T},n} \ln\left(1 + \frac{\text{I}_{\text{ph,n}}{}^{-I}}{I_{0,n}}\right) - R_{\text{s}}I \tag{11.14}$$

Thus, Eq. (11.14) is considered as modest electricity energy generation model that is transformed from the biogas from bioreactor, and it can be explained as simply as the following equation:

$$V = V_{\text{oc},n} + V_{\text{T},n} \ln\left(1 + \frac{I}{\text{I}_{\text{ph,n}}}\right) - R_{\text{s}}I \tag{11.15}$$

Results and Discussion

Since the anaerobic *co-digestion* of domestic biowaste including human feces was lead into an anaerobic bioreactor, thus, the *methanogenesis* process began to produce biogas into the bioreactor right way. Naturally, the formation of biogas from the biowaste is being examined by computerized gas chromatograph [19, 27, 40] (Figs. 11.5 and 11.6).

Therefore, a model of bioreactor module described the generation of maximum bioenergy from domestic waste considering protective anaerobic detention chamber (Fig. 11.1a). Naturally, the model of the bioreactor module is being simplified by the determination of accurate form of the current-voltage (*I–V*) curb considering the mode of single-diode electricity circuit [28, 33, 53].

The next step is to calculate the electricity energy generation I_{pv} from biogas production by the calculation from the mode of current flow into the diode panel (Fig. 11.2a), accounting for *I–V–R* relationship (Fig. 11.2b), and biogas received by the diode to convert to alternating current (AC) for using domestic energy demand (Fig. 11.2c).

The below equation represents the electricity energy output from biogas (CH4):

$$P_{pv} = \eta_{pvg} A_{pvg} G_t \tag{11.16}$$

where η_{pvg} represents the methane generation efficiency, A_{pvg} represents the electricity energy generation, and G_t represents the current flow in the circuit cell. Thus, η_{pvg} can be rewritten as follows:

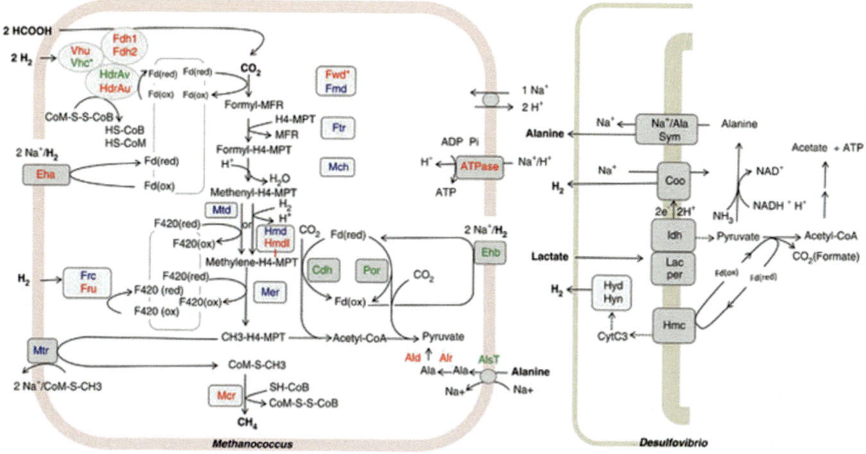

Fig. 11.5 The pathway of the methanogenesis mechanism depicting the biosynthesis of Methanococcus maripaludis and Desulfovibrio vulgaris to conduct bioenergy generation by consuming sludge

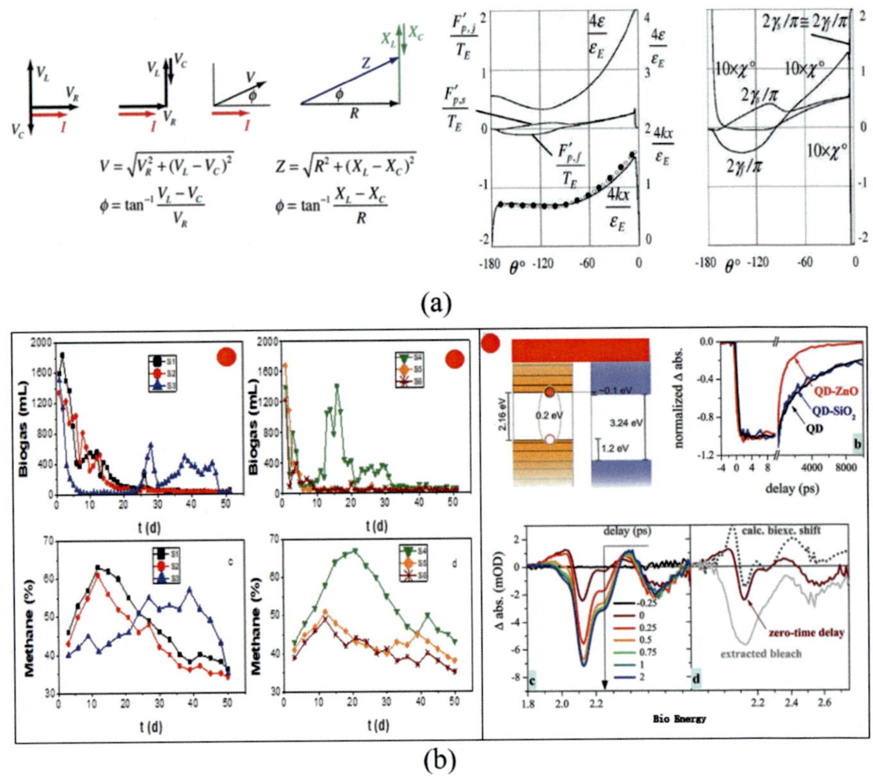

Fig. 11.6 (**a**) The biowaste transformation rate into the bioreactor in different directions and angles and (**b**) the production rate of biogas and the bioenergy considering the bioreactor methane content of the biowaste

$$\eta_{\text{pvg}} = \eta_{\text{r}}\eta_{\text{pc}}[1 - \beta(T_{\text{c}} - T_{\text{c ref}})] \tag{11.17}$$

η_{pc} represents the power factor effectiveness once it is equal to 1; β represents the energy cofactor (0.004–0.006 per °C); η_{r} represents the mode of energy production; and $T_{\text{c ref}}$ is the cell temperature in °C which can be obtained from the following equation:

$$T_{\text{c}} = T_{\text{a}} + \left(\frac{\text{NOCT} - 20}{800}\right)\text{G}_{\text{t}} \tag{11.18}$$

Here, T_{a} represents the ambient temperature in °C, G_{t} represents the current flow in a circuit cell (W/s), and NOCT represents the standard operating cell temperature in Celsius (°C) degree. The total electricity energy production in the circuit panel is estimated by the following equation:

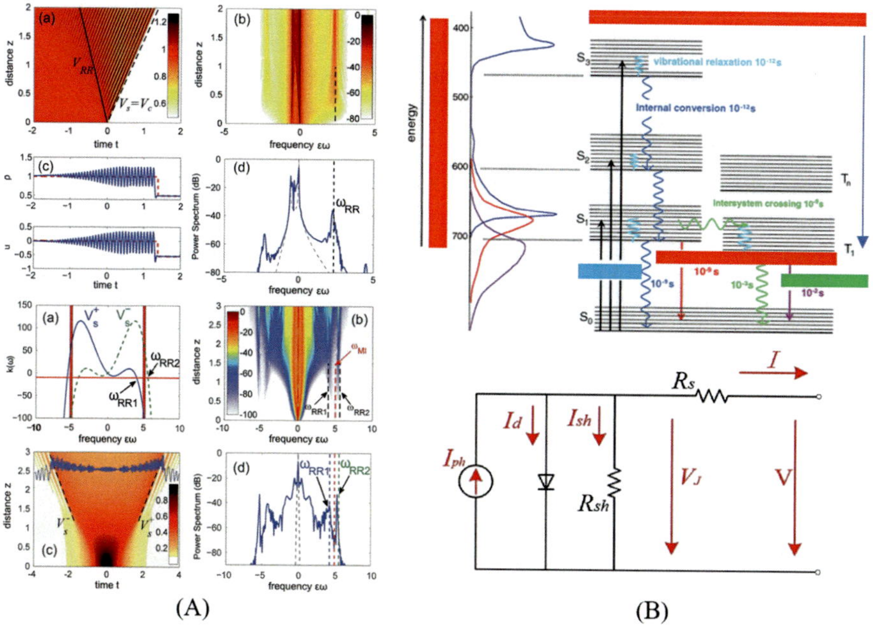

Fig. 11.7 (**a**) MATLAB simulation calculating the electricity energy generation from bioenergy that shows various frequencies and distances of the electric charges into the single-diode circuit cell. (**b**) The conversion mechanism of electricity energy DC into AC for the use as the prime source of power supply for a building

$$I_{\mathrm{t}} = I_{\mathrm{b}}R_{\mathrm{b}} + I_{\mathrm{d}}R_{\mathrm{d}} + (I_{\mathrm{b}} + I_{\mathrm{d}})R_{\mathrm{r}} \tag{11.19}$$

The current flow into the circuit cells is determined by the functional mode of its P–N junction that is able to produce electricity by conducting the interconnection of series-parallel configuration of the circuit cell [32, 45, 50] (Fig. 11.7).

Implementation of the standard single-diode circuit cell and the function of N_{s} series and N_{p} parallel connection in relation to current generation can be expressed as

$$I = N_{\mathrm{p}}\left[I_{\mathrm{ph}} - I_{\mathrm{rs}}\left[\exp\left(\frac{q(V + IR_{\mathrm{s}})}{AKTN_{\mathrm{s}}} - 1\right)\right]\right] \tag{11.20}$$

where

$$I_{\mathrm{rs}} = I_{\mathrm{rr}}\left(\frac{T}{T_{\mathrm{r}}}\right)^3 \exp\left[\frac{E_{\mathrm{G}}}{AK}\left(\frac{1}{T_{\mathrm{r}}} - \frac{1}{T}\right)\right] \tag{11.21}$$

Here, in Eqs. (11.20) and (11.21), q represents the generation of electron charge (1.6×10^{-19} C), K is the Boltzmann constant, A represents the cell standard cofactor, and T represents the cell temperature (K). I_{rs} represents the cell reverse current at

T, T_r represents the cell referred temperature, I_{rr} represents the reverse current at T_r, and E_G represents the band-gap energy flow into the circuit cell. The electric current I_{ph} formation conforming the circuit cell's temperature can be simplified as follows:

$$I_{ph} = \left[I_{SCR} + k_i(T - T_r)\frac{S}{100} \right] \tag{11.22}$$

I_{SCR} represents the cell short-circuit current and electricity energy generation, k_i represents the short-circuit current temperature coefficient, and S represents the electricity energy (kW). Thus, the I–V relationship into the circuit cell can be expressed simply as

$$I = I_{ph} - I_D \tag{11.23}$$

$$I = I_{ph} - I_0 \left[\exp\left(\frac{q(V + R_s I)}{AKT} - 1 \right) - \frac{V + R_s I}{R_{sh}} \right] \tag{11.24}$$

I_{ph} represents the electricity current (A), I_D represents the functional current (A), I_0 represents the inverse current (A), A represents the functional constant, q represents the charge of the electron (1.6×10^{-19} C), K is the Boltzmann constant, T represents the cell temperature (°C), R_s represents the series resistance (ohm), R_{sh} represents the shunt resistance (ohm), I represents the cell current (A), and V represents the circuit cell voltage (V). Thus, the output electricity current into the circuit panel is thus described as follows:

$$I = I_{PV} - I_{D1} - \left(\frac{V + IR_S}{R_{SH}} \right) \tag{11.25}$$

where

$$I_{D1} = I_{01} \left[\exp\left(\frac{V + IR_s}{a_1 V_{T1}} \right) - 1 \right] \tag{11.26}$$

$$I_{D2} = I_{01} \left[\exp\left(\frac{V + IR_s}{a_2 V_{T2}} \right) - 1 \right] \tag{11.27}$$

I_{01} and I_{02} represent the reverse currents of cell, respectively, and V_{T1} and V_{T2} represent the thermal voltages of the respective cell. The cell idealist constants are denoted as a_1 and a_2. Then, the simplified equation of the cell mode is described as

$$v_{\mathrm{oc}} = \frac{V_{\mathrm{oc}}}{cK\,T/q} \tag{11.28}$$

$$P_{\max} = \frac{\frac{V_{\mathrm{oc}}}{cK\,T/q} - \ln\left(\frac{V_{\mathrm{oc}}}{cK\,T/q} + 0.72\right)}{\left(1 + \frac{V_{\mathrm{oc}}}{K\,T/q}\right)}\left(1 - \frac{V_{\mathrm{oc}}}{\frac{V_{\mathrm{oc}}}{I_{\mathrm{SC}}}}\right) \times \left(\frac{V_{\mathrm{oc0}}}{1 + \beta \ln \frac{G_0}{G}}\right)\left(\frac{T_{\mathrm{o}}}{T}\right)^{\gamma} I_{\mathrm{sc0}}\left(\frac{G}{G_o}\right)^{\alpha} \tag{11.29}$$

where v_{oc} represents the normal value of the open-circuit voltage, V_{oc} represents the thermal voltage $V_{\mathrm{t}} = nkT/q$, c represents the constant current flow, K is the Boltzmann constant, T represents the temperature in Kelvin, α represents the nonlinear cofactor, q represents the electron charge, γ represents the factor representing all the nonlinear temperature-voltage function, and β represents the cell module coefficient. Since Eq. (11.29) depicted the tip energy generation by the circuit cell, therefore, the equation of total power output for an array with N_{s} cells connected in series and N_{p} cells connected in parallel with power P_{M} for each mode can be expressed as

$$P_{\mathrm{array}} = N_{\mathrm{s}} N_{\mathrm{p}} P_{\mathrm{M}} \tag{11.30}$$

Conversely, the derivative of the power with respect to current will equate to peak electricity energy production

$$\left.\frac{\mathrm{d}P}{\mathrm{d}I}\right|_{\mathrm{mpp}} = \left.\frac{\mathrm{d}(VI)}{\mathrm{d}I}\right|_{\mathrm{mpp}} = V_{\mathrm{mpp}} + I_{\mathrm{mpp}}\left.\frac{\mathrm{d}V}{\mathrm{d}I}\right|_{\mathrm{mpp}} \tag{11.31}$$

and

$$V = V_{\mathrm{oc},n} + V_{\mathrm{T,n}} \ln\left(1 - \frac{I}{\mathrm{I}_{\mathrm{ph},n}}\right) - R_{\mathrm{s}} I \tag{11.32}$$

Thus, the net electricity energy production volt (V) from biogas is finally computed as $V = V_{\mathrm{oc},n} + V_{\mathrm{T,n}} \ln\left(1 - \frac{I}{\mathrm{I}_{ph,\mathrm{n}}}\right) - R_{\mathrm{s}} I$, where the total amount of power has been determined using the equation

$$P_{\max}=\frac{\frac{V_{oc}}{cK\,T/q}-\ln\left(\frac{V_{oc}}{cK\,T/q}+0.72\right)}{\left(1+\frac{V_{oc}}{K\,T/q}\right)}\left(1-\frac{V_{oc}}{\frac{V_{oc}}{I_{SC}}}\right)\times\left(\frac{V_{oc0}}{1+\beta\ln\frac{G_0}{G}}\right)\left(\frac{T_o}{T}\right)^{\gamma}I_{sc0}\left(\frac{G}{G_o}\right)^{\alpha} \tag{11.33}$$

considering the parameter of

$$P_{\text{array}}=N_sN_pP_M \tag{11.34}$$

The electricity energy production is therefore accomplished per mole biogas production, which is equivalent to 1.4 eV/moles. Since 0.4 kg biowaste can produce 81 mole biogas, thus, the total electricity generation from 0.4 kg biowaste is equivalent to 1.4 × 81 = 113.4 eV (cc). Since 1.4 eV is equal to 27.77 kWT, thus, the total electricity energy production would be (27.77 kW × 81) = 2249.37 kW eV per day [34, 36]. If a commercial building and an office consumption is roughly 2200 kWh/day for a building with a 20 m × 20 m footprint and height of 20 m, 0.4 kg/day biowaste is sufficient enough to meet the total energy demand for this building, which is environmentally friendly.

Conclusion

The advancement of building construction in both urban and suburban regions around the globe has been quickening tremendously for the past 50 years. Thus, environmental change is expanding quickly because of the traditional utilization of fossil fuel by building sectors throughout the world. Conventional domestic waste and wastewater treatment processes are causing serious ecological contamination, making harm to human well-being and hampering the animal and plant kingdom. Here, "Sustainable Sanitation Technology," an inventive technology for the building sector, would be the front-line science to mitigate the complete energy need for a building without using any utility service connection. Because this innovation "Sustainable Sanitation Technology" could deliver sustainable energy by utilizing the building's cellar as the acting bioreactor to create biogas from the household biowaste and then converting it into electricity energy to meet the total indispensable energy need for a building, which is also environmentally friendly.

Acknowledgments This research was conducted solely by Md. Faruque Hossain, and the author confirms that it does not have any financial interest by any means. Any discoveries, conclusions, and recommendations expressed in this paper are exclusively those of the author, who affirms that the article has no conflict of interest for publication in a suitable journal and/or publishers.

References

1. Alfredo, M., et al. (2017). In focus: Biotechnology and chemical technology for biorefineries and biofuel production. *Journal of Chemical Technology & Biotechnology, 92*(5), 897–898.
2. Ashley, B., et al. (2017). Accelerating net terrestrial carbon uptake during the warming hiatus due to reduced respiration. *Nature Climate Change*. https://doi.org/10.1038/NCLIMATE3204
3. Achard, F., et al. (2014). Determination of tropical deforestation rates and related carbon losses from 1990 to 2010. *Global Change Biology, 20*, 2540–2554.
4. Alessandro, A., Friedlingstein, P., et al. (2015). Spatio-temporal patterns of terrestrial gross primary production: A review: GPP Spatio-temporal patterns. *Reviews of Geophysics*. https://doi.org/10.1002/2015RG000483
5. Andres, R., Boden, T., & Higdon, D. (2014). A new evaluation of the uncertainty associated with CDIAC estimates of fossil fuel carbon dioxide emission. *Tellus B, 66*, 23616. https://doi.org/10.3402/tellusb.v66.23616
6. Arnell, N., et al. (1996). *Climate change 1995: Impacts, adaptations, and mitigation of climate change* (R. T. Watson, et al., Eds.) (pp. 325–363). Cambridge University Press.
7. Canadell, J. G., et al. (2007). Contributions to accelerating atmospheric CO2 growth from economic activity, carbon intensity, and efficiency of natural sinks. *Proceedings of the National Academy of Sciences of the USA, 104*, 18866–18870.
8. Chen, J., Dimitrov, D., Dimitrova, T., Timans, P., et al. (2008). *Carrier density profiling of ultra-shallow junction layer through corrected C-V plotting*. Extended Abstracts – 2008 8th international workshop on junction technology (IWJT'08).
9. Chen, Y., et al. (2017). A pan-tropical cascade of fire driven by El Niño/southern oscillation. *Nature Climate Change, 7*, 906–911.
10. Chevallier, F. (2015). On the statistical optimality of CO2 atmospheric inversions assimilating CO2 column retrievals. *Atmospheric Chemistry and Physics, 15*, 11133–11145. https://doi.org/10.5194/acp-15-11133-2015
11. Ciais, P., et al. (2013). Chapter 6: Carbon and other biogeochemical cycles. In T. Stocker, D. Qin, & G.-K. Platner (Eds.), *Climate Change 2013 the physical science basis*. Cambridge University Press.
12. Landig, A. J., Koski, J. V., Scarlino, P., Mendes, U. C., Blais, A., Reichl, C., Wegscheider, W., Wallraff, A., Ensslin, K., & Ihn, T. (2018). Coherent spin–photon coupling using a resonant exchange qubit. *Nature, 560*, 179–184.
13. Colonna, P., Casati, E., Trapp, C., Mathijssen, T., Larjola, J., Turunen-Saaresti, T., & Uusitalo, A. (2015). Organic Rankine cycle power systems: From the concept to current technology, applications and an outlook to the future. *Journal of Engineering for Gas Turbines and Power, 137*(10), 100801-1–100801-19.
14. Le Querel, C., Andrew, R. M., et al. (2016). Global carbon budget 2016. *Earth System Science Data, 8*, 605–649. https://doi.org/10.5194/essd-8-605-2016
15. Vo"ro" smarty, C. J., Fekete, B., Meybeck, M., & Lammers, R. (2000). Global biogeochem. *Cycles, 14*, 599.
16. Bauer, J. E., Cai, W.-J., Raymond, P. A., Bianchi, T. S., Hopkinson, C. S., & Regnier, P. A. G. (2013). The changing carbon cycle of the coastal ocean. *Nature, 504*, 61–70.
17. Davis, S. J., & Calderia, K. (2010). Consumption-based accounting of CO2 emissions. *Proceedings of the National Academy of Sciences of the USA, 107*, 5687–5692.
18. Dietzenbacher, E., Pei, J. S., & Yang, C. H. (2012). Trade, Production fragmentation, and China's carbon dioxide emissions. *Journal of Environmental Economics and Management, 64*, 88–101.
19. Duce, R. A., et al. (2008). Impacts of atmospheric anthropogenic nitrogen on the open ocean. *Science, 320*, 893–897.
20. Earles, J. M., Yeh, S., & Skog, K. E. (2012). Timing of carbon emissions from global forest clearance. *Nature Climate Change, 2*, 682–685.

21. Erb, K.-H., et al. (2013). Bias in the attribution of forest carbon sinks. *Nature Climate Change, 3*, 854–856.
22. Feely, R. A. (2008). Global nitrogen deposition and carbon sinks. *Nature Geoscience, 1*, 430–437.
23. Gelfand, I., Sahajpal, R., Zhang, X., Izaurralde, R., Gross, K., & Robertson, G. (2013). Sustainable bioenergy production from marginal lands in the US Midwest. *Nature, 493*(7433), 514–517.
24. Gonzalez-Gaya, B., et al. (2016). High atmospheric–ocean exchange of semivolatile aromatic hydrocarbons. *Nature Geoscience, 9*, 438–442. https://doi.org/10.1038/ngeo2714
25. Hossain, M. F. (2019). Sustainable technology for energy and environmental benign building design. *Journal of Building Engineering, 22*, 130–139. (Elsevier).
26. Hossain, M. F. (2019). Breakdown of Bose-Einstein photonic structure to produce sustainable energy. *Energy Report, 5*, 202–209. (Elsevier). 2018.
27. Hossain, M. F. (2018). Green science: Advanced building design technology to mitigate energy and environment. *Renewable and Sustainable Energy Reviews, 81*(2), 3051–3060. (Elsevier).
28. Hossain, M. F. (2018). Bose-Einstein (B-E) photonic energy structure reformation for cooling and heating the premises naturally. *Advanced Thermal Engineering, 142*, 100–109. https://doi.org/10.1016/j.applthermaleng.2018.06.057. (Elsevier).
29. Hossain, M. F. (2018). Global environmental vulnerability and the survival period of all living beings on earth. *International Journal of Environmental Science and Technology, 16*, 755–762.
30. Hossain, M. F. (2018). Photonic thermal energy control to naturally cool and heat the building. *Applied Thermal Engineering, 131*, 576–586. (Elsevier).
31. Hossain, M. F. (2018). Green science: Decoding dark photon structure to produce clean energy. *Energy Reports, 4*, 41–48. (Elsevier).
32. Hossain, M. F. (2018). Photon energy amplification for the design of a micro PV panel. *International Journal of Energy Research*. https://doi.org/10.1002/er.4118.(Wiley).2017
33. Hossain, M. F. (2017). Green science: Independent building technology to mitigate energy, environment, and climate change. *Renewable and Sustainable Energy Reviews, 73*, 695–705. (Elsevier).
34. Hossain, M. F. (2016). Solar energy integration into advanced building design for meeting energy demand. *International Journal of Energy Research, 40*, 1293–1300. (Wiley).
35. Hossain, M. F. (2016). Production of clean energy from cyanobacterial biochemical products. *Strategic Planning for Energy and the Environment, 3*, 6–23. (Taylor and Francis).
36. Houghton, R. A. (2017). Balancing the global carbon budget. *Annual Review of Earth and Planetary Sciences, 35*, 313–347.
37. Karl, D. M., & Church, M. J. (2014). Microbial oceanography and the Hawaii Ocean Time-series programme. *Nature Review Microbiology, 12*, 699–713.
38. Kenneth, G., et al. (2018). Unravelling the link between global rubber price and tropical deforestation in Cambodia. *Nature Plants, 5*, 47–53.
39. Li, W., Ciais, P., Wang, Y., Peng, S., et al. (2016). Reducing uncertainties in decadal variability of the global carbon budget with multiple datasets. *Proceedings of the National Academy of Sciences of the USA*. https://doi.org/10.1073/pnas.1603956113
40. Liu, Z., Guan, D., et al. (2015). Reduced carbon emission estimates from fossil fuel combustion and cement production in China. *Nature, 524*, 335–338.
41. van Dam, J. C. (Ed.). (1999). *Impacts of climate change and climate variability on hydrological regimes*. Cambridge University Press.
42. Milliman, J., & Mei-e, R. (1995). *Climate change: Impact on coastal habitation* (pp. 57–83) (D. Eisma, Ed.). CRC Press.
43. Ballantyne, A. P., Alden, C. B., Miller, J. B., Tans, P. P., & White, J. W. C. (2012). Increase in observed net carbon dioxide uptake by land and oceans during the last 50 years. *Nature, 488*, 70–72.
44. Betts, R. A., Jones, C. D., Knight, J. R., Keeling, R. F., & Kennedy, J. J. (2016). El Nino and a record CO2 rise. *Nature Climate Change, 6*, 806–810.

45. Grätzel, M. (2001). Photoelectrochemical cells. *Nature, 414*(6861), 338–344.
46. Izadyar, N., Ong, H., Chong, W., & Leong, K. (2016). Resource assessment of the renewable energy potential for a remote area: A review. *Renewable and Sustainable Energy Reviews, 62*, 908–923.
47. Kane, M. (2003). Small hybrid solar power system. *Energy, 28*(14), 1427–1443.
48. Liu, Y., Xie, J., Ong, C., Vecitis, C., & Zhou, Z. (2015). Electrochemical wastewater treatment with carbon nanotube filters coupled with in situ generated H_2O_2. *Environmental Science: Water Research & Technology, 1*(6), 769–778.
49. Pierre, R., Lauerwald, R., & Ciais, P. (2014). Carbon leakage through the terrestrial-aquatic Interface: Implications for the anthropogenic CO_2 budget. *Procedia Earth and Planetary Science, 10*, 319–324.
50. Miller, P. P., et al. (2015). Audit of the global carbon budget: Estimate errors and their impact on uptake uncertainty. *Biogeosciences, 12*, 2565–2584. https://doi.org/10.5194/bg-12-2565-2015
51. Prietzel, J., Zimmermann, L., Schubert, A., & Christophel, D. (2016). Organic matter losses in German Alps forest soils since the 1970s most likely caused by warming. *Nature Geoscience, 9*, 543–548.
52. Romero-García, J., Sanchez, A., Rendón-Acosta, G., Martínez-Patiño, J., Ruiz, E., Magaña, G., & Castro, E. (2016). An olive tree pruning biorefinery for co-producing high value-added bioproducts and biofuels: Economic and energy efficiency analysis. *Bioenergy Research, 9*(4), 1070–1086.
53. Ruiz, H., Martínez, A., & Vermerris, W. (2016). Bioenergy potential, energy crops, and biofuel production in Mexico. *Bioenergy Research, 9*(4), 981–984.
54. Schwietzke, S., et al. (2016). Upward revision of global fossil fuel methane emissions based on isotope database. *Nature, 538*, 88–91.
55. Stephens, B. B., et al. (2007). Weak northern and strong tropical land carbon uptake from vertical profiles of atmospheric CO_2. *Science, 316*, 1732–1735.
56. Van der Werf, G. R., Dempewolf, J., et al. (2008). Climate regulation of fire emissions and deforestation in equatorial Asia. *Proceedings of the National Academy of Sciences of the USA, 15*, 20350–20355.
57. Weiland, P. (2009). Biogas production: Current state and perspectives. *Applied Microbiology and Biotechnology, 85*(4), 849–860.
58. Liu, Y., Wu, P., Liu, F., Li, F., An, X., Liu, J., Wang, Z., Shen, C., & Sand, W. (2019). Electroactive modified carbon nanotube filter for simultaneous detoxification and sequestration of Sb(III). *Environmental Science & Technology, 53*(3), 1527–1535.
59. Yin, Y., Ciais, P., Chevallier, F., et al. (2016). Variability of fire carbon emissions in equatorial Asia and its non-linear sensitivity to El Niño. *Geophysical Research Letters, 43*, 10472–10479.

Chapter 12
Sustainable Infrastructure: Invisible Roads and Transportation Engineering

Introduction

Urban and suburban areas massively depend on transportation infrastructure networks which are primarily constructional with concrete and asphalt, and it does not have enough vegetation to absorb heat caused by these asphalt and concrete [1]. Recent research found that transportation infrastructure on earth is approximately 0.9% of the total planetary surface area of 196.9 million mi^2, which is equivalent to 1.77 million miles squared infrastructure on earth, causing nearly 6% of global warming by reflecting heat (albedo) back to the space [2, 3]. On the other hand, conventional energy utilization for the transportation sectors is not only costly but also causes adverse environmental impact [4, 5]. A variety of studies have been performed to understand long-term climate variations by conventional energy utilization by the transportation sectors that is casing nearly 28% global energy consumption, which is equivalent to megaton CO_2 and responsible for 28% of global warming, and thus infrastructure and transportation fuel cases total 34% global warming [6, 7]. To mitigate transportation infrastructure crisis and its adverse environmental impact, I, therefore, proposed a new technology of maglev transportation infrastructure system for building better transportation infrastructure system.

A recent study by Cai and Chen described the dynamic characteristics, magnetic suspension systems, vehicle stability, and suspension control laws of maglev/guideway coupling systems about the maglev transportation system [8], but the fact is commercial application of this research modeling considering life cycle cost analysis, technology implementation, and infrastructure development did not show any possibility to apply it commercially [1, 9]. Therefore, the approach of this research is to apply the maglev transportation infrastructure commercially for confirming a greener and cleaner transportation infrastructure system where all vehicles shall run just over 2 ft above the earth surface at flying stage by the act of propulsive and impulsive superconducting force. Since the vehicle will run by electromagnetic force, it will not require any energy while running over the maglev. To mitigate

M. F. Hossain, *Global Sustainability*, https://doi.org/10.1007/978-3-031-34575-3_12

energy consumption when vehicles need to run on maglev area, additional technology has also been proposed to implement wind energy into the vehicles while it is in motion as a backup energy source. Thus, a detailed mathematical modeling using MATLAB Simulink software has been implemented for this wind energy utilization for the vehicles by performing turbine and drive train modeling [10–12]. A concerted research effort has been performed recently on climate science and found that currently 402 ppm CO_2 is present in the atmosphere causing global warming, which is required to be cut down to 300 ppm CO_2 to confirm global cooling at comfortable stage [13, 14]. Once maglev transportation infrastructure system is implemented throughout the world, it will reduce 34% CO_2 per year. Thus, it will take only $\left\{\int_{300}^{402}(1-0.34)\mathrm{d}x\right\} = 67.32$ years to cool the atmosphere, resulting in no more climate change after 68 years. It will simply be the most innovative technology in modern science for sustainable infrastructure to mitigate the cost and global warming dramatically.

Simulations and Methods

To model this underground sustainable maglev transportation infrastructure, I have formulated the following calculation by using MATLAB software in terms of (1) guideway model system by adopting Bernoulli-Euler beam equation of series of simply supported beams and (2) calculation of magnetic forces for uplift levitation and lateral guidance with allowable levitation and guidance distance considering lateral vibration control LQR algorithm, tuning parameters, and maglev dynamics.

Guideway Model

To prepare the guideway modeling considering a free body diagram (Fig. 12.1), I have considered multiple magnets with equal intervals (d) that are to be traveling at various level speeds of speed v, where m is the beam weight, c the damping coefficient, EI_y the flexural rigidity in the y direction, EI_z the flexural rigidity in the z direction, l the car length, m_w the lumped mass of magnetic wheel, m_v the distributed mass of the rigid car body, and $\theta_{i\,=\,x,\,y,z}$ the midpoint rotation components of the rigid car body. Considering these, I have formulated the equations of motion for the jth guideway girder carrying a moving maglev vehicle suspended by multiple magnetic forces as follows:

$$m\ddot{u}_{z,j} + c_z\dot{u}_{z,j} + EI_z u''''_{z,j} = p_0 - \sum_{k=1}^{K}\left[G_{z,k}(i_k, h_{z,k})\varphi_j(x_k, t)\right] \tag{12.1}$$

and

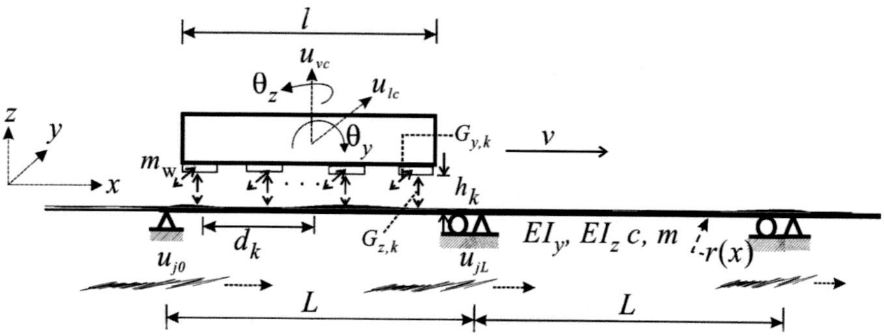

Fig. 12.1 A free body diagram showing the maglev guideway vs vehicle force considering weight and motion where the superconducting guideway is below the vehicle body. It is functioned by a series of equal-distant concentrated masses to levitate the vehicle up to the superconducting guideway beam; the maglev bar gets stimulated by the lateral multi-support motion which is induced by the superconducting force to allow traveling on longitudinal direction

$$\varphi_j(x_k, t) = \delta(x - x_k)\left[H\left(t - t_k - \frac{(j-1)L}{v}\right) - H\left(t - t_k - \frac{jL}{v}\right)\right] \tag{12.2}$$

together with the following boundary conditions with lateral (y-direction) support movements:

$$\begin{aligned} &u_{y,j}(0,t) = u_{yj0}(t), u_{y,j}(L,t) = u_{yjL}(t), \\ &EI_z u''_{z,j}(0,t) = EI_z u''_{z,j}(L,t) = 0 \end{aligned} \tag{12.3}$$

$$\begin{aligned} &u_{z,j}(0,t) = u_{z,j}(L,t) = 0 \\ &EI_y u''_{y,j}(0,t) = EI_y u''_{y,j}(L,t) = 0 \end{aligned} \tag{12.4}$$

where $(x)' = \partial(x)/\partial x$; $(x) = \partial(x)/\partial t$; $u_{z,j}(x,t)$ is the vertical deflection of the jth span; $u_{y,j}(x,t)$ is the lateral deflection of the jth span; L is the span length, K is the number of magnets attached to the rigid levitation frame; $\delta(x)$ is Dirac's delta function; $H(t)$ is the unit step function; $k = 1, 2, 3,\ldots,$ Kth moving magnetic wheel on the beam; $t_k = (k-1)d/v$ is the arrival time of the kth magnetic wheel into the beam; x_k is the position of the kth magnetic wheel on the guideway; and ($G_{y,k}$, $G_{z,k}$) are lateral guidance and uplift levitation forces of the kth lumped magnet in the vertical and lateral directions [15].

Magnetic Forces of Uplift Levitation and Lateral Guidance

Since the maglev vehicle will run over guideway by superconducting force with lateral ground motion (as shown in Fig. 12.1), thus, guidance forces tuned by the maglev system need to be controlled by the lateral motion of the moving maglev vehicle. Therefore, this study adopts the lateral guidance force ($G_{y,k}$) and the uplift levitation force ($G_{z,k}$) to keep and guide the kth magnet of the vehicle; those could be expressed as

$$G_{y,k} = K_0 \left(\frac{i_k(t)}{h_{z,k(t)}} \right)^2 K_{k,z} \tag{12.5}$$

$$G_{y,k} = K_0 \left(\frac{i_k(t)}{h_{z,k(t)}} \right)^2 (1 - K_{y,k}) \tag{12.6}$$

where $K_{y,k}$ and $K_{z,k}$ represent induced guidance factors, and they are given by

$$K_{y,k} = \frac{\chi_k \times h_{y,k}}{W(1+\chi_k)}, \qquad K_{z,k} = \frac{\chi_k \times h_{y,k}}{W(1+\chi_k)} \tag{12.7}$$

In Eqs. (12.6) and (12.7), $K_0 = \mu_0 N_0^2 A_0/4$ is the coupling factor, $\chi_k = \pi h_{y,k}/4h_{z,k}$, W the pole width, μ_0 the vacuum permeability, N_0 the number of turns of the magnet windings, A_0 the pole face area, $i_n(t) = i_0 + \imath_n(t)$ the electric current, $\imath_n(t)$ the deviation of current, and (i_0, h_{y0}, h_{z0}) the desired current and air gaps around a specified nominal operating point of the maglev wheels at *static* equilibrium. The uplift levitation ($h_{y,k}$) and lateral guidance ($h_{z,k}$) gaps are, respectively, given by

$$h_{y,k}(t) = h_{y0} + u_{1,k}(t) - u_{y,j}(x_k), u_{1,k}(t) = u_{1c}(t) + d_k\theta_z \tag{12.8}$$

$$h_{z,k}(t) = h_{z0} + u_{v,k}(t) - u_{z,j}(x_k) + r(x_k), u_{v,k}(t) = u_{vc}(t) + d_k\theta_y \tag{12.9}$$

where ($u_{1,k}$, $u_{v,k}$) are the displacements of the kth magnetic wheel in the y and z directions, (u_{1c}, u_{vc}) are the midpoint displacements of the rigid car, (θ_y,θ_z) are the midpoint rotations of the rigid car, $r(x)$ is the irregularity of guideway, and d_k is the location of the kth magnetic wheel to the midpoint of the rigid beam. As indicated in Eqs. (12.6), (12.7), and (12.8), the motion-dependent nature and guidance factors ($\kappa_{y,k}$,$\kappa_{z,k}$) dominate the control forces of the maglev vehicle-guideway system. Thus, the equations of motion of the 4-DOF rigid maglev vehicle (see Fig. 12.1) are written as

$$M_0\ddot{u}_{1c} = g(t) + \sum\nolimits_{k=1}^{K} G_{y,k}, \quad I_T\ddot{\theta}_Z = g(t) \times l + \sum\nolimits_{k=1}^{K} \left[G_{y,k}d_k\right] \tag{12.10}$$

$$M_0\ddot{u}_{vc} = p_0 + \sum_{k=1}^{K} G_{z,k}, \quad I_T\ddot{\theta}_y = -\sum_{k=1}^{K} [G_{z,k}d_k] \tag{12.11}$$

in which $M_0 = m_v l + Km_w$ is the lumped mass of the vehicle, $g(t)$ the control force to tune the lateral response of the maglev vehicle, I_T the total mass moment of inertia of the rigid car, and $p_0 = M_0 g$ the lumped weight of the maglev vehicle.

Wind Energy Modeling for the Vehicles

Though the vehicle will run by electromagnetic force, a wind turbine generator is to be used for powering vehicle as the additional source of energy to exit vehicle from road and park where maglev system is not available. Thus, the model is developed by doubly fed induction generator (DFIG) for producing electricity for transportation vehicles [11, 12, 16]. The fundamental equation governing the mechanical power of the wind turbine is

$$P_w = \frac{1}{2} C_p(\lambda, \beta)\rho A V^3 \tag{12.12}$$

where ρ is the air density (kg/m^3), C_p is the power coefficient, A is the intercepting area of the rotor blades (m^2), V is the average wind speed (m/s), and λ is the tip speed ratio [17]. The theoretical maximum value of the power coefficient C_p is 0.593; C_p is also known as Betz's coefficient. Mathematically,

$$\lambda = \frac{R\omega}{V} \tag{12.13}$$

where R is the radius of the turbine (m), ω is the angular speed (rad/s), and V is the average wind speed (m/s). The energy generated by wind can be obtained by

$$Q_w = P \times (\text{Time})[\text{kWh}] \tag{12.14}$$

It is well known that wind velocity cannot be obtained by a direct measurement from any particular motion [15, 18]. In data taken from any reference, the motion needs to be determined for that particular motion; then, the velocity needs to be measured at a lower motion.

$$v(z) \ln\left(\frac{Z_r}{Z_o}\right) = v(Z_r) \ln\left(\frac{Z}{Z_0}\right) \tag{12.15}$$

where Z_r is the reference height (m), Z is the height at which the wind speed is to be determined, Z_0 is the measure of surface roughness (0.1–0.25 for cropland), $v(z)$ is the wind speed at height z (m/s), and $v(z_r)$ is the wind speed at the reference height

z (m/s). The power output in terms of the wind speed shall be estimated using the following equation:

$$P_w(v) = \begin{cases} \dfrac{v^k - v_C^k}{v_R^k - v_C^k} \cdot P_R & v_C \leq v \leq v_R \\ P_R & v_R \leq v \leq v_F \\ 0 & v \leq v_C \text{ and } v \geq v_F \end{cases} \quad (12.16)$$

where P_R is rated power, v_C is the cut-in wind speed, v_R is the rated wind speed, v_F is the rated cut-out speed, and k is the Weibull shape factor [19]. When the blade pitch angle is zero, the power coefficient is maximized for an optimal TSR [2]. The optimal rotor speed is to be calculated by

$$\omega_{opt} = \frac{\lambda_{opt}}{R} V_{wn} \quad (12.17)$$

which will give

$$V_{wn} = \frac{R\omega_{opt}}{\lambda_{opt}} \quad (12.18)$$

where ω_{opt} is the optimal rotor angular speed in rad/s, λ_{opt} is the optimal tip speed ratio, R is the radius of the turbine in meters, and V_{wn} is the wind speed in m/s.

The turbine speed and mechanical powers are depicted in the following graph (Fig. 12.2) with increasing and decreasing rates of wind speed while the vehicle is in motion. When the wind is steady, the persistence forecasts yield good results. When the wind speed is increased rapidly, sudden "ramps" in power output are generated, which is a tremendous benefit for capturing the energy.

Wind Energy Storage in Battery System

Standard Simulink/SimPowerSystems has been calculated by using MATLAB Simulink for the wind energy conversion that is to be stored in circuit-implemented inverter as a storage buffer, and all the electricity is to be supplied through the battery according to Peukert's law to start the engine and to be used when the vehicle is not in motion.

Design of Traffic Control

Though underground maglev system has the capability to allow run up to 580 kph, the vehicles' high speed shall be calculated based on traffic flow, composition,

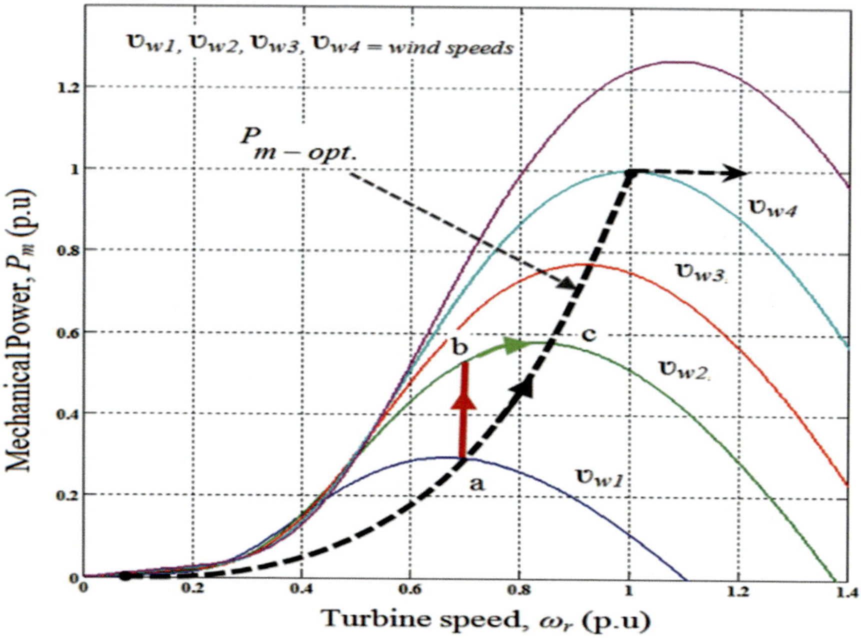

Fig. 12.2 Relationship between mechanical power generation and turbine speeds at different wind speeds for an implementation in a car

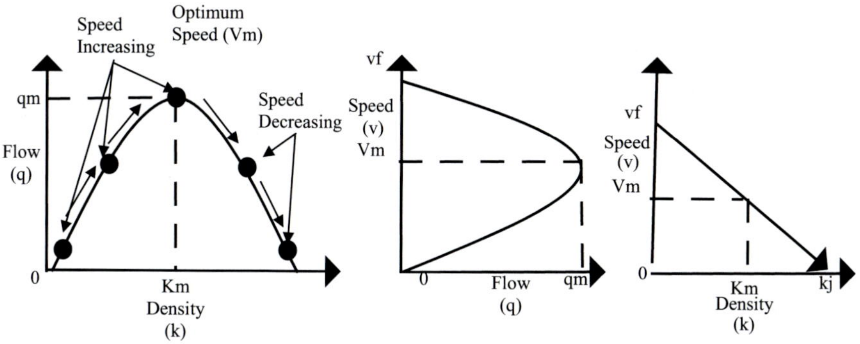

Fig. 12.3 Greenshield's fundamental diagrams: (**a**) Speed vs vehicle density, (**b**) flow vs vehicle density, and (**c**) speed vs flow analysis

volume, number and location of access points, and local environment importantly allotting sufficient number of lanes considering Greenshield's following road and highway capacity analysis (Fig. 12.3).

Since the maglev technology is invisible, thus, to alert the drivers and pedestrian, the maglev roads, highways, and its exits have to be constructed by landscaping by covering the guideway by herb (green grass), and in between lanes, at least 2 ft should be left blank (no landscaping) in order to differentiate the lanes.

Results and Discussion

Based on the mathematical modeling described above, I have performed load resistance factor design (LRFD) calculation considering the following equation and selected W24 × 84 beam which is the continuous maglev underground runs (metal track guideway) that need to be structurally sound to carry enough current, load, and levitation force of the vehicles:

$$Fy \propto \frac{nl^2}{h} \tag{12.19}$$

$$Fx \propto \frac{-1}{ktvx} \; Fx \propto \frac{-nl^2}{h} \tag{12.20}$$

where Fy is the vehicle weight, n total number of coils in maglev, l current on each coil, h height of levitation, t thickness of conduction track, and k conductivity of track.

To construct under maglev guideway just 2 ft below the earth surface, it will be needed to have a U-shaped cross section to fix the pole position [63, 64]. Naturally heavy-duty waterproofing membrane is to be used to protect the maglev underground runs for avoiding floods and moisture. It is well researched that the propulsion coils run in elliptical loops along both walls of the guideway, generating magnetic force when electricity runs through them [65, 68]. So, levitation and guidance coils will be formed that will create their own magnetic force once the applied superconducting magnets pass on it where propulsion and levitation are the key factor to run the vehicle. In propulsion, as the direction of the current charges back and forth in the propulsion coils above the wall of the guideway, the north and south poles will reserve repeatedly, propelling the vehicle by alternating force of attracting and repulsion (Fig. 12.4). In levitation, as the vehicle passes, an electric current is induced in the coil along the guideway, and the vehicle will be levitated by the force of attraction, which will pull up on the magnet in the vehicle, as well as by repulsion, which will push up on the magnet [20].

To create levitation and lateral balance in the vehicle, an electromagnetic induction is to be used. To confirm the most efficient and economical way to produce the powerful magnetic field by using the superconducting coils, I have assumed the permanent currents of about 700,000 amperes go through these superconducting coils [21], hence creating a strong magnetic field of almost 5 teslas, i.e., 100,000 times stronger than the earth magnetic field by implementing the following block diagram (Fig. 12.5).

Simply, it can be explained that when an electric current flows through the propulsion coils, a magnetic field is produced. The forces of attraction and repulsion between the coils and the superconducting magnets on the vehicle propel the vehicle forward in a flying stage up to 4 ft height where 2 ft shall be considered underground

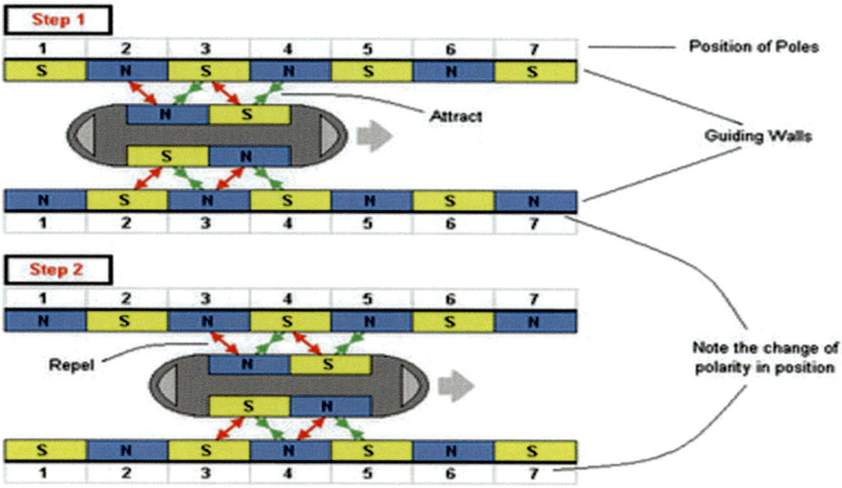

(a)

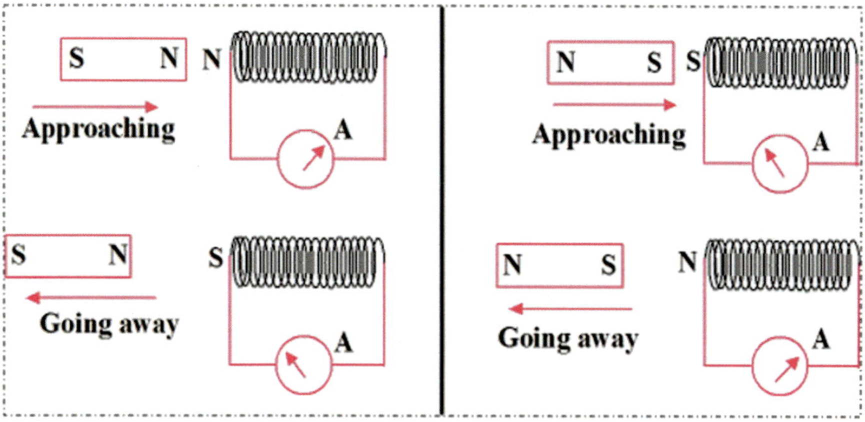

(b)

Fig. 12.4 The above figures indicate the polarization of the coil in different cases. (**a**) Schematic diagram of the director of the running vehicle (must be construction with magnet as shown on this diagram) on maglev propulsion via propulsion coils. (**b**) Near the receding S-pole becomes an N-pole to oppose the going away of the bar magnet's S-pole

cover and the other 2 ft just over the earth surface (Fig. 12.6). The vehicle's speed is to be adjusted by altering the timing of the polarity shift in the propulsion coils' magnetic field between north and south with the possibility of maximum speed of 580 kph. As the vehicle passes just 2 ft above the guideway (1 ft from the earth surface), an electric current is induced in the levitation and guidance coils, creating

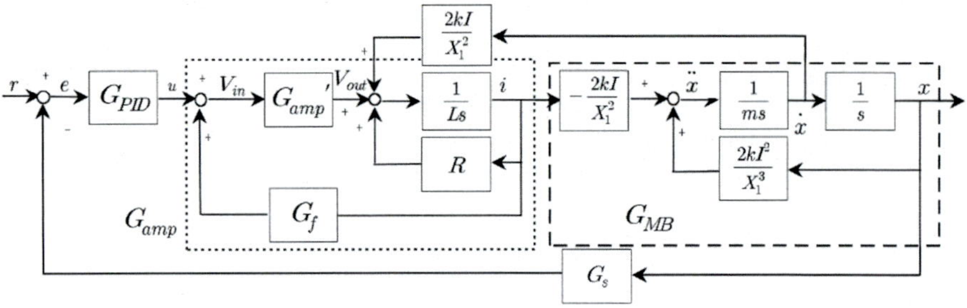

Fig. 12.5 Block diagram to control the mathematically modeled magnetic bearing system, a process to design the driver to operate the electromagnet. Here, the method is to determine the peripheral device values of the linear amplifier circuit that has the desired output by applying a generic algorithm and identify the magnetic bearing system

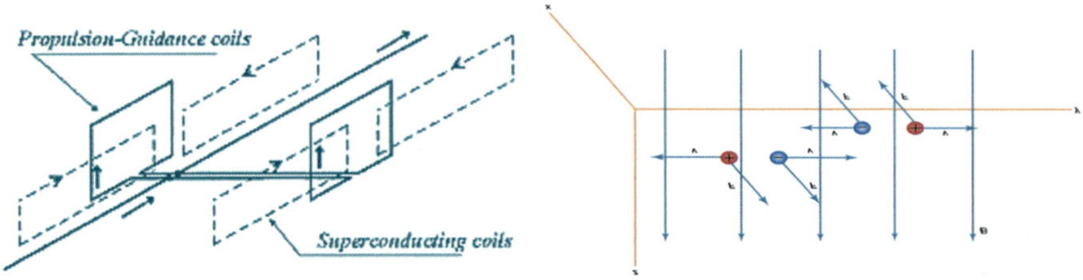

Fig. 12.6 The maglev vehicle's force and directional diagram as shown by propulsion guidance coils and superconducting coils

opposite magnetic poles in the upper and lower loops. The upper loops become the polar opposite of the vehicle's magnets, producing attraction, which pulls the vehicle up. The lower loops have the same pole as the magnets. This generates repulsion, which pushes the vehicle in the same direction up. The two forces combine to levitate the vehicle while maintaining its lateral balance between the walls of the guideway.

Subsequently, a niobium-titanium alloy should be used to create superconducting magnets for maglev, but to reach superconductivity, they must be kept cold. In order to keep the alloy cool, liquid helium should be used at a temperature of −269 °C since alloy retains superconductivity at temperatures up to −263 °C, though the maglev system can operate better at 6 °C to produce sufficient magnetic force.

In addition to underground maglev, the wind turbine generation system for the backup energy source is to be implemented by the optimal operation of the whole system and based on a robustness test, performed by adding a wind speed signal and power coefficient.

These conditions permit the application of the wind profile considered to be a wind speed signal with a mean value of 8 m/s and a rated wind speed of 10 m/s; the whole system is tested under standard conditions with a stator voltage of approximately 50% for 0.5 s between 4 and 4.5 s, approximately 25% between 6 and 6.5 s,

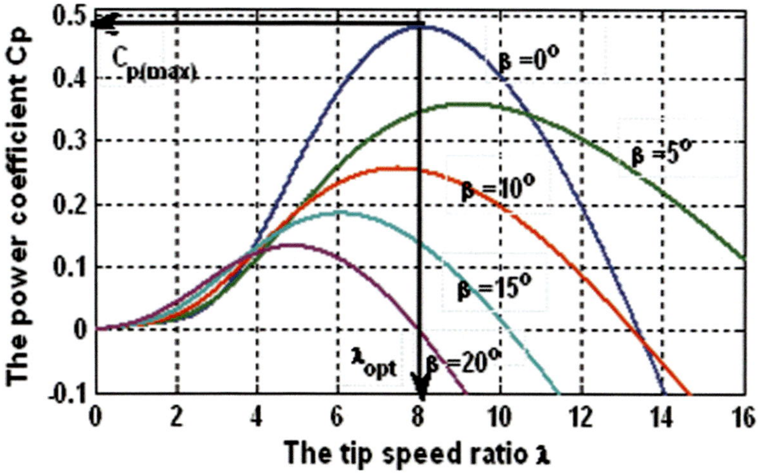

Fig. 12.7 The maximum values of C_p are achieved for the curve associated with $\beta = 2°$. From this curve, the maximum value of C_p ($C_{p,max} = 0.5$) is obtained for $\lambda_{opt} = 0.91$. This value (λ_{opt}) represents the optimal speed ratio

and 50% between 8 and 8.5 s (Fig. 12.7). Thus, the machine is considered to be functioning in ideal conditions (no perturbations and no parameter variations). Moreover, to guarantee a unity power factor at the stator side, the reference for the reactive power is to be set to zero [22]. As a result of increasing wind speed, the generator shaft speed achieved maximum angular speed by tracking the maximum power point speed. Thus, the wind turbine always works optimally since the pole placement technique is to be used to design the tracking control [14]. Consequently, decoupling among the components of the rotor current was also performed to confirm that the control system worked effectively. The bidirectional active and reactive power transfer between the rotor and power system is exchanged by the generator according to the super synchronous operation, achieving the nominal stator power, and the reactive power can be controlled by the load-side converter to obtain the unit's power factor to generate energy for powering vehicles.

Cost Engineering

The order of magnitude cost estimate was performed by using HCSS (HeavyBid) software standard union rate of New York State locals with a project of 10% general condition, 10% overhead and profit, and 3% contingency over the hard cost of labor, materials, and equipment comparing between maglev infrastructure and traditional infrastructure system for a sample of 100 miles long and 128 ft wide (12 ft wide of four lanes in each direction, two-sided 10 ft service space, and 6 ft median in the center of the road). In order to determine that the underground guideway (w24 × 84) can last long, I have calculated again the LRFD to provide the shoring for both sides for the entire 100-miles-long and 128-ft-wide (12 ft wide of four lanes for each direction, two-sided 10 ft service space, and 6 ft median in the center of the road) construction cost considering standard excavation up to 6 ft deep, with appropriate shoring with minimum embedment depth L4 of 5 ft and standard soil pressure $\Upsilon_s = 120$ lbf/ft^3, angle of pressure $\Phi = 21°$, and the soil pressure coefficient $c = 800$ lbf/ft^2. To prepare the conceptual estimate, we need to determine the length of soldier piles. I have counted 6′ OC (on center) soldier piles at both sides by illustration and using the following LRFD method that soldier piles must be set at to support the necessary excavation and/or earth pressure against collapse:

$$\text{Active earth pressure } K_a = \tan^2\left(45° - \frac{\Phi}{2}\right) \tag{12.21}$$

$$\text{Passive earth pressure } K_b = \tan^2\left(45° + \frac{\Phi}{2}\right) \tag{12.23}$$

Use Eqs. (12.22) and (12.23) to find the lateral earth pressure the solid piles must support.

$$\begin{aligned} P_{EM} &= \Upsilon_s\, h\, k_{a,\text{piles}} \\ &= \left(120\frac{\text{lbf}}{\text{ft}^3}\right)(6.0)\tan^2\left(45° - \frac{21°}{2}\right) \\ &= 340.128\ \text{lbf}/\text{ft}^2 \end{aligned}$$

To determine the type of steel beams required for the soldier piles, we have taken the bending moments about the tributary area of the piles.

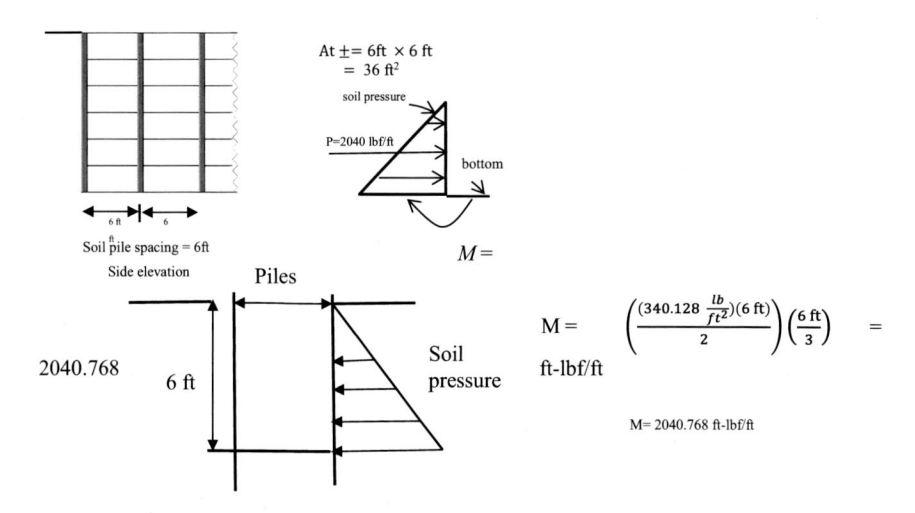

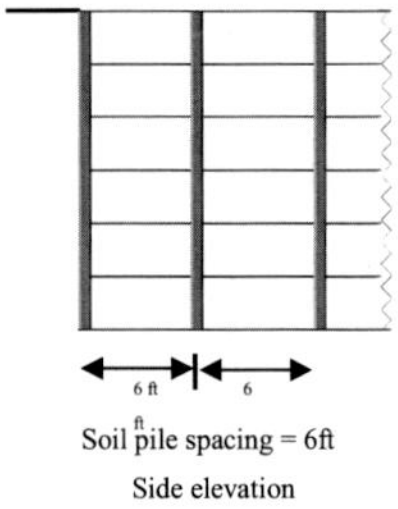

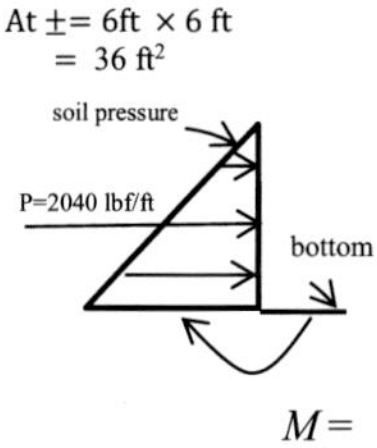

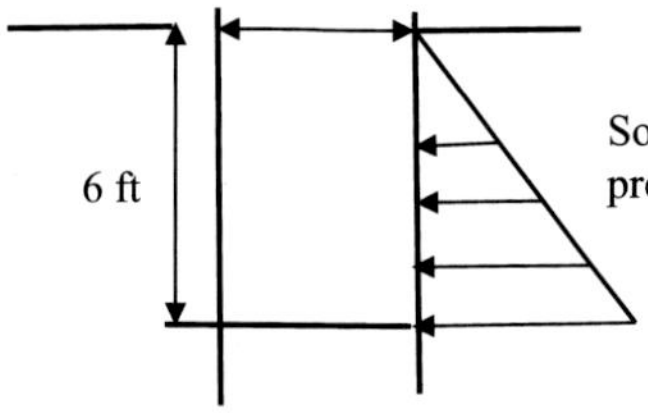

$$M = \left(\frac{(340.128\ \frac{lb}{ft^2})(6\ \text{ft})}{2}\right)\left(\frac{6\ \text{ft}}{3}\right) = 2040.768\ \text{ft-lbf/ft}$$

M= 2040.768 ft-lbf/ft

The moment is a distributed moment applied to the base of the tributary area of each soldier pile. Therefore, the moment is 2040.768 ft-lbf per foot. The total moment on the soldier pile (at the base) is

$$\begin{aligned} M0 &= M(6\text{ ft}) \\ &= \left(2040.768\ \frac{\text{ft-lbf}}{\text{ft}}\right)(6\text{ ft}) \\ &= 12{,}244.61\text{ ft-lbf} \end{aligned}$$

Now,

$$\begin{aligned} Z_{\text{req}} &= \frac{M0}{\Phi b\ Fy} = \frac{\left(12\frac{\text{in}}{\text{ft}}\right)(12{,}244.61\text{ ft-lbf})}{(0.9)\left(50{,}000\frac{\text{lbf}}{\text{in}^2}\right)} \\ &= 3.27\text{ in}^3 \end{aligned}$$

From AISC tables, the soldier piles have been selected to be W12 × 26, and the perpendicular support w8 × 12 members are 6 ft long.

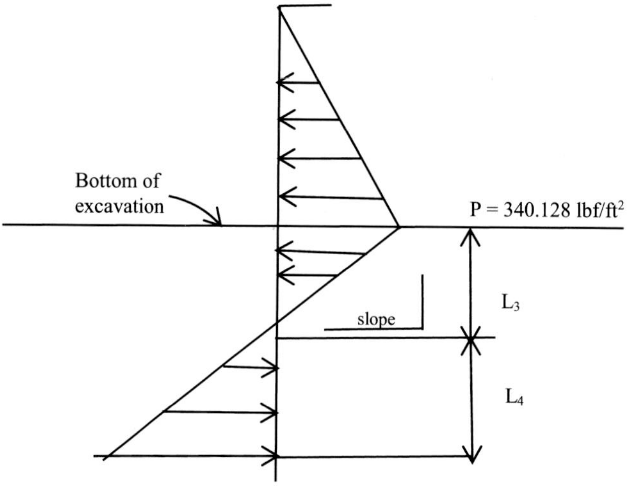

Then, we have determined the depth required below subgrade by calculating the passive earth pressure coefficient using Eq. (12.23):

$$\begin{aligned} K_{\text{p}} &= \tan^2\left(45° + \frac{\Phi}{2}\right) \\ &= \tan^2\left(45° + \frac{21°}{2}\right) \\ &= 2.12 \end{aligned}$$

Then, we have calculated the active earth pressure coefficient using Eq. (12.22):

$$\begin{aligned} K_a &= \tan^2\left(45° - \frac{\Phi}{2}\right) \\ &= \tan^2\left(45° - \frac{21°}{2}\right) \\ &= 0.4724 \end{aligned}$$

In order to determine the slopes of the excavation, depth is required. Since below the bottom of the excavation, the pressure is both considered to be passive and have the same slope. The slope of the pressure profile above the reversal point is calculated from the standard equation for the slope, using L_3 as the rise and Υhk_a as the run (a value equal to the lateral earth pressure, expressed this way for the purposes of cancelation). Thus, the slope of the pressure profile below the reversal point can be calculated similarly, using L_4 as the rise and the product of $\Upsilon L_4 k_p$ as the run. Because the slopes are the same, the two equations can be equated. Rearranging to solve for L_3,

$$\frac{L_3}{\Upsilon hk_a} = \frac{L_4}{\Upsilon L_4 k_p}$$

$$\begin{aligned} L_3 &= \frac{hk_a}{k_p} = \frac{(6\ \text{ft})(0.4724)}{2.12} \\ &= 1.337\ \text{ft} \end{aligned}$$

The necessary embedment depth is

$$1.337\ \text{ft} + 5\ \text{ft} = 6.337\ \text{ft}$$

The total required soldier pile length is

$$6.337\ \text{ft} + 6\ \text{ft} = 12.337\ \text{ft}\ (13\ \text{ft assumed})$$

So, I have determined that the solder pile (W12 × 26) should be 13 ft long, and the perpendicular support (w8 × 12) should be 6 ft long as the support for structurally sound maglev construction.

Construction Cost Estimate Comparison

The order of magnitude cost estimate was performed by using HCSS (HeavyBid) software standard union rate of New York State locals with a project of 10% general condition, 10% overhead and profit, and 3% contingency over the hard cost of labor, materials, and equipment comparing between maglev infrastructure and traditional infrastructure system for a sample of 100 miles long and 128 ft wide (12 ft wide of four lanes in each direction, two-sided 10 ft service space, and 6 ft median in the center of the road). To construct the long-lasting and sophisticated underground maglev, I have performed load resistance factor design (LRFD) calculation and selected w24 × 84 beam that the continuous maglev underground runs (structural beam) are structurally sound. Then I have calculated the required shoring concept for 100-miles-long and 128-ft-wide construction cost considering standard excavation up to 6 ft deep, with appropriate shoring with minimum embedment depth L_4 of 5 ft and standard soil pressure $\Upsilon_s = 120$ lbf/ft^3, angle of pressure $\Phi = 21°$, and the soil pressure coefficient c = 800 lbf/ft^2 in order to determine the length of soldier piles. So, I have calculated by using LRFD methods again that selected that the solder pile (W12 × 26) should be 13 ft long, and the perpendicular support (w8 × 12) should be 6 ft long as the support maglev construction.

Cost of Maglev Infrastructure

The proposed maglev infrastructure, therefore, requires shoring, excavation, structural steel, and concrete operation, and thus, I have calculated the estimate considering the following components:

Shoring at 13′ deep with w24 × 26 steel soldier piles at 6′ OC both sides at $2/lf; top rail w8 × 12 both sides at $2/lf; 6′ length w8 × 12 perpendicular support 20 OC at $2/lf; and protection board 1,372,800 ft^2 both sides at $4/ft^2. Thus, the total cost would be $23,724,800.

Excavation (52,800’$_{\text{length}}$ × 128$_{\text{width}}$ × 6$_{\text{deep}}$ × 1.3$_{\text{fluff factor}}$)/27 of 19,524,266.67 yd^3 at $56/yd^3 cost for digging, stock piling, and backfilling. The total cost would be $1,093,358,933.

Cost of Materials: 100 miles maglev system with structural steel (w24 × 84) support for 8 lanes of $354,816,000; 2 × 2 structural concrete strip footing at $150/yd^3 of $93,866,666; reinforcement bar at 100 lb/yd^3 of $62,577,778; and concrete form at $2/ft^2 of $16,896,000. Thus, the total cost of material is $528,156,445.

Cost of Labor: 200 iron workers for 2704 working days at $100/h; 100 concrete cement workers for 2704 working days at $90/h; 100 laborers for 2704 working days at $70/h; and 50 equipment operators for 2704 working days at $100/h. Thus, the total labor cost is $886,912,000 considering the standard 8 h a day.

Equipment Cost: 10 small renting at $1000/day; 10 small tool renting at $250/day; and 271 concrete pumps at $2000 each. Thus, the total equipment cost is $34,342,000.

Other Costs: Engineering service at $5/ft^2 and survey team at $4400/day for each working day; thus, the total cost is $349,817,600. The net construction cost by adding 10% general condition, 10% overhead and profit, and 3% contingency into the excavation, material, labor, equipment, and other costs would be $3,587,063,487.

Cost of Traditional Road Infrastructure

A typical highway consists of 8″ asphalt surface course, 4″ binder course, 4″ base course, and 12″ aggregate with standard wire mesh or framing, and thus, we have calculated the estimate considering the following components:

Excavation ($52,800_{length} \times 128_{width} \times 2.33_{deep} \times 1.3_{fluff\ factor}$)/27 of 7,581,924 yd^3 at \$56/yd^3 cost for digging, stock piling, and backfilling. The total cost would be \$424,587,744.

Cost of Materials: \$50/yd^3; 4″ base course of 834,370 yd^3 at \$50/yd^3; wire mesh or framing of (528,000 × 128) at \$1/ft^2, and 12″ subbase aggregate of 2,503,111 yd^3 at \$25/yd^3. Thus, the total cost of material is \$380,472,775.

Cost of Labor: 200 asphalt cement workers for 2704 working days at \$100/h; 200 labor foremen for 2704 working days at \$100/h; 200 laborers for 2704 working days at \$70/h; 200 equipment operators for 2704 working days at \$100/h; 100 truck drivers for 2704 working days at \$100/h; and 200 small roller engineers for 2704 working days at \$100/h. Thus, the total cost is \$2,249,728,000.

Equipment Cost: 200 roller renting at \$1000/week; 200 milling renting at \$10,000/week; and 100 truck renting at \$500/week. Thus, the total cost is \$502,171,429.

Other Costs: Detailing and shop drawing at \$10/ft^2; engineering service at \$5/ft^2; survey team at \$4400/day for each working day; banking service of 301,037 yd^3 at \$1000/yd^3; and maiden concrete divider of 106,468 yd^3 at \$818/yd^3. Thus, the total cost is \$1,326,694,600. The net construction cost by adding 10% general condition, 10% overhead and profit, and 3% contingency into the excavation, material, labor, equipment, and other costs would be \$6,805,115,863.

Cost Saving

In this article, I have calculated cost saving by using standard 100 miles highway of 128 ft wide (12 ft wide of four lanes in each direction, two-sided 10 ft service space, and 6 ft median in the center of the road) as an experimental tool to compare construction cost in between conventional and maglev infrastructure system. The total cost estimate for traditional infrastructure is \$6,805,115,863, and the maglev infrastructure system cost is only \$3,587,063,487 for the same 100 miles highways, and the net cost saving is \$3,218,052,377 (Table 12.1). Consequently, it is will reduce neatly 50% cost once maglev infrastructure system is used for the construction of invisible infrastructure, which is also benign to the environment.

Table 12.1 This cost comparison is prepared by using HCSS cost data 2019 for material by utilizing selective manufacturers and labor rate in accordance with international of union wage of each specified trade workers considering US location. The equipment rental cost is estimated as current rental market in conjunction with the standard practice of construction of the production

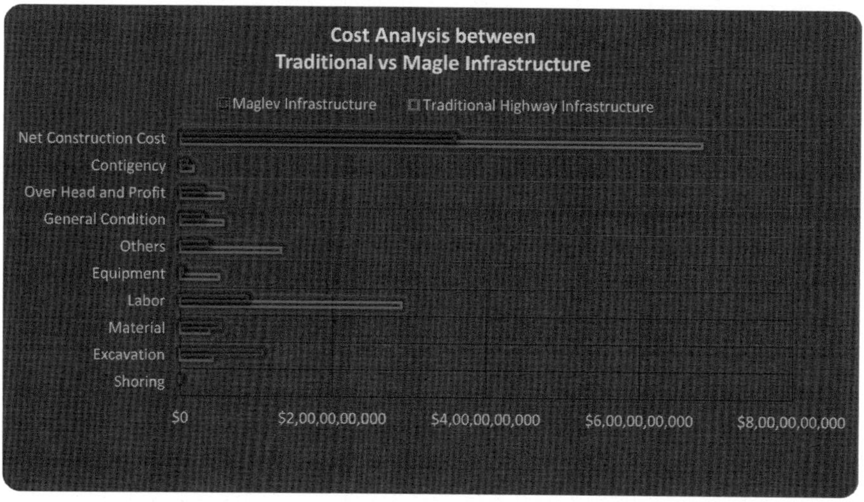

Conclusions

Traditional transportation infrastructure construction and maintenance throughout the world are not only expensive but it is also consuming 5.6×10^{20} J/y (560 EJ/y) of fossil fuel, which is indeed dangerous of a cliché when discussing about climate [16, 23]. In order to mitigate this issue, better infrastructure transportation planning needs to be achieved where environmental sustainability and climate adaptation have been confirmed to create more resilient and vibrant communities. Interestingly, this *sustainable infrastructure* technology proposed for running urban transportation system will be the emergent technology in modern science since it will run by repulsive force and attractive force at the levitated (flying) stage over the maglev system and will be powered by the wind without consuming fossil fuel. Simply, this sustainable infrastructure technology to run the transportation system would be the innovative technology ever to console infrastructure, transportation, energy, and global warming crisis dramatically.

Acknowledgments The author, Md. Faruque Hossain, declares that any findings, predictions, and conclusions described in this article are solely performed by the author, and it is confirmed that there is no conflict of interest for publishing this research paper in a suitable journal and/or publisher.

References

1. Chang, S. (2003). Evaluating disaster mitigations: Methodology for urban infrastructure systems. *Natural Hazards Review*, 186–196. https://doi.org/10.1061/(ASCE)1527-6988(2003)4:4(186)
2. Thomas, A., Pugh, M., MacKenzie, A. R., Whyatt, J. D., & Hewitt, C. N. (2012). Effectiveness of green infrastructure for improvement of air quality in urban street canyons. *Environmental Science and Technology, 46*(14), 7692–7699. https://doi.org/10.1021/es300826w
3. El-Anwar, O., Ye, J., & Orabi, W. (2015). Efficient optimization of post-disaster reconstruction of transportation networks. *Journal of Computing in Civil Engineering, 30*, 04015047.
4. Yau, J. D. (2012). Lateral vibration control of a low-speed maglev vehicle in cross winds. *Wind and Structures, 15*(3), 263–283.
5. O'Neill, B. C., & Oppenheimer, M. (2004). Climate change impacts sensitive to path to stabilization. *Proceedings of the National Academy of Sciences, 101*, 16411–16416.
6. Mann, M. E., Bradley, R. S., & Hughes, M. K. (1998). Global-scale temperature patterns and climate forcing over the past six centuries. *Nature, 392*, 779–787.
7. Website: http://www.supraconductivite.fr/en/index.php?p=applications-trains-maglev-more
8. Cai, Y., Chen, S. S., Rote, D. M., & Coffey, H. T. (1996). Vehicle/guideway dynamic interaction in maglev systems. *Journal of Dynamic Systems, Measurement, and Control, 118*, 526–530.
9. Cai, Y., & Chen, S. S. (1997). Dynamic characteristics of magnetically-levitated vehicle systems. *Applied Mechanics Reviews, 50*(11), 647–670.
10. Chakib, R., Essadki, A., & Cherkaoui, M. (2014). Modeling and control of a wind system based on a DFIG by active disturbance rejection control. *International Review on Modelling and Simulations (IREMOS), 7*(4), 626.
11. Jeong, S.-Y., Nguyen, T. H., Le, Q. A., & Lee, D.-C. (2016). High-performance control of three-phase four-wire DVR systems using feedback linearization. *Journal of Power Electronics, 16*, 351.
12. He, P., Wen, F., Ledwich, G., & Xue, Y. (2013). Small signal stability analysis of power systems with high penetration of wind power. *Journal of Modern Power Systems and Clean Energy, 1*, 241.
13. Hossain, M. F. (2016). Solar energy integration into advanced building design for meeting energy demand and environment problem. *International Journal of Energy Research*. https://doi.org/10.1002/er.3525
14. Werner, J. P., Luterbacher, J., & Smerdon, J. E. (2012). A pseudoproxy evaluation of Bayesian hierarchical modelling and canonical correlation analysis for climate field reconstructions over Europe. *Journal of Climate, 26*, 851.
15. Khodakarami, J., & Ghobadi, P. (2016). Urban pollution and solar radiation impacts. *Renewable and Sustainable Energy Reviews, 57*, 965–976.
16. Hossain, M. F. (2016). Theory of global cooling. *Energy Sustainability, and Society, 6*, 24.
17. Zheng, X. J., Wu, J. J., & Zhou, Y. H. (1997). Numerical analyses on dynamic control of five-degree-of-freedom maglev vehicle moving on flexible guideways. *Journal of Sound and Vibration, 235*, 43–61.
18. Elmansouri, A., El Mhamdi, J., & Boualouch, A. (2016). *Wind energy conversion system using DFIG controlled by back-stepping and RST controller*. 2016 international conference on electrical and information technologies (ICEIT).
19. Zhao, C. F., & Zhai, W. M. (2002). Maglev vehicle/guideway vertical random response and ride quality. *Vehicle System Dynamics, 38*(3), 185–210.
20. Lala, H., & Karmakar, S. (2015). *Continuous wavelet transform and artificial neural network based fault diagnosis in 52 bus hybrid distributed generation system*. 2015 IEEE students conference on engineering and systems (SCES).

21. Abdelmalek, S., Barazane, L., Larabi, A., & Belmili, H. (2015). *Contributions to diagnosis and fault tolerant control based on proportional integral observer: Application to a doubly-fed induction generator*. 2015 4th international conference on electrical engineering (ICEE).
22. Zheng, X. J., Wu, J. J., & Zhou, Y. H. (2005). Effect of spring non-linearity on dynamic stability of a controlled maglev vehicle and its guideway system. *Journal of Sound and Vibration, 279*, 201–215.
23. Soong, T. T. (1990). *Active structural control: Theory and practice*. Longman Scientific & Technical.
24. Aldo, D., & Alfred, R. (1999). *Design of an integrated electromagnetic levitation and guidance system for Swiss Metro*. EPE'99, Lausanne, Swiss.
25. Astrom, K. J., & Hagglund, T. (1988). *Automatic tuning of PID controllers*. Instrument Society of America.
26. Bhandari, B., Poudel, S. R., Lee, K.-T., & Ahn, S.-H. (2014). Mathematical modeling of hybrid renewable energy system: A review on small hydro-solar-wind power generation. *International Journal of Precision Engineering and Manufacturing-Green Technology, 1*, 157–173.
27. Kamili, H., & Riffi, M. E. (2016). Portfolio optimization using the bat algorithm. *International Review on Computers and Software (IRECOS), 11*(3), 277.
28. Kerrouche, K., Mezouar, A., Boumedien, L. (2013). *A simple and efficient maximized power control of DFIG variable speed wind turbine*. 3rd international conference on systems and control.
29. Loucif, M., & Boumediene, A. (2015). *Modeling and direct power control for a DFIG under wind speed variation*. 2015 3rd international conference on control engineering & information technology (CEIT).
30. Lu, L. (2002). Investigation on wind power potential on Hong Kong islands—An analysis of wind power and wind turbine characteristics. *Renewable Energy, 27*, 1–12.

Chapter 13
Sustainable Transportation Technology: Application of Hybrid Wind and Solar Energy in the Transportation Sector

Introduction

The current magnitude of wind turbine installations for producing electricity seriously impacts global habitats and ecosystems [1–3]. In addition, possible radar interference is also a concern for the construction of wind turbines to produce electricity [4, 5]. Prominent concerns about the production of solar energy by using solar panels installed over vast land areas include visibility and the misuse of land, which affects the harmonium of the landscape [7, 8]. In addition, the technological application for the conversion of wind and solar power into electricity supply through the national grid, including cost concerns, and engineering challenges along with transmission and operational mechanisms are likely to restrict the global commercial use of wind and solar energy [1, 3, 6].

Several studies in the past conducted on wind energy and solar energy have been carried out to use renewable energy as an alternative source to mitigate global energy crisis. Boumassata et al. [5] suggested that grid power control based on a wind energy conversion system is quite applicable throughout the world. Another study by Elmansouri et al. [7] suggested that a wind energy conversion system using a doubly fed induction generator (DFIG) controlled by back-stepping and an RST controller may be the most interesting technology for utilizing this energy in the urban transportation sector to reduce energy and environmental crises. Recently, Hossain and Fara [16] revealed that solar energy has a tremendous capability to power vehicles using such a natural system implementation on transportation vehicles, and another study by Hossain [14, 15] revealed that wind energy has tremendous potential to meet the net energy demand of running a vehicle for the development of sustainable transportation system. Although their research on both wind and solar energies hypothetically suggested analyzing the possibility of utilizing this renewable energy for many sectors, applications of these renewable hybrid energy conversion technologies have not yet been attempted in the transportation sector [6, 9, 10].

M. F. Hossain, *Global Sustainability*, https://doi.org/10.1007/978-3-031-34575-3_13

Simply, the knowledge gap and the novelty of their research for the application of hybrid wind and solar power have led to its being ignored despite the tremendous potential it has to meet the near-term and long-term energy mitigation needs in the transportation sector globally.

Thus, to utilize these renewable energies, in this study, a hybrid model of wind turbines and solar panels was installed on a vehicle to capture wind and solar energy and create naturally the required energy to power the transportation vehicle. Subsequently, the drive train model, the energy conversion mechanism, and the process of electricity energy generation through the main subsystems were mathematically calculated using MATLAB Simulink [7, 11, 12].

Simply, this sustainable transportation technology would be the innovative mechanism of implementation of hybrid wind and solar energy that will enable transportation vehicles to utilize this abundant energy resource, which indeed would be a cutting-edge technology to reduce the cost of energy and environmental vulnerability worldwide in the transportation sector.

Material and Methods

Modeling of Wind Turbine

Wind energy is measured here in volume from the air as the kinetic force [8, 14–16], which can be installed in front of the car in order to form energy by the rotation of wind turbine to convert kinetic wind into electric energy [13–15]. The process is such that the wind turbines rotate in the clockwise direction, and the main shaft connected to the gearbox in the nacelle causes the spinning process and then transfers energy to the generator, which eventually converts the kinetic energy into electric energy to power the transportation vehicle [17, 18]. Thus, in this research, the model is used through the application of MATLAB Simulink to develop wind turbines to install on transportation vehicles, which will be driven by airflow due to the motion of the vehicle [19–21]. Consequently, a series of mathematical analyses have been integrated to perform simulations in a doubly fed induction generator (DFIG) to convert energy into electricity [20–22]. Subsequently, the conversion of wind energy is also analyzed using DC and AC converter circuits to power vehicles [11, 23]. The active stator is thus applied here in regulating the whole mechanism of wind conversion, which is represented by the following equation:

$$P_{\mathrm{w}} = \frac{1}{2} C_{\mathrm{p}}(\lambda, \beta) \rho A V^3 \tag{13.1}$$

where ρ is the density of air in kg/m^3, C_{p} is the power coefficient, V is the average wind speed in m/s, λ is the tip speed ratio, and A is the intercepting area of wind rotor blades in m^2. C_{p} is 0.593 at its peak [24, 25]. The average tip speed ratio (TSR) is given in terms of velocity ratios, as shown:

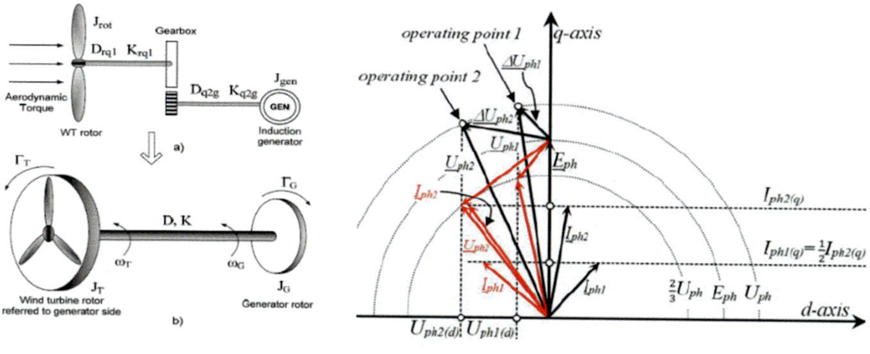

(a) (b)

Fig. 13.1 (**a**) The representation of a schematic of a wind turbine installed on a transportation vehicle. (**b**) The different orientation of d-axis (Z) and q-axis (Z_r) at maximum rotor speed ($z = I_{ph}$)

$$\lambda = \frac{R\omega}{V} \tag{13.2}$$

where R is the radius of the turbine (m), ω the angular velocity (rad/s), and v the mean wind velocity (m/s). The power in the turbine is given by:

$$Q_w = P \times T \tag{13.3}$$

Some of the factors that may have been interfered here with wind flow and velocity that are used as the factored in the equation below and take into account the error proneness due to obstacles, which is expressed as:

$$v(z)\ln\left(\frac{z_r}{z_o}\right) = v(z_r)\ln\left(\frac{z_r}{z_o}\right) \tag{13.4}$$

where Z_r is the height of the vehicle (m), Z_0 the surface roughness (0.1–0.25), Z the wind speed, $v(z)$ the wind velocity at height z (m/s), and $v(Z_r)$ the wind velocity at height Z_r (m/s) (Fig. 13.1).

Furthermore, the velocity of wind is calculated by integrating the turbine rotation that helps in establishing the attainment of equilibrium wind velocity in a steady way within the turbine positioned on the transportation vehicle. The equation for calculation is as shown below:

$$P_w(v) = \begin{cases} \dfrac{v^k - v_C^k}{v_R^k - v_C^k} . P_R & v_C \le v \le v_R \\ P_R & v_R \le v \le v_F \\ 0 & v \le v_C \text{ and } v \ge v_F \end{cases} \tag{13.5}$$

where P_R is the rated power, v_C the cut-in wind velocity, v_R the rated wind velocity, v_F the rated cut-out velocity, and k the Weibull shape cofactor.

The maximum power is obtained from the angular velocity of the generator. Therefore, the maximum power point tracking (MPPT) is derived from the optimum rotor velocity using the equation below:

$$\omega_{\text{opt}} = \frac{\lambda_{\text{opt}}}{R} V_{\text{wn}} \tag{13.6}$$

The equation is converted to:

$$V_{\text{wn}} = \frac{R\omega_{\text{opt}}}{\lambda_{\text{opt}}} \tag{13.7}$$

where ω_{opt} is the maximum rotor angular velocity (rad/s), λ_{opt} the maximum tip velocity factor, R the radius of the turbine (m), and V_{wn} the wind velocity in m/s.

The entry of turbine force into the gearbox is being modeled by applying the torsional multibody dynamic, which is expressed by the following equation [26, 27]:

$$\begin{bmatrix} \dot{\omega}_l \\ \dot{\omega}_g \\ \dot{T}_{l_2} \end{bmatrix} = \begin{bmatrix} -\frac{K_l}{J_l} & 0 & -\frac{1}{J_l} \\ 0 & -\frac{K_g}{J_g} & \frac{1}{n_g J_g} \\ \left(B_{l_2} - \frac{K_{l_2} K_r}{J_r}\right) & \frac{1}{n_g}\left(\frac{K_{l_2} K_r}{J_g} - B_{l_2}\right) & -K_{l_2}\left(\frac{J_r + n_g^2 J_g}{n_g^2 J_g J_r}\right) \end{bmatrix} \begin{bmatrix} \omega_l \\ \omega_g \\ T_{l_2} \end{bmatrix} + \begin{bmatrix} \frac{1}{J_r} \\ 0 \\ \frac{K_{l_2}}{J_r} \end{bmatrix} T_m + \begin{bmatrix} 0 \\ -\frac{1}{J_g} \\ \frac{K_{l_2}}{n_g J_g} \end{bmatrix} T_g \tag{13.8}$$

The relationship is being analyzed here to obtain the moment of inertia, which is calculated using the following equation and represents a combined mass of the hub and blades:

$$J_t \dot{\omega}_t = T_a - K_t \omega_t - T_g \tag{13.9}$$

and

$$K_t = K_r + n_g^2 K_g \tag{13.10}$$

where J_t is the turbine rotor moment of inertia (kg m^2), ω_t the angular velocity of the rotor (rad/sec^2), K_t the turbine cofactor (Nm rad^{-1} s^{-1}), and K_g the generator factor (NM/rad/s).

The other factor that is modeling the drive train, which is a component of the torque at the rotor shaft and passes through the gearbox, has been analyzed to reduce the complexity associated with the turbine mass. Thus, the one-mass model derived from the torsional model is being considered using the following matrix [24, 29]:

$$\begin{bmatrix} \dot{\omega}_l \\ \dot{\omega}_g \\ \dot{T}_{l_2} \end{bmatrix} = \begin{bmatrix} -\frac{K_l}{J_l} & 0 & -\frac{1}{J_l} \\ 0 & -\frac{K_g}{J_g} & \frac{1}{n_g J_g} \\ \left(B_{l_2} - \frac{K_{l_2} K_r}{J_r}\right) & \frac{1}{n_g}\left(\frac{K_{l_2} K_r}{J_g} - B_{l_2}\right) & -K_{l_2}\left(\frac{J_r + n_g^2 J_g}{n_g^2 J_g J_r}\right) \end{bmatrix} \begin{bmatrix} \omega_l \\ \omega_g \\ T_{l_2} \end{bmatrix} + \begin{bmatrix} \frac{1}{J_r} \\ 0 \\ \frac{K_{l_2}}{J_r} \end{bmatrix} T_m + \begin{bmatrix} 0 \\ -\frac{1}{J_g} \\ \frac{K_{l_2}}{n_g J_g} \end{bmatrix} T_g \tag{13.11}$$

Consequently, a combination of mutual damping from one mass and the weight of the blades is used in calculating the turbine inertia [30, 31]. The blades are designed to distribute weight equally and provide the best outcome in calculating the torque values as the sum of all the components. The one-mass model is used in the calculation by factoring the mass moment of inertia of the turbines, as shown below:

$$J_t \dot{\omega}_t = T_a - K_t \omega_t - T_g \tag{13.12}$$

and

$$K_t = K_r + n_g^2 K_g \tag{13.13}$$

where J_t is the rotor moment of inertia for turbine (kg m^2); ω_t the angular speed of low shaft (rad/s^2); K_t the damping coefficient of the turbine (Nm/rad/s), which is linked to the aerodynamic resistance; and K_g the coefficient of the generator damping (Nm/rad/s), which shows the mechanical friction and windage.

The one-mass model is thus used in the analysis of wind speed based on the air mass flow, followed by the calculation of the mechanism used in converting energy and finally the analysis of the whole process of conversion [20, 21, 32].

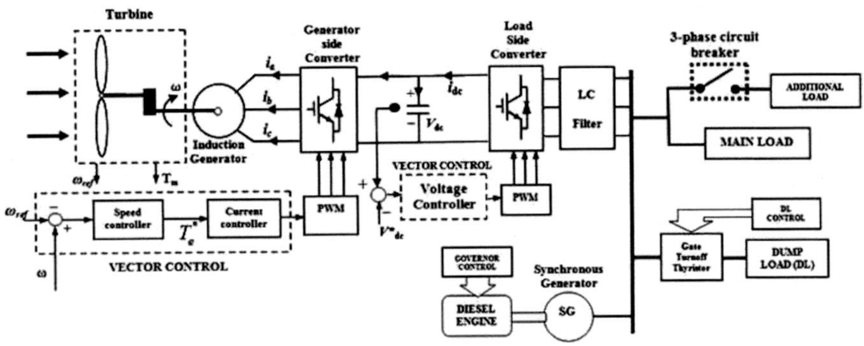

Fig. 13.2 A block diagram of the wind turbine control system, and the main components involved are the generator-side convertor, turbine, as well as load-side convertor

Conversion of Wind Energy

Since the wind is transferred to the electrical power system through the DFIG, thus, the conversion process is coupled by several stages that occur in the aerodynamic system, thereafter releasing electric energy (Fig. 13.2). The real log determination of velocity in the DFIG systems is applied in the analysis of the velocity systems and controlled through the aerodynamic system [17, 33]. The deterministic and stochastic elements are calculated using the equation below, and it is a component of wind speed V transferred in the rotor.

$$V(t) = V_0 + \sum_{i=1}^{n} A_i \sin(\omega_i t + \varphi_i) \tag{13.14}$$

In the above equation, V_0 is the average component, A_i the magnitude, ω_i the pulsation, and ψ_i the initial phase for every turbulence.

Consequently, the conversion energy from kinetic to electrical is determined by taking into consideration the turbulence function, which is a component of the rotational blades. The kinetic energy is dependent on the speed of the rotor and power coefficient C_p, as shown in the equation below [9, 27]:

$$P_{\mathrm{aer}} = \frac{1}{2} C_\mathrm{p}(\lambda, \beta)\rho R\pi^2 V^3, \tag{13.15}$$

where ρ is the air density, R the blade length, and V the wind speed. The captured wind coefficient $C_\mathrm{p}(\lambda)$ is calculated by considering the pitch angle and the wind speed (Q_t), as shown in the equation below [26, 34]:

$$\lambda = \frac{\Omega_T * R}{V}. \tag{13.16}$$

Modeling of Solar Energy

Since the sun produces electromagnetic waves that propagate in the bulk of photons, thus, it is proposed to be trapped by using solar panels installed on transportation vehicle [3, 19]. The modeling process of the transferred energy in this research has been achieved by employing the concept of classical statistical physics on the photon wave frequency (Fig. 13.3). Here, the components used in modeling the maximum solar energy are photon excitation (1.4 eV) and energy value (27.77 MW/m^2) in order to confirm the use of classical statistical physics to show the solar radiation frequencies (Fig. 13.3a) and the representation of maximum solar energy formation (Fig. 13.3b).

Subsequently, the maximum energy intake in the solar panels installed in the transportation vehicle has also been modeled using the photovoltaic module model and used to describe how photovoltaics are generated using the thermocouple, radiation shield, and temperature component on the solar panel (Fig. 13.3). The maximum power output simplified the photovoltaic solar radiance model embedded in the photovoltaic module and attached to the solar radiation logarithmic and temperature module [6, 20, 21, 31]. Naturally, the configuration model in a single-diode circuit is being applied in the calculation of the parameter characteristics of current-voltage (I-V) and considering their numbers [13, 17, 35]. The solar panel collects the solar radiance that can be used in deriving the active solar volt parameter (Iv+), which is expressed as a single-diode circuit, as shown in Fig. 13.3.

Solar Energy Conversion

Solar energy conversion is implemented by determining the current production by calculating the Ipv from a one-diode model, which is derived from the I-V-R relationship by employing the photovoltaic array collected and converting from DC to AC for use on the transportation vehicle (Fig. 13.4).

The equation shown below is derived from calculating the energy output in the solar panel and linked to the solar radiation:

$$P_{\mathrm{pv}} = \eta_{\mathrm{pvg}} A_{\mathrm{pvg}} G_{\mathrm{t}} \tag{13.17}$$

where η_{pvg} is the efficiency of solar panel generation, A_{pvg} represents the solar panel generator area (m^2), and G_{t} is the solar radiation in a tiled module plane (W/m^2). The solar panel generation efficiency can be written as:

$$\eta_{\mathrm{pvg}} = \eta_{\mathrm{r}} \eta_{\mathrm{pc}} [1 - \beta(T_{\mathrm{c}} - T_{\mathrm{c\ ref}})] \tag{13.18}$$

where η_{pc} is the efficiency of power conditioning, and during the application of MPPT, it is equivalent to 1:β, which is also 0.004–0.006 per °C (temperature

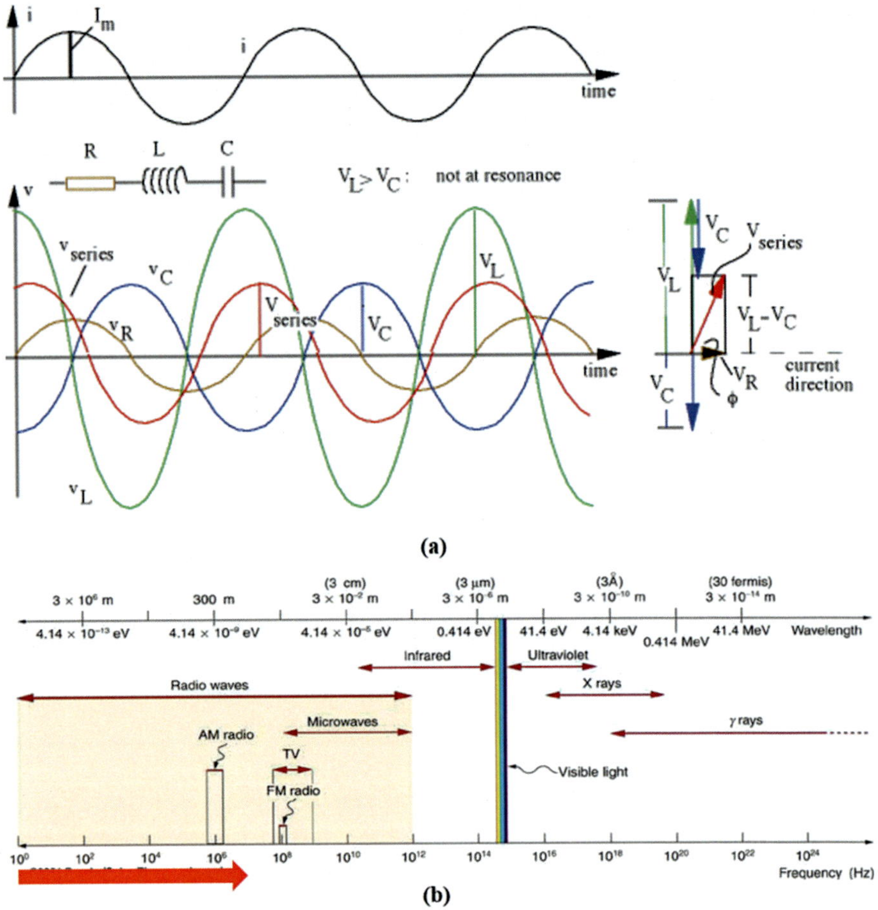

Fig. 13.3 Representation of the PV system model: (**a**) the photovoltaic solar radiance in the photovoltaic mode and having temperature, thermocouple, and radiation as the variations and (**b**) Simulink block that constitutes the active solar volt parameter into the diode circuit as the current direction

coefficient); η_r is the module efficiency; and $T_{c\ ref}$ is the reference cell temperature measured in °C. The quantity can be calculated using the equation below:

$$T_c = T_a + \left(\frac{\text{NOCT} - 20}{800}\right) G_t \tag{13.19}$$

The variables in the equations are defined as follows: T_a, the ambient temperature (°C); G_t, the solar radiance in the tilted module plane (W/m^2); and final NOCT, the standard operating temperature (°C). The equation below is utilized to obtain the total radiance in the solar cell, which is a variable of the diffused solar irradiance:

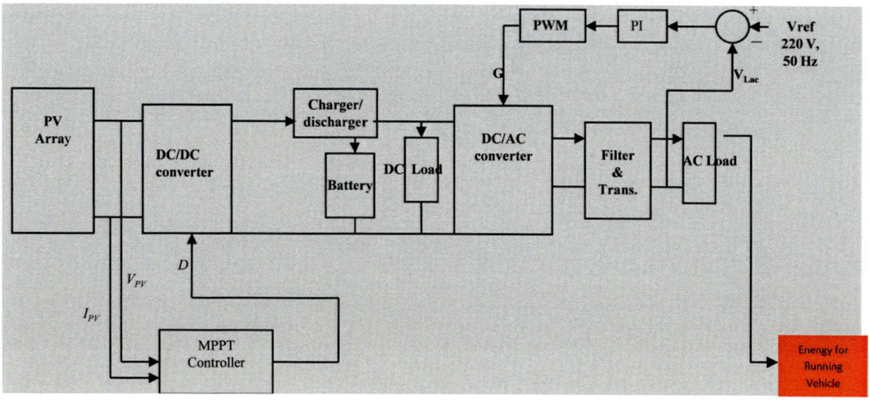

Fig. 13.4 MATLAB is used for modeling the single-diode circuit for the photovoltaic cell (PV), resulting in the generation of current followed by the conversion of DC to AC by using the I-V-R relationships

$$I_t = I_b R_b + I_d R_d + (I_b + I_d) R_r \tag{13.20}$$

Here, the energy is generated in solar cells because of the P-N junction semiconductor are interconnections of solar panel cells in series which eventually form electricity generation [2, 6, 33, 37].

Electrical Subsystem

After the analysis of the wind and solar energy conversion system, using mathematical models, the result is then integrated into the hybrid energy system to introduce the driving force through the electrical subsystem [34, 36]. The analysis of the DFIG system is utilized in the analysis of the hybrid system that is used to establish the electric system that can generate electrical energy. The process is summarized using the equation below [16, 34, 36].

$$\phi_s = \phi_{ds} \Rightarrow \phi_{qs} = 0. \tag{13.21}$$

The above equation is used in calculating the stator constant flux ϕ_s and functional voltage V_s, as shown below:

$$\begin{aligned} v_{ds} &= 0 \\ v_{qs} &= \omega_s \times \phi_s = V_s \end{aligned} \tag{13.22}$$

Therefore, the stator voltage vectors are a function of the direction of quadrate advances and explained using the equations below to derive the rotor voltage:

$$\begin{cases} v_{\mathrm{dr}} = \sigma L_{\mathrm{r}} \dfrac{di_{\mathrm{dr}}}{dt} + R_{\mathrm{r}} i_{\mathrm{dr}} + fem_d \\ v_{\mathrm{qr}} = \sigma L_{\mathrm{r}} \dfrac{di_{\mathrm{qr}}}{dt} + R_{\mathrm{r}} i_{\mathrm{qr}} + fem_q \end{cases} \tag{13.23}$$

where $\mathrm{fem_d}$ represents the coupling term in the d-axis and $\mathrm{fem_q}$ represents the coupling term in the q-axis. The two can be represented using the following equation:

$$\begin{cases} \mathrm{fem_d} = -\sigma L_{\mathrm{r}} L_{\mathrm{r}} i_{\mathrm{qr}} \\ \mathrm{fem_q} = \sigma L_{\mathrm{r}} \omega_{\mathrm{r}} i_{\mathrm{dr}} + s \dfrac{M}{L_{\mathrm{s}}} V_{\mathrm{s}}. \end{cases} \tag{13.24}$$

Equations (13.10) and (13.12) are used to obtain the current fluxes, as shown below:

$$\begin{cases} \phi_{\mathrm{ds}} = L_{\mathrm{s}} i_{\mathrm{ds}} + M i_{\mathrm{dr}} \\ 0 = L_{\mathrm{s}} i_{\mathrm{qs}} + M I_{\mathrm{qr}} \end{cases} \tag{13.25}$$

Equation (13.15) is used to calculate the currents, as shown below:

$$\begin{cases} i_{\mathrm{ds}} = \dfrac{\phi_{\mathrm{ds}} - M i_{\mathrm{dr}}}{L_{\mathrm{s}}} \\ i_{\mathrm{qs}} = -\dfrac{M}{L_{\mathrm{s}}} i_{\mathrm{qr}}. \end{cases} \tag{13.26}$$

The stator is associated with rotor currents in the equations below:

$$\begin{cases} P_{\mathrm{s}} = -V_{\mathrm{s}} * \dfrac{M}{L_{\mathrm{s}}} i_{\mathrm{qr}} \\ Q_{\mathrm{s}} = -V_{\mathrm{s}} * \dfrac{M}{L_{\mathrm{s}}} \left(i_{\mathrm{dr}} - \dfrac{\phi_{\mathrm{ds}}}{M} \right) \end{cases}, \tag{13.27}$$

where i_{qr} represents the stator active and i_{dr} shows the reactive powers. The integration of DFIGs on the transportation vehicle together with the control variables from the power system is an important approach to creating hybrid renewable energy to power the transportation sector naturally.

Results and Discussions

The Wind Turbine Model

The model of wind turbine system that is being proposed to be installed on the transportation vehicle to enhance the operation of the whole system in both the MPPT control and stator flux to generate wind energy. MATLAB Simulink is subsequently applied in the mathematical calculation of all mechanisms involved in rotor converter control (Fig. 13.5). Since the performance of the wind turbine operations at different values of wind speeds is related to the pitch angles, the degrees of angular speed are obtained by performing a robustness test by incorporating both the voltage dips and wind speed signals [34, 36].

The maximum angular speed acquired by the shaft in the generator is due to the increased speed of the wind, through the process of tracking the maximum power point speed through the functionalities of the generator, supported by stator active power. Thus, the load-side converter controls the power to derive the power factor that results in energy output [38, 39].

Then the determination of MPPT control was calculated through the electromagnetic torque and integrated to confirm the stator active power; therefore, the process was facilitated by the power speed ratio that helped in tracking the maximum power point during the process of energy conversion [28, 40]. The maximum power cofactor is here achieved by the use of wind turbine shaft speed, which indicates that the optimum electromagnetic torque is obtained from MPPT control [1, 35]. The determination of rotor dynamics was from the active and reactive powers, and this tool was placed when the MPPT control deduced the wind turbine efficiency, which is expressed as:

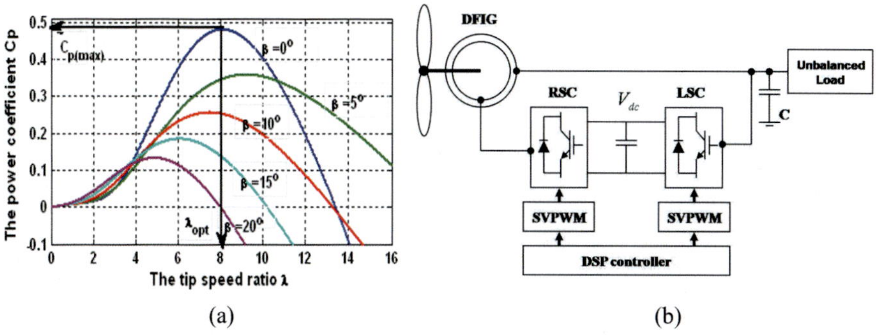

Fig. 13.5 (**a**) How the maximum values of C_p are attained from the diagram when $\beta = 2°$. The maximum value of $C_p = 0.5$ at $\lambda_{opt} = 0.91$. The λ_{opt} stands for the optimal speed ratio and has a rated speed of wind of 10 m/s and the mean value rated at 8 m/s. The standard conditions for testing are such that the stator voltage is estimated at 50% for 4–5 s, 25% for 6–6.5 s, and 50% for 8–8.5 s. (**b**). The DFIG is a factor that was obtained from the relationships between the wind signals that allowed the use of DSP control and the wind turbine that resulted in V_{dc} energy in the turbine

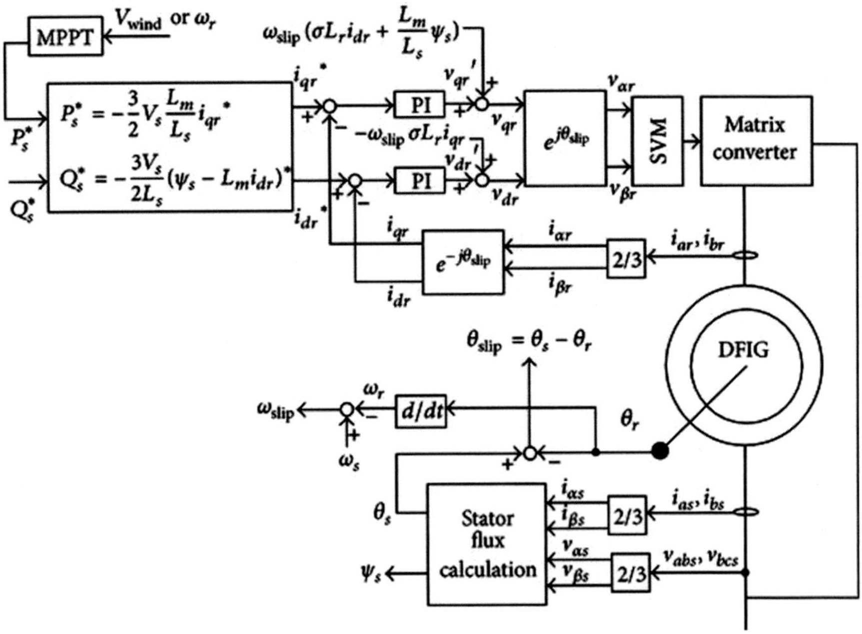

Fig. 13.6 The block diagram of the MPPT (V_{wind}) showing the control system of the DFIG velocity in the aerodynamic subsystem, which power the vehicles using the stator flux rotor

$$\begin{cases} i_{\text{qr}}^{*} = -\dfrac{L_{\text{s}}}{MV_{\text{s}}} P_{\text{s}}^{*} \\ i_{\text{dr}}^{*} = -\dfrac{L_{\text{s}}}{MV_{\text{s}}} \left(Q_{\text{s}}^{*} - \dfrac{V_{\text{s}}^{2}}{\omega_{\text{s}} L_{\text{s}}} \right) \end{cases}. \tag{13.28}$$

Here, the assessment of this mechanism revealed an accurate speed airflow that was taken in by the wind turbine, and the tracking of wind predicted velocity was made effective through the use of MPPT control, which prompted adjustment of the electromagnetic torque in the DFIG, producing the energy in a consistent way (Fig. 13.6).

Subsequently, the drive train model is formed following the synchronization of d-q comments and is being calculated by the equation below:

$$V_{\text{q}} = -R_{\text{s}} i_{\text{q}} - L_{\text{q}} \frac{di_{\text{q}}}{dt} - \omega L_{\text{d}} i_{\text{d}} + \omega \lambda_{\text{m}} \tag{13.29}$$

The electronic torque is obtained using the equation below:

$$T_{\text{e}} = 1.5p\left[\lambda i_{\text{q}} + \left(L_{\text{d}} - L_{\text{q}}\right) i_{\text{d}} i_{\text{q}}\right] \tag{13.30}$$

where L_q is the resistance along the q-axis, L_d is the inductance resistance along the d-axis, i_q is the current along the q-axis, i_d is the current along the d-axis, V_q is the voltage along the q-axis, V_d is the voltage along the d-axis, ω_r represents the angular velocity of the rotor, λ is the amplitude of the induced influx, and p is the number of pairs of poles. The following equations of the q-q frame are applicable for the squirrel cage induction generator (SCIG):

$$\begin{bmatrix} V_{qs} \\ V_{ds} \\ V_{qr} \\ V_{dr} \end{bmatrix} = \begin{bmatrix} R_s + pL_s & 0 & pL_m & 0 \\ 0 & R_s + pL_s & 0 & pL_m \\ pL_m & -\omega_r L_m & R_r + pL_r & -\omega_r L_r \\ \omega_r L_m & pL_m & \omega_r L_r & R_r + pL_r \end{bmatrix} \begin{bmatrix} i_{qs} \\ i_{ds} \\ i_{qr} \\ i_{dr} \end{bmatrix} \tag{13.31}$$

The equation used for the stator side is as shown below:

$$V_{qs} = R_s i_{qs} + \frac{d}{dt}\lambda_{qs} \tag{13.32}$$

The equation used for the rotor side is as shown below:

$$V_{qr} = R_r i_{qr} + \frac{d}{dt}\lambda_{qr} - \omega_r \lambda_{dr} \tag{13.33}$$

The equation for gap flux linkage is as shown below:

$$\lambda_{qr} = L_m \left(i_{qr} + i_{qs}\right) \tag{13.34}$$

R_s represents the stator winding resistance, R_r is the motor winding resistance, L_m represents the magnetizing inductance, L_{ls} represents the stator leakage inductance, L_{ir} is the rotor leakage inductance, ω_r is the electrical rotor angular speed, i_d is the current, V_q and V_d represent the voltages, and λ_d and λ_q show the fluxes of the d-q model [17, 39].

Equation (13.37) can be substituted in Eq. (13.38) to generate the output power and torque of the turbine (T_t), as shown below:

$$P_w = \frac{1}{2}\rho A C_p(\lambda, \beta)\left(\frac{R\omega_{opt}}{\lambda_{opt}}\right)^3 \tag{13.35}$$

$$T_t = \frac{1}{2}\rho A C_p(\lambda, \beta)\left(\frac{R}{\lambda_{opt}}\right)^3 \omega_{opt} \tag{13.36}$$

The following equation is used to calculate the power coefficient C_p, which is a nonlinear function:

$$C_p(\lambda, \beta) = c_1\left(c_2\frac{1}{\lambda} - c_3\beta - c_4\right)e^{-c_5\frac{1}{\lambda_i}} + c_6\lambda \tag{13.37}$$

with

$$\frac{1}{\lambda_{\mathrm{i}}} = \frac{1}{\lambda + 0.08\beta} - \frac{0.035}{\beta^3 + 1} \tag{13.38}$$

The constants C1–C6 are outlined in the subsequent sections.

Therefore, the energy conversion circuit diagram that was proposed to be implemented through Simulink played an important role in the generation of energy that could be collected through the wind.

Conversion of Wind Energy

The conversion of wind energy is then proposed to implement through nonlinear procedures and characteristics that are integrated to facilitate the conversion from wind energy to electric energy [17]. The components of the control systems are the fuzzifications that determine the net energy output after the conversion process through the use of a fuzzy inference system (FIS) [8, 28, 32]. Subsequently, the FLC is used in calculating the wind turbine and the DFIG subsystems, which is helpful in solving several issues during the transformation process, as shown by the following equation:

$$\begin{cases} e_{\Omega_{\mathrm{g}}(n)} = \Omega_{\mathrm{g}}^{*}(n) - n_2(n-1) \\ \Delta e_{\Omega_{\mathrm{g}}(n)} = \Omega_{\mathrm{g}}^{*}(n) - \Omega_{\mathrm{g}}(n-1) \end{cases} \tag{13.39}$$

Here, the fuzzy controller is used in the determination of the net realistic energy variants. The triangular, trapezoidal, and symmetrical components of the wind turbine generator were being proposed to be used to confirm the wind energy production. Variables such as the net current formation were used in calculating the FLC and the DFIG of the model using the equations below:

$$\begin{cases} e_{i_{\mathrm{dr}}(n)} = i_{\mathrm{dr}}^{*}(n) - i_{\mathrm{dr}}(n) \\ e_{i_{\mathrm{dr}}}(n) = i_{\mathrm{qr}}^{*}(n) - i_{\mathrm{qr}}(n) \end{cases} \tag{13.40}$$

The simplified version of the equation is in the form below:

$$\begin{cases} e_{i_{\mathrm{dr}}(n)} = i_{\mathrm{dr}}^{*}(n) - i_{\mathrm{dr}}(n) \\ e_{i_{\mathrm{dr}}}(n) = i_{\mathrm{qr}}^{*}(n) - i_{\mathrm{qr}}(n) \end{cases} \tag{13.41}$$

Here, the quantification process of the fuzzy controller has been achieved successfully using both the input and output variables of the energy systems. After quantification, the wind energy transformation rate was determined by the behavior of the wind speed rotor of the wind turbine [6, 14, 15]. The determination of the accuracy and complexity of the wind energy conversion is achieved through the computation of the changes that take place in the output power. Here, the amount of power output from a wind energy electric system (WEES) depends upon the peak power points; thus, the model on a sedan car is being utilized for extracting maximum power from the WEES. Since the main MPPT control method is being presented in this model, thus, the MPPT controllers are used for extracting maximum possible power in WEES [28, 54]. The power generation is interestingly related to the speed of wind, which is coming from the speed of car running. The analysis is, therefore, clarified as per the following figure, which depicts the relationship between wind speed (due to the motion of a car) miles per hour (MPH) and kWh power production. The results show that at the tip a mean 8 kWh at MPH for 10 MPH average where energy starts to produce 2 MPH wind speed just after the engine got started by battery. Since a standard car required to get fully energized 20 kWh, thus, if only the car run at 10 MPH for 2 hours, it will get fully charged to run for an average 200 miles; consequently, if the car run at 60 MPH, it will take only 20 minutes to get fully charged to run for the same millage (Fig. 13.7). Simply, the results confirm that generation of electrical energy in the wind turbine depends on the wind speed and that the required power output of the system is sufficient to run the car.

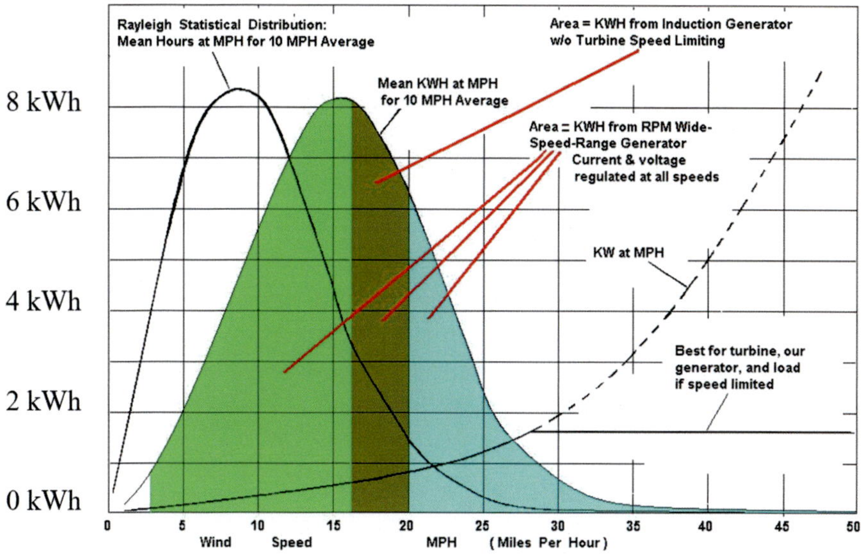

Fig. 13.7 Shows the transformation of wind power into the car in relation to the wind speed vs energy production with the mean hours (10 MPH average), kWh from induction generator, and kWh from RPM of the speed range generator current and voltage

Solar Energy Modeling

Then the dynamic photon proliferation is being calculated to help demonstrate the quantity of solar energy captured by the panels. Therefore, the calculation is performed using Eqs. (13.43) and (13.44). To confirm the net capture of the photon energy, then the real condition *j*(GD), which is the solar panel transportation vehicle, is calculated. The quantity is used to define the density of states (DOS) through the release of the magnitude $V(\omega)$ in the active solar panel cell [4, 11, 26]. As a result, the solar energy is captured by considering different dimensions of the solar panel transportation vehicle, such as 1D, 2D, and 3D [31], as shown in Table 13.1.

Here, the 1D, 2D, and 3D transportation PV panels require the sharp frequency cut-off at Ω_d, which is used to enhance the photon energy generation of the DOS, as shown in Fig. 13.8. Therefore, the two major variables that are dilogarithmic are $\mathrm{Li}_2(x)$ and $e_{rfc}(x)$. The eigenfunctions of Maxwell's rule and photon eigenfrequencies are used in determining the DOS values of the other panels, which are denoted as $\varrho_{PC}(\omega)$ [27, 38]. The DOS is also utilized in 1D solar cells, and it is given by the equation $\varrho_{PC}(\omega) \propto \frac{1}{\sqrt{\omega - \omega_e}}\Theta(\omega - \omega_e)$, where $\Theta(\omega - \omega_e)$ is known as the Heaviside step function. The frequency of the PBE in a DOS is represented by the quantity ω_e.

Here, the DOS is a critical factor to accurately calculate the qualitative state of the photon energy of the photon cell in the vehicle solar panel theoretically; thus, the DOS is being used classically next to an anisotropic PBE on the 1D, 2D, and 3D solar panel [11, 35]. Thus, the DOS of the photon energy is derived here as $\varrho_{PC}(\omega) \propto \frac{1}{\sqrt{\omega - \omega_e}}\Theta(\omega - \omega_e)$, which is further clarified in relation to the electromagnetic field factor [4, 7, 19, 38]. The photon energy is approximated as $\varrho_{PC}(\omega) \propto -[\ln|(\omega - \omega_0)/\omega_0| - 1]\Theta(\omega - \omega_e)$ in 1D and 2D solar panels, where ω_e is the center of the DOS distribution.

Simply, here, $J(\omega)$ is clarified as the DOS forces generated in the cell of the solar panel; thus, here, the net solar energy release magnitude $V(\omega)$ within the acting PV cell of the vehicle solar panel is calculated as [4, 11, 26]

$$J(\omega) = \varrho(\omega)|V(\omega)|^2. \tag{13.42}$$

Hereafter, this paper defines the proliferative photon dynamics $\langle a(t)\rangle = u(t, t_0)\langle a(t_0)\rangle$ and the PB frequency ω_c, where the function $u(t, t_0)$ describes the photon structure. $u(t, t_0)$ is derived by the dissipative integral-differential equation as in Eq. (13.18):

$$u(t, t_0) = \frac{1}{1 - \Sigma'(\omega_b)} e^{-i\omega(t - t_0)} + \int_{\omega_e}^{\infty} d\omega \frac{J(\omega)e^{-i\omega(t - t_0)}}{[\omega - \omega_c - \Delta(\omega)]^2 + \pi^2 J^2(\omega)}, \tag{13.43}$$

where $\Sigma'(\omega_b) = [\partial\Sigma(\omega)/\partial\omega]_{\omega = \omega_b}$ and $\Sigma(\omega)$ is the self-energy correction of the PB photon introduced in the reservoirs.

Table 13.1 Different photon structures of the density of states (DOS) in different dimensions of the solar panel. The photon dynamics are used in determining the quantity that shows the unit area (ω) as well as the self-energy induction in the reservoir between include C, η, and χ [4, 38]

Photon	Unit area $J(\omega)$ for different DOS*	Reservoir-induced self-energy correction $\Sigma(\omega)$*
1D	$\frac{C}{\pi}\frac{1}{\sqrt{\omega-\omega_e}}\Theta(\omega-\omega_e)$	$\varrho_{PC}(\omega)\propto\frac{1}{\sqrt{\omega-\omega_e}}\Theta(\omega-\omega_e),-\frac{C}{\sqrt{\omega_e-\omega}}$
2D	$\Omega_d-\eta\left[\ln\left\|\frac{\omega-\omega_0}{\omega_0}\right\|-1\right]\Theta(\omega-\omega_e)\Theta(\Omega_d-\omega)$	$\eta\left[\mathrm{Li}_2\left(\frac{\Omega_d-\omega_0}{\omega-\omega_0}\right)-\mathrm{Li}_2\left(\frac{\omega_0-\omega_e}{\omega_0-\omega}\right)-\varrho_{PC}(\omega)\propto-[\ln\|(\omega-\omega_0)/\omega_0\|-1]\Theta(\omega-\omega_e)\right]$
3D	$\chi\sqrt{\frac{\omega-\omega_e}{\Omega_C}}\exp\left(-\frac{\omega-\omega_e}{\Omega_C}\right)\Theta(\omega-\omega_e)$	$\chi\left[\pi\sqrt{\frac{\omega_e-\omega}{\Omega_C}}\exp\left(-\frac{\omega-\omega_e}{\Omega_C}\right)e_{rfc}\sqrt{\frac{\omega_e-\omega}{\Omega_C}}-\sqrt{\pi}\right]$

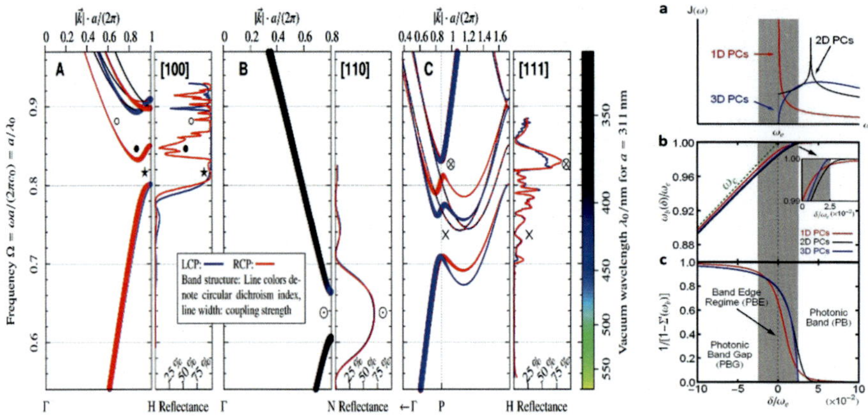

Fig. 13.8 (1) The energy conversion modes and the photonic band structure. (2) The diagram showing the relationship between the units and the frequency at several DOS such as 1D, 2D, and 3D on the solar panel in subpanel (2a). In subpanel (2b), there is a representation of the frequencies of the photovoltaic modes for functional tuning. In subpanel (2c), there are different levels of photonic modes for the release of energy, which is related to the transformation from the PBG to the photonic band (PB) area for creating net solar energy [7, 37]

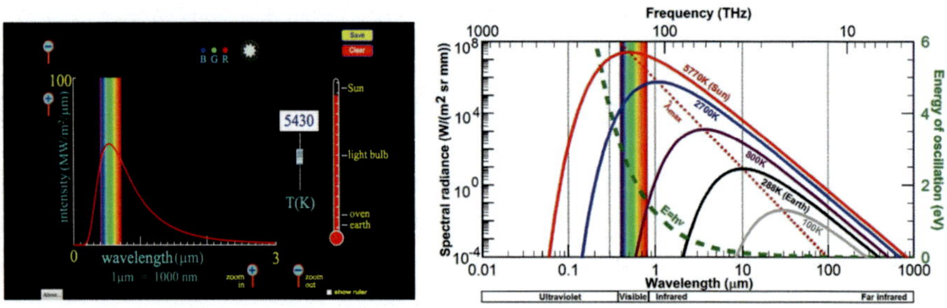

Fig. 13.9 An illustration of blackbody solar irradiation in different temperature at 5770 K. The power is 6.31 × 107 (W/m^2), peak E is 1.410 (eV), peak λ is 0.88 (μm), and peak μ is 2.81 × 107 (W/m^2·eV)

$$\Sigma(\omega) = \int_{\omega_e}^{\infty} d\omega' \frac{J(\omega')}{\omega - \omega'}. \tag{13.44}$$

The solar energy frequency mode in the PBG ($0 < \omega_b < \omega_e$) is being represented by the frequency ω_b in Eq. (13.47). This energy can be calculated under the pole condition $\omega_b - \omega_c - \Delta(\omega_b) = 0$, where the principal-value integral is given by $\lesssim\Delta(\omega) = \mathcal{P}\left[\int d\omega' \frac{J(\omega')}{\omega - \omega'}\right]$.

Hence, Fig. 13.9a shows the cooling photonic dynamics of the proliferation energy $|u(t, t_0)|$, and the energy is determined in 3D, 2D, and 1D photon cells with different measures of the detuning parameter δ [8, 17, 20, 21]. Thus, the proliferation magnitude is then integrated into the PB to the PBG area and shows a plot of the rates of solar energy generation dynamics $\kappa(t)$. From the results, it is shown that dynamic photons are generated rapidly once ω_c has crossed PB from the PBG area. Since the $u(t, t_0)$ range is $1 \geq |u(t, t_0)| \geq 0$, the crossover area is defined to satisfy $0.9 \gtrsim |u(t \rightarrow \infty, t_0)| \geq 0$. This represents $-0.025\omega_e \lesssim \delta \lesssim 0.025\omega_e$, with solar energy generation rate $\kappa(t)$ within PBG ($\delta < -0.025\omega_e$) and next to PBE ($-0.025\omega_e \lesssim \delta \lesssim 0.025\omega_e$).

More specifically, the PB was first being considered as the solar energy determination n_0, i. e. , $\rho(t_0) = |n_0\rangle\langle n_0|$, derived theoretically by real-time quantum feedback control [36, 38], solving Eq. (13.47), in consideration of the state of energy state photon generation at time t:

$$\rho(t) = \sum_{n=0}^{\infty} \mathcal{P}_n^{(n_0)}(t)|n_0\rangle\langle n_0| \tag{13.45}$$

$$\mathcal{P}_n^{(n_0)}(t) = \frac{[v(t,t)]^n}{[1 + v(t,t)]^{n+1}}[1 - \Omega(t)]^{n_0} \times \sum_{k=0}^{\min\{n_0, n\}} \binom{n_0}{k}\binom{n}{k}\left[\frac{1}{v(t,t)}\frac{\Omega(t)}{1 - \Omega(t)}\right]^k, \tag{13.46}$$

where $\Omega(t) = \frac{|u(t,t_0)|^2}{1+v(t,t)}$. The findings show that energy state solar photons are introduced into dynamic states $\mathcal{P}_n^{(n_0)}(t)$ of $|n_0\rangle$ and graphs of the proliferation of photon dissipation $\mathcal{P}_n^{(n_0)}(t)$ in the primary state $|n_0 = 5\rangle$ as well as in the limit of steady states $\mathcal{P}_n^{(n_0)}(t \to \infty)$. Therefore, the proliferation of the produced energy state photons will ultimately reach a non-equilibrium state of energy that supplies enough energy to run the transportation vehicle.

Solar Energy Conversion

Since the magnitude of solar irradiance is usually considered as the solar energy transformed into electrical energy for the transportation vehicle, thus, the static light quanta are being counted in relation to the range of ν_r to $\nu_r + d\nu_r$ [14, 15, 38]. The optimal solar irradiation is being attained here at 1.4 eV, and thus the energy value of 27.77 mW/m^2 eV has been configured mathematically, and thus, a solar panel installed in a transportation vehicle shall produce 277,700 kW annually or 760.08 kW daily on an average of 5 hours of solar irradiance during peak levels [19, 30]. The amount of energy that is simply produced by a solar panel of approximately 1 m^2 on a peak level day is equivalent to 23 gallons of gas that can be used to run any standard transportation vehicle for up to 900 km. Considering the waste factor based on physical principles, the energy is lost during the conversion of solar energy to direct current (DC) power and finally to alternating current (AC), which is approximately 0.8 [8, 11, 20, 21]. Since solar panels have an efficiency of up to 80%, the transformation of energy by a solar panel may be up to 125% higher, and thus, the net energy will remain the same, which is equivalent to (277,700 × 1.25 × 0.8) = 277,700 kW annually or 760 kW daily (Fig. 13.9).

Electrical Subsystem

Once both wind and solar energy conversion modeling have been completed, then these two renewable energy systems have been integrated using MATLAB 9.0 in order to form hybrid energy. The utilization of this integrated hybrid renewable energy has been configured via the electrical subsystem. Hence, the functional mechanism of the electrical subsystem is important in transforming the electric energy produced into DC before it is finally converted into AC. The functional mechanism is quantified by the d-q synchronous voltage equation below:

$$V_q = -R_s i_q - L_q \frac{di_q}{dt} - \omega L_d i_d + \omega \lambda_m \tag{13.47}$$

The electrical energy produced can therefore be simplified as an equation that demonstrates the dynamic modeling of active energy, as shown below:

$$\begin{bmatrix} V_{qs} \\ V_{ds} \\ V_{qr} \\ V_{dr} \end{bmatrix} = \begin{bmatrix} R_s + pL_s & 0 & pL_m & 0 \\ 0 & R_s + pL_s & 0 & pL_m \\ pL_m & -\omega_r L_m & R_r + pL_r & -\omega_r L_r \\ \omega_r L_m & pL_m & \omega_r L_r & R_r + pL_r \end{bmatrix} \begin{bmatrix} i_{qs} \\ i_{ds} \\ i_{qr} \\ i_{dr} \end{bmatrix} \tag{13.48}$$

Here, electricity energy generation is précised by further analysis of stator control and expressed as:

$$V_{qs} = R_s i_{qs} + \frac{d}{dt} \lambda_{qs} \tag{13.49}$$

which can be further simplified as:

$$V_{qr} = R_r i_{qr} + \frac{d}{dt} \lambda_{qr} - \omega_r \lambda_{dr} \tag{13.50}$$

In case an air gap flux leakage occurs in generation, the equations are further corrected to be:

$$\lambda_{qr} = L_m \left(i_{qr} + i_{qs} \right) \tag{13.51}$$

where R_s, R_r, L_m, L_{ls}, L_{lr}, ω_r, i_d, i_q, V_d, V_q, λ_d, and λ_q represent the resistance of the stator, current resistance, current inductance, stator leakage inductance, current leakage inductance, electrical conductance, current, voltage, and fluxes of the d-q model, respectively [39]. Then, the net hybrid electricity energy (T_t) produced is calculated as:

$$P_w = \frac{1}{2}\rho A C_p(\lambda,\beta)\left(\frac{R\omega_{opt}}{\lambda_{opt}}\right)^3 \tag{13.52}$$

$$T_t = \frac{1}{2}\rho A C_p(\lambda,\beta)\left(\frac{R}{\lambda_{opt}}\right)^3 \omega_{opt} \tag{13.53}$$

The equation below is used in determining the power coefficient (Cp) as a nuclear function:

$$C_p(\lambda,\beta) = c_1\left(c_2\frac{1}{\lambda} - c_3\beta - c_4\right)e^{-c_5\frac{1}{\lambda_i}} + c_6\lambda \tag{13.54}$$

where:

$$\frac{1}{\lambda_i} = \frac{1}{\lambda + 0.08\beta} - \frac{0.035}{\beta^3 + 1} \tag{13.55}$$

The net hybrid solar and wind energy generation module produced was then determined using an energy conversion circuit and a converter available in MATLAB 9.0 used in powering the transportation vehicle (Fig. 13.10).

The formation of wind energy by a running car at 10 kph can produce 8 kWh, for an average wind speed of 2 kph. As a standard car requires 20 kWh to get fully energized, the results indicate that it is possible to fully charge the car and run it for 200 km if the car runs at 10 kph for 2 hours during charging. Consequently, if the car runs at 60 kph, it will take 20 min to get fully charged and can run 900 km in only 5 hours at speed of 60 kph. Subsequently, the amount of energy is produced by the acting solar panel of a car of approximately 1 m^2 on a peak level is 760 kW daily on an average of 5 hours, which is equivalent to 23 gallons of gas that can be used to run any standard sedan car for up to 900 km. Thus, the hybrid energy of wind and solar power can run 1800 km in 5 hours running the vehicle at a speed of 60 kph.

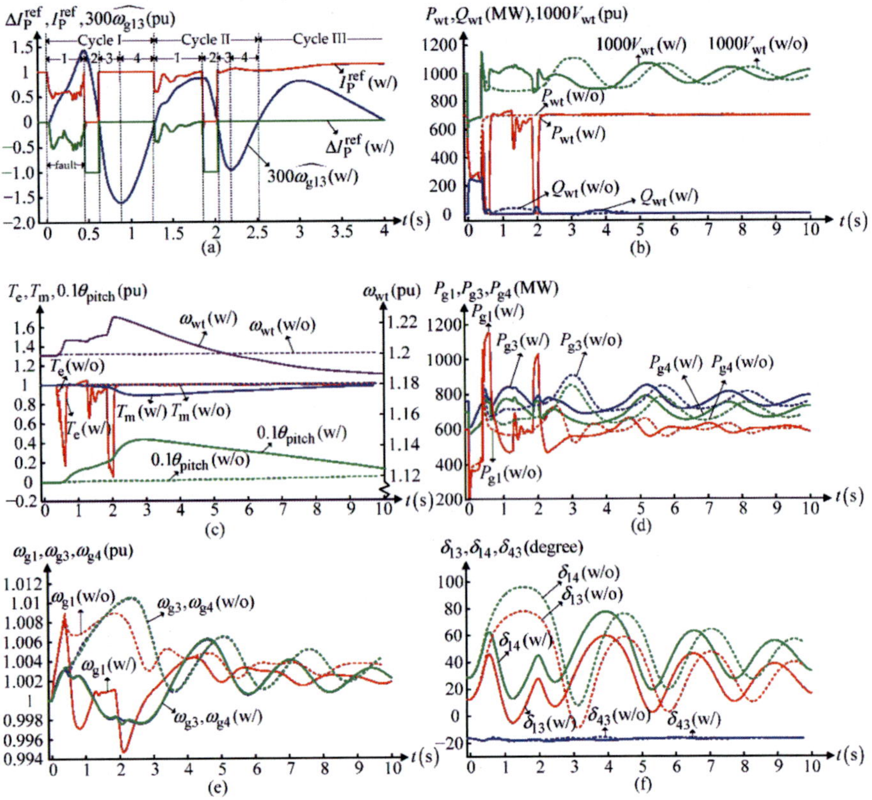

Fig. 13.10 An illustration shows of how the hybrid current output (MW) can be formed from wind turbine and solar panel in order to transform it into the electrical subsystem to power the vehicle

Battery Modeling

Eventually, a battery modeling is being calculated to store the energy to characterize the starting power to ignite the vehicle and the backup power source while the vehicle is not on the windy or non-solar hours by incorporating Peukert's law of battery charge (Fig. 13.11), which is:

$$t_{\text{discharge}} = H\left(\frac{C}{IH}\right)^{k}, \tag{13.56}$$

where t represents the battery charge time, C represents the battery capacity, I represents the current flow, H represents the rated discharge time, and k is Peukert's coefficient, which is calculated as follows:

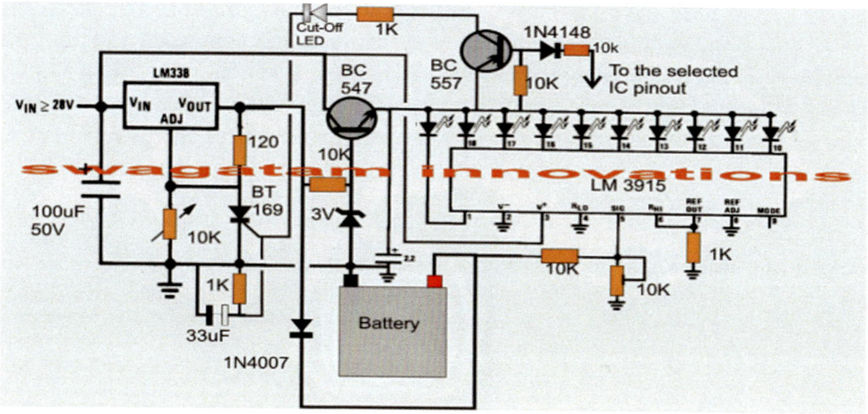

Fig. 13.11 Block diagram of the battery model to control its voltage source to start the engine and run the car when it is not in motion

$$k = \frac{\log T_2 - \log T_1}{\log I_1 - \log I_2}, \tag{13.57}$$

where I_1 and I_2 represent the variance of the charge current flow rates and T_1 and T_2 represent the corresponding time for completely charging the battery, which is defined as:

$$t_{\text{charging}} = \frac{\text{Ampere hour of battery}}{\text{Charging current}}$$

Conclusions

The conventional development of the transportation sector all over the world has increased and, thus, the consumption of fossils does, resulting in deadly depletion of global energy and an environmental crisis. To meet the total energy needs in the global transportation sector, which is environmentally friendly, thus, in this work, a sustainable transportation technology considering the utilization of hybrid renewable energy of wind and solar power has been conducted. The mathematical result of this hybrid energy formation from both wind and solar power confirms that it can supply the energy to 1800 km in just 5 hours functions of wind turbine and solar panel. The deliberation of wind energy by a running car at 60 kph produces enough energy to run 900 km and the solar energy is produced by the acting PV panel of 1 m^2 function on the car only 5 hours in daylight can produce enough energy to run the car 900 km. Subsequently, the implementation of these two energies through the electrical subsystem to form hybrid energy in just 5 hours which can power the car to run 1800 km. Besides, the battery modeling system is being conducted to ignite the engine and supply the energy when the car is not in a running condition. Consequently, the findings of this research suggest that the large-scale adoption of wind and solar energy to run any vehicles with calculative measure of energy requirement for each vehicle can be an innovative solution to meet the energy demand in the transportation sector globally.

Acknowledgments The author, Md. Faruque Hossain, declares that any findings, predictions, and conclusions described in this article are solely performed by the author, and it is confirmed that there is no conflict of interest for publishing this research paper in a suitable journal and/or publisher.

References

1. Abulizi, M., Peng, L., Francois, B., & Li, Y. (2014). Performance analysis of a controller for doubly-fed induction generators based wind turbines against parameter variations. *International Review of Electrical Engineering, 9*, 264–269. https://doi.org/10.15866/iree.v9i2.1797
2. Bahri, N., & Ouled Amor, W. (2019). Intelligent power supply management of an autonomous hybrid energy generator. *International Journal of Sustainable Engineering, 12*, 312–332. https://doi.org/10.1080/19397038.2019.1581852
3. Bakhsh, F. I., & Khatod, D. K. (2014). A novel method for grid integration of synchronous generator based wind energy generation system. In *2014 IEEE international conference on power electronics, drives and energy systems (PEDES)*. IEEE, Mumbai, India, pp. 1–6.
4. Bento, F., & Cardoso, A. J. M. (2018). A comprehensive survey on fault diagnosis and fault tolerance of DC-DC converters. *Chinese Journal of Electrical Engineering, 4*, 1–12. https://doi.org/10.23919/CJEE.2018.8471284
5. Boumassata, A., Kerdoun, D., & Madaci, M. (2015). Grid power control based on a wind energy conversion system and a flywheel energy storage system. In *IEEE EUROCON 2015 – international conference on computer as a tool (EUROCON)*. IEEE, Salamanca, Spain, pp. 1–6.
6. Darvish Falehi, A. (2018). Augment dynamic and transient capability of DFIG using optimal design of NIOPID based DPC strategy. *Environmental Progress & Sustainable Energy, 37*, 1491–1502. https://doi.org/10.1002/ep.12811
7. Elmansouri, A., El-mhamdi, J. E., & Boualouch, A. (2016). Wind energy conversion system using DFIG controlled by back-stepping and RST controller. In *2016 international conference on electrical and information technologies (ICEIT)*. IEEE, Tangiers, Morocco, pp. 312–318.
8. Elyaalaoui, K., Ouassaid, M., & Cherkaoui, M. (2016). Supervision system of a wind farm based on squirrel cage asynchronous generator. In *2016 international renewable and sustainable energy conference (IRSEC)*. IEEE, Marrakech, Morocco, pp. 403–408.
9. Gaillard, A., Poure, P., & Saadate, S. (2009). Reactive power compensation and active filtering capability of WECS with DFIG without any over-rating. *Wind Energy, 13*, 603–614. https://doi.org/10.1002/we.381
10. Ghedamsi, K., & Aouzellag, D. (2010). Improvement of the performances for wind energy conversions systems. *International Journal of Electrical Power & Energy Systems, 32*, 936–945. https://doi.org/10.1016/j.ijepes.2010.02.012
11. Hamoud, F., Doumbia, M. L., & Cheriti, A. (2014). Hybrid PI-sliding mode control of a voltage source converter based STATCOM. In *2014 16th international power electronics and motion control conference and exposition*. IEEE, Antalya, Turkey, pp. 661–666.

12. Heydari, M., & Smedley, K. (2015). Comparison of maximum power point tracking methods for medium to high power wind energy systems. In *2015 20th conference on electrical power distribution networks conference (EPDC)*. IEEE, Zahedan, Iran, pp. 184–189.
13. Hossain, M. F. (2017). Design and construction of ultra-relativistic collision PV panel and its application into building sector to mitigate total energy demand. *Journal of Building Engineering, 9*, 147–154. https://doi.org/10.1016/j.jobe.2016.12.005
14. Hossain, M. F. (2019). Flying transportation technology. In M. F. Hossain (Ed.), *Sustainable design and build: Building, energy, roads, bridges, water and sewer systems* (pp. 282–300). Butterworth-Heinemann.
15. Hossain, M. F. (2019). Sustainable technology for energy and environmental benign building design. *Journal of Building Engineering, 22*, 130–139. https://doi.org/10.1016/j.jobe.2018.12.001
16. Hossain, M. F., & Fara, N. (2016). Integration of wind into running vehicles to meet its total energy demand. *Energy, Ecology and Environment, 2*, 35–48. https://doi.org/10.1007/s40974-016-0048-1
17. Junyent-Ferré, A., & Gomis-Bellmunt, O. (2011). Wind turbine generation systems modeling for integration in power systems. In A. Zobaa & R. Bansal (Eds.), *Handbook of renewable energy technology* (pp. 53–68). World Scientific.
18. Junyent-Ferré, A., Gomis-Bellmunt, O., Sumper, A., Sala, M., & Mata, M. (2010). Modeling and control of the doubly fed induction generator wind turbine. *Simulation Modelling Practice and Theory, 18*, 1365–1381. https://doi.org/10.1016/j.simpat.2010.05.018
19. Karthikeyan, R., & Parvathy, A. K. (2015). Peak load reduction in micro smart grid using non-intrusive load monitoring and hierarchical load scheduling. In *2015 international conference on smart sensors and systems (IC-SSS)*. IEEE, Bangalore, India, pp. 1–6.
20. Kerrouche, K., Mezouar, A., & Belgacem, K. (2013). Decoupled control of doubly fed induction generator by vector control for wind energy conversion system. *Energy Procedia, 42*, 239–248. https://doi.org/10.1016/j.egypro.2013.11.024
21. Kerrouche, K., Mezouar, A., & Boumedien, L. (2013). A simple and efficient maximized power control of DFIG variable speed wind turbine. In *3rd international conference on systems and control* (pp. 894–899). IEEE.
22. Lap-Arparat, P., & Leephakpreeda, T. (2019). Real-time maximized power generation of vertical axis wind turbines based on characteristic curves of power coefficients via fuzzy pulse width modulation load regulation. *Energy, 182*, 975–987. https://doi.org/10.1016/j.energy.2019.06.098
23. Ligang, H., Xiangdong, W., & Kang, Y. (2015). Optimal speed tracking for double fed wind generator via switching control. In *The 27th Chinese control and decision conference (2015 CCDC)*. IEEE, Qingdao, China, pp. 2638–2643.
24. Ling, Y. (2018). A fault ride through scheme for doubly fed induction generator wind turbine. *Australian Journal of Electrical and Electronics Engineering, 15*, 71–79. https://doi.org/10.1080/1448837X.2018.1525172
25. Loucif, M., & Boumédiène, A. (2015). Modeling and direct power control for a DFIG under wind speed variation. In *2015 3rd international conference on control, Engineering & Information Technology (CEIT)*. IEEE, Tlemcen, Algeria, pp. 1–6.
26. Mahato, S. N., Singh, S. P., & Sharma, M. P. (2013). Dynamic behavior of a single-phase self-excited induction generator using a three-phase machine feeding single-phase dynamic load. *International Journal of Electrical Power & Energy Systems, 47*, 1–12. https://doi.org/10.1016/j.ijepes.2012.10.067
27. Mani, P., Lee, J. H., Kang, K. W., & Joo, Y. H. (2020). Digital controller design via LMIs for direct-driven surface mounted PMSG-based wind energy conversion system. *IEEE Transactions on Cybernetics, 50*, 3056–3067. https://doi.org/10.1109/TCYB.2019.2923775
28. Mohd Zin, A. A. B., Pesaran, H. A. M., Khairuddin, A. B., Jahanshaloo, L., & Shariati, O. (2013). An overview on doubly fed induction generators′ controls and contributions to

wind based electricity generation. *Renewable and Sustainable Energy Reviews, 27*, 692–708. https://doi.org/10.1016/j.rser.2013.07.010
29. Mwaniki, J., Lin, H., & Dai, Z. (2017). A condensed introduction to the doubly fed induction generator wind energy conversion systems. *Journal of Engineering, 2017*, 1–18. https://doi.org/10.1155/2017/2918281
30. Ouassaid, M., Elyaalaoui, K., & Cherkaoui, M. (2015). Reactive power capability of squirrel cage asynchronous generator connected to the grid. In *2015 3rd international renewable and sustainable energy conference (IRSEC)*. IEEE, Marrakech, Morocco, pp. 1–4.
31. Pugh, T. A. M., MacKenzie, A. R., Whyatt, J. D., & Hewitt, C. N. (2012). Effectiveness of green infrastructure for improvement of air quality in urban street canyons. *Environmental Science & Technology, 46*, 7692–7699. https://doi.org/10.1021/es300826w
32. Rad, M. A. V., Ghasempour, R., Rahdan, P., Mousavi, S., & Arastounia, M. (2020). Techno-economic analysis of a hybrid power system based on the cost-effective hydrogen production method for rural electrification, a case study in Iran. *Energy, 190*, 116421. https://doi.org/10.1016/j.energy.2019.116421
33. Rani, M. D., & Kumar, M. S. (2017). Development of doubly fed induction generator equivalent circuit and stability analysis applicable for wind energy conversion system. In *2017 international conference on recent advances in electronics and communication technology (ICRAECT)*. IEEE, Bangalore, India, pp. 55–60.
34. Ribeiro, E., Monteiro, A., Cardoso, A.J.M., Boccaletti, C., 2014. Fault tolerant small wind power system for telecommunications with maximum power extraction, in: 2014 IEEE 36th International Telecommunications Energy Conference (INTELEC). IEEE, Vancouver, BC, Canada, pp. 1–6.
35. Saeed, M. S. R., Mohamed, E. E. M., & Sayed, M. A. (2016). Design and analysis of dual rotor multi-tooth flux switching machine for wind power generation. In *2016 eighteenth international middle east power systems conference (MEPCON)*. IEEE, Cairo, Egypt, pp. 499–505.
36. Touaiti, B., Azza, H. B., & Jemli, M. (2016). A MRAS observer for sensorless control of wind-driven doubly fed induction generator in remote areas. In *2016 17th international conference on sciences and techniques of automatic control and computer engineering (STA)*. IEEE, Sousse, Tunisia, pp. 526–531.
37. Touaiti, B., Azza, H. B., & Jemli, M. (2019). Control scheme for stand-alone DFIG feeding an isolated load. In *2019 10th international renewable energy congress (IREC)*. IEEE, Sousse, Tunisia, pp. 1–6.
38. Upadhyay, V.C., Sandhu, K.S., 2018. Reactive power management of wind farm using STATCOM. In *2018 international conference on emerging trends and innovations in engineering and technological research (ICETIETR)*. IEEE, Ernakulam, India, pp. 1–5.
39. Venkatesan, C., Sundararaman, K., & Gopalakrishnan, M. (2017). Grid integration of PMSG based wind energy conversion system using variable frequency transformer. In *2017 international conference on intelligent computing, instrumentation and control technologies (ICICICT)*. IEEE, Kannur, India, pp. 1451–1456.
40. Zohoori, A., Vahedi, A., Noroozi, M. A., & Meo, S. (2017). A new outer-rotor flux switching permanent magnet generator for wind farm applications. *Wind Energy, 20*, 3–17. https://doi.org/10.1002/we.1986

Chapter 14
Theory of Mind Machine

Introduction

Since the imagination power has the extreme influence on mind energy, thus, in this research, a novel mathematical transformation modeling is being conducted primarily on clarifying the imagination power and mind energy and its transformation mechanisms into the energy state travel machine to travel anywhere based on imagination. Simply, in this work, a bridge of the imagination and mind has been computed in order to clarify how the imagination can be decoded into activated mind energy and then transformed into energy state travel machine to travel anywhere. Naturally, the energy momentum into the imagination power is being modeled to release its activated energy considering its nano-point dynamics of QED to form activated mind energy [18, 29]. Subsequently, this activated mind energy is also being modeled to convert it into energy state travel machine by implementing its quantum Higgs boson ($H \rightarrow \gamma\gamma^{-}$) electromagnetic fields to travel anywhere based on imagination [10, 12]. This energy conversion mechanism from imagination to mind and then into the energy state travel machine shall indeed be a noble scientific discovery ever to confirm an ultimate transportation system for mankind.

Methods and Simulation

Activation of Imagination Power ($\mathbf{I}_p$)

For the activation of imagination power considering the brain's sphericity, a mathematical modeling is being performed considering the quantum electrodynamics (QED) of the human brain and expressed by the following law of cosines:

M. F. Hossain, *Global Sustainability*, https://doi.org/10.1007/978-3-031-34575-3_14

$$\rho\cos(c) = \rho\cos(a)\cos(b) + \rho\sin(a)\sin(b)\cos(C) \tag{14.1}$$

Here a, b, and c represent the arc lengths in radians corresponding to the sides of a spherical horizon of the mind, and C represents the angle of the vertex corresponding to the brain of *arc* with its length c [8, 23, 25]. Subsequently, with the implementation of its zenith angle Θ, the following terms are being written as the laws of cosines: $C = h = r$; $c = \Theta$; $a = \frac{1}{2}\pi - \phi$; $b = \frac{1}{2}\pi - \delta$. Thus, the equation is being expressed as:

$$\rho\cos(\Theta) = \sin(\phi)\sin(\delta) + \cos(\phi)\cos(\delta)\cos(r) \tag{14.2}$$

To confirm this calculation, a further clarification has been conducted considering a general derivative as below:

$$\begin{aligned}\rho\cos(\theta) = {} & \rho\sin(\phi)\sin(\delta)\cos(\beta) + \sin(\delta)\cos(\phi)\sin(\beta)\cos(\gamma) \\ & + \cos(\phi)\cos(\delta)\cos(\beta)\cos(h) - \cos(\delta)\sin(\phi)\sin(\beta)\cos(\gamma)\cos(h) \\ & - \cos(\delta)\sin(\beta)\sin(\gamma)\sin(h)\end{aligned} \tag{14.3}$$

Here, β represents the angle corresponding to the horizon of the imagination and γ corresponding to the azimuth angle of the brain, and thus, the sphere of the imagination parameter, which originated from the brain dynamics, is denoted here as R_E, and its mean distance is being denoted here as R_0 considering the average distance corresponding to the mind dynamical unit (*m.u.*), and the energy constant is denoted here as S_0 [7, 13, 19]. Thus, the imagination power flux onto the imagination of the sphere is calculated as:

$$Q = \begin{cases} S_o \dfrac{R_0^2}{R_E^2}\cos(\theta) & \cos(\theta) > 0 \\ 0 & \cos(\theta) \le 0 \end{cases} \tag{14.4}$$

Here, the mean Q over a time is the mean value of Q in relation to its rotation within the time of $h = \pi$ to $h = -\pi$, and thus, the calculation can be expressed as:

$$Q^{-\text{time}} = -\frac{1}{2\pi}\int_{\pi}^{-\pi} Q dh \tag{14.5}$$

Since h_0 is the time related mind sphere once Q is the positive, thus, it will be feasible during the imagination power presence in the brain when $\Theta = 1/2\,\pi$, or for h_0 considering following equation by solving as:

$$\rho\sin(\phi)\sin(\delta) + \cos(\phi)\cos(\delta)\cos(h_{0)} = 0 \tag{14.6}$$

or

$$\rho\cos(h_0) = -\tan(\phi)\tan(\delta) \tag{14.7}$$

Since tan(φ)tan(δ) > 1, thus, the imagination power will confirm its existence at $h = \pi$, so $h_0 = \pi$ in a $Q^{-\text{time}} = 0$; $\frac{R_0^2}{R_E^2}$ would remain constant into the imagination power by presenting an imagination sphere, which can be expressed by the following integral:

$$\begin{aligned}\int_{\pi}^{-\pi} Q\mathrm{d}h &= \int_{h_0}^{-h_0} Q\mathrm{d}h = S_0\frac{R_0^2}{R_E^2}\int_{h_0}^{-h_0}\cos(\theta)\mathrm{d}h \\ &= S_0\frac{R_0^2}{R_E^2}\left[h\sin(\phi)\sin(\delta) + \cos(\phi)\cos(\delta)\sin(h)\right]_{h=h_0}^{h=-h_0} \\ &= -2S_0\frac{R_0^2}{R_E^2}\left[h_0\sin(\phi)\sin(\delta) + \cos(\phi)\cos(\delta)\sin(h_0)\right]\end{aligned} \tag{14.8}$$

Therefore:

$$Q^{-\text{time}} = \frac{S_0}{\pi}\frac{R_0^2}{R_E^2}\left[h_0\sin(\phi)\sin(\delta) + \cos(\phi)\cos(\delta)\sin(h_0)\right] \tag{14.9}$$

Since θ is representing here as mainstream imagination that describes the orbit of the mind, thus, $\theta = 0$ corresponding to its vernal equinox of δ at its spherical position that can be written as:

$$\delta = \varepsilon\sin(\theta) \tag{14.10}$$

Here, ε represents the mainstream longitude of sphere ϖ that is related to the vernal equinox, and thus, the equation can be written as:

$$R_E = \frac{R_0}{1 + e\cos(\theta - \omega)} \tag{14.11}$$

or

$$\frac{R_0}{R_E} = 1 + e\cos(\theta - \omega) \tag{14.12}$$

Given the above clarification of ϖ, ε, s_0, and e from mind dynamical computation, mind energy L_e can be computed for any latitude φ and θ of the imaginative sphere (Fig. 14.1). Here, $\theta = 0°$ is representing the vernal equinox, and $\theta = \infty$ is

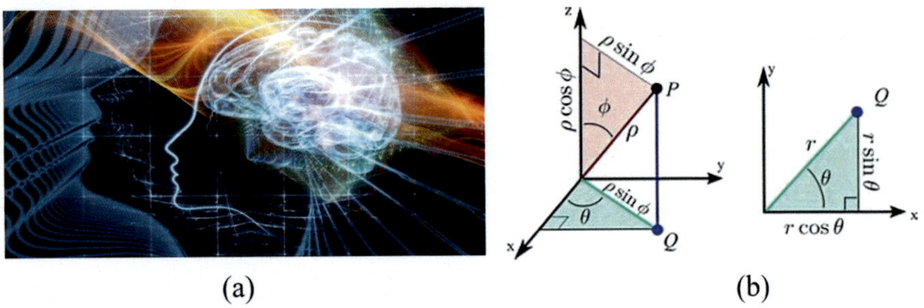

Fig. 14.1 The distribution of imagination power: (**a**) schematic diagram of the imagination sphericity of the mind and (**b**) the imagination power with respect to the law of cosines into the brain

representing the sphere of the imagination, and thus, the equation can be simplified for imagination power activation in the brain and can be expressed as:

$$M_e = S_0\left(1 + \cos\left(2\pi\frac{n}{x}\right)\right) \tag{14.13}$$

Here, S_0 is an energy constant, n is the unit of imagination, x is the imagination, and π is the sphere of the imagination that optimized the generation of imagination power (I_p) in the brain in order to pave the formation of the mind energy (M_e).

Computation of Mind Energy ($\mathbf{M}_e$)

Since the mind energy is the hidden particle, thus, mind energy (M_e) is being clarified considering its quantum electrodynamics (QED) by analyzing its minimal way of new U(1) gauge field considering its vectorial (V) dynamics of energy of its tangential (T) field at Z dimension of the brain [1, 8, 9]. Simply, here, the new U (1) gauge symmetry and its kinetic energy proliferation corresponding to the mind energy field, the *standard model hypercharge field*, are being computed to confirm the determination of striking electron flow from the mind energy (Fig. 14.2). Here, the striking electron of the mind energy (M_e) will be scattered at an angle θ with speed of c, which is being expressed as the real fermion of the energy momentum of vectors of constituent particles of the body:

$$\overrightarrow{De} = \left|\frac{Edp}{c}, \overrightarrow{Pd}\right| + \overrightarrow{Xef}\left[\frac{Eef}{c}, \overrightarrow{Pef}\right] \tag{14.14}$$

Since mind energy momentum is conserved and, thus, $\Delta P = 0$, therefore, the sum of the initial momentum must be equal to the sum of the final momentum [22, 38, 39]. Thus, it can be rewritten as the sum of the vectors as:

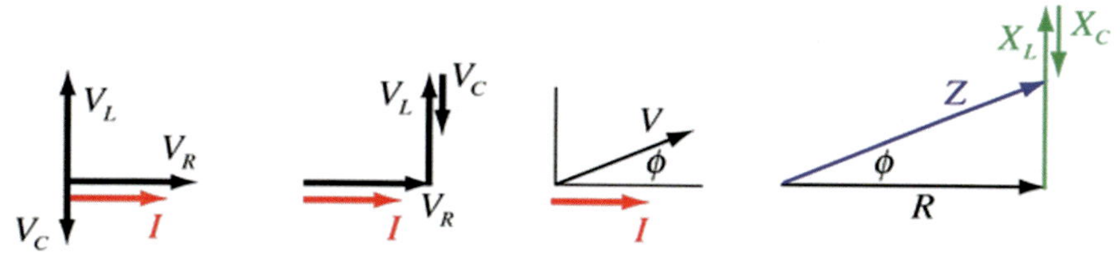

$$V = \sqrt{V_R^2 + (V_L - V_C)^2} \qquad Z = \sqrt{R^2 + (X_L - X_C)^2}$$

$$\phi = \tan^{-1}\frac{V_L - V_C}{V_R} \qquad \phi = \tan^{-1}\frac{X_L - X_C}{R}$$

Fig. 14.2 The conceptual model of mind energy U(1) gauge symmetry corresponding to its vectorial dynamics of tangential striking electron in order to pave it to transform into an energy state travel machine considering the kinetic energy field parameter of the body

$$\overrightarrow{De} + \overrightarrow{Xei} = \overrightarrow{Xp} + \overrightarrow{Xef} \tag{14.15}$$

Since the isolated four momentums of the released electron are being calculated, the equation can be rewritten as:

$$\left(\overrightarrow{\overrightarrow{De}} + \overrightarrow{Xei} - \overrightarrow{Xp}\right)^2 = \overrightarrow{Xef}^2 \tag{14.16}$$

Multiplying the mind energy momentum of the vectors together with the cross term, here it can be rewritten as:

$$2\,\overrightarrow{Xd}\,.\overrightarrow{Xei} = 2 * \left[\frac{Ed}{c}, \overrightarrow{Pd}\right].\left[m_e c, \vec{0}\right] = 2Edm_e \tag{14.17}$$

Subsequently, multiplying together the momentums of the initial stationary electron considering the generation of mind energy, the equation can be rewritten as the following:

$$m_{\mathrm{d}}^{2}c^{2} - m_{\mathrm{e}}^{2}c^{2} + - 2m_{\mathrm{e}}E_{\mathrm{p}} - 2\left(\frac{E_{\mathrm{d}}E_{\mathrm{p}}}{c^{2}} - \frac{P_{\mathrm{d}}E_{\mathrm{p}}\cos(\theta)}{c}\right) + 2E_{\mathrm{d}}m_{\mathrm{e}} = m_{\mathrm{e}}^{2}c^{2} \tag{14.18}$$

which can be simplified as:

$$m_{\mathrm{d}}^{2}c^{2} - 2\left(m_{\mathrm{e}}E_{\mathrm{p}} + \left(\frac{E_{\mathrm{d}}E_{\mathrm{p}}}{c^{2}} - \frac{P_{\mathrm{d}}E_{\mathrm{p}}}{c}\cos(\theta)\right) - E_{\mathrm{d}}m_{\mathrm{e}}\right) = 0 \tag{14.19}$$

Here, to confirm the energy generation from mind of the momentums, the vector energy momentum has been calculated by conducting the following mathematical probe:

$$2\left(m_{\mathrm{e}}E_{\mathrm{p}} + \left(\frac{E_{\mathrm{d}}E_{\mathrm{p}}}{c^{2}}\right) - \frac{P_{\mathrm{d}}E_{\mathrm{p}}}{c}\cos(\theta)\ )\right) = m_{\mathrm{d}}^{2}c^{2} - 2E_{\mathrm{d}}m_{\mathrm{e}} \tag{14.20}$$

Here, the expression for energy generation is related to the vectorial momentum of mind energy; thus, the limit $m_{\mathrm{d}} \rightarrow 0$ can be expressed in the case of scattering in order to transform it into an energy state travel machine. Since $m_{\mathrm{d}} \rightarrow 0$, $P_{\mathrm{d}} \rightarrow E_{\mathrm{c}}$, and thus, the mind energy generation can be expressed by adding the differential scattering cross section in order to transform it into the energy state travel machine by the configuration of this mind energy transform dynamic mechanism as the following equation:

$$De = \frac{\partial\sigma_{\mathrm{kn}}}{\partial\Omega} = \frac{r_0^2}{2}\,\frac{\left(P_{\mathrm{d}}^{2}c^{2} + m_{\mathrm{d}}^{2}c^{4}\right)}{\left(\dfrac{-\left(\frac{m_{\mathrm{d}}^{2}c^{2}}{2m_{\mathrm{e}}} - E_{\mathrm{d}}\right)}{\left(1+\left(\frac{E_{\mathrm{d}}}{m_{\mathrm{e}}c^{2}} - \frac{P_d}{\overset{c}{m_{\mathrm{e}}}}\cos(\theta)\right)\right)}\right)^{2}}$$

$$\times\left[\left(\frac{\left(P_{\mathrm{d}}^{2}c^{2} + m_{\mathrm{d}}^{2}c^{4}\right)^{1/2}}{\dfrac{-\left(\frac{m_{\mathrm{d}}^{2}c^{2}}{2m_{\mathrm{e}}} - E_{\mathrm{d}}\right)}{\left(1+\left(\frac{E_{\mathrm{d}}}{m_{\mathrm{e}}c^{2}} - \frac{P_{\mathrm{d}}}{m_{\mathrm{e}}c}\cos(\theta)\right)\right)}}\right) + \frac{\dfrac{-\left(\frac{m_{\mathrm{d}}^{2}}{2m_{\mathrm{e}}} - E_{\mathrm{d}}\right)}{\left(1+\left(\frac{E_{\mathrm{d}}}{m_{\mathrm{e}}c^{2}} - \frac{P_{\mathrm{d}}}{m_{\mathrm{e}}c}\cos(\theta)\right)\right)}}{\left(P^{2}C^{2}+m_{\mathrm{d}}^{2}C^{4}\right)^{1/2}} - \sin(\theta)^{2}\right] \tag{14.21}$$

Computation of Energy State Travel Machine (T_m)

Since the energy state travel machine (T_m) is the force particle of the mind energy, thus, the nanoscopic numbers of body particles are being calculated here in relation to its dynamic force generation by implementation of Higgs boson ($H \rightarrow \gamma\gamma^{-}$) electromagnetic fields of the human body [2, 6, 20]. Simply, the energy content within the energy state travel machine (T_m) will be released from the mind energy considering its standard matter (H^*) on its free quantum fields (n) as an activated state of machine energy (Fig. 14.3).

Subsequently, the quantum field of this energy state travel machine (T_m) is to be the components of all free quantum momentum of the body; thus, its components (t, x) are being clarified into a vector $x = (ct, x)$ considering its energy momentum of $\eta_{\mu v} =$ diag $(\pm 1, \mp 1, \mp 1, \mp 1)$ of the Klein-Gordon equations, expressed by the following Table 14.1:

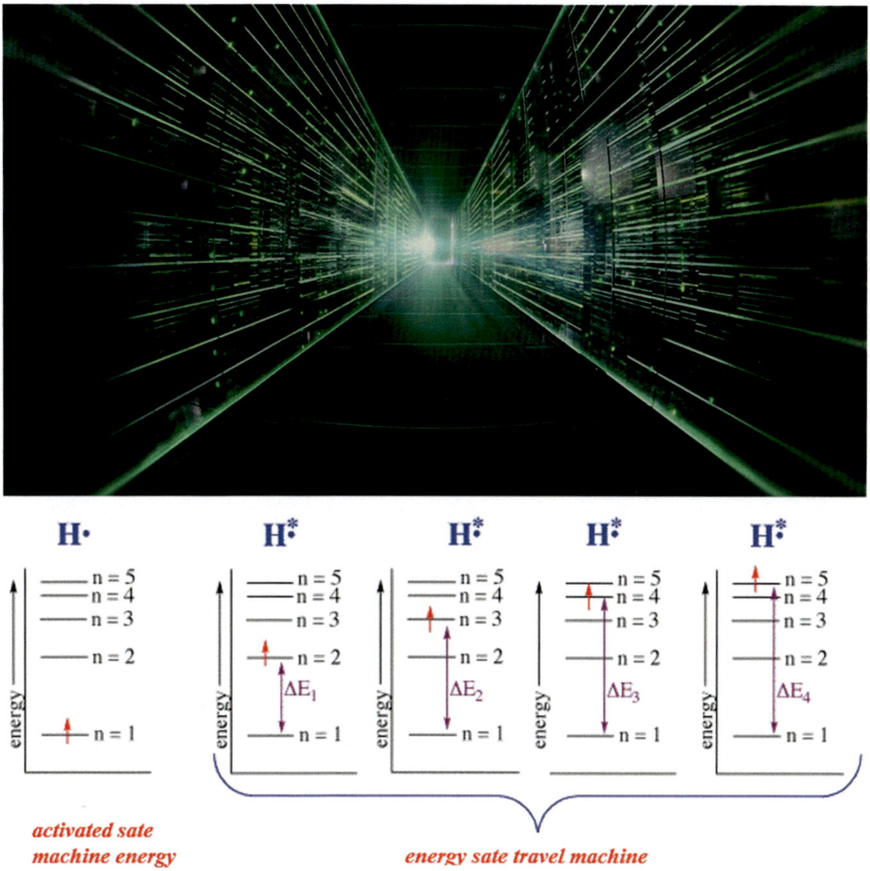

Fig. 14.3 The concept of energy state mind machine distribution (H^*) due to the quantum mechanical functions of the human body considering free quantum fields (n) corresponding to its vectorial state

Table 14.1 The Klein-Gordon equation in normal units with metric of $\eta_{\mu\nu} =$ diag $(\pm 1, \mp 1, \mp 1, \mp 1)$

	Position space $x = (ct, x)$	Fourier transformation $\omega = \frac{E}{\hbar},\ k = p/\hbar$	Momentum space $p = \left(, \frac{E}{c}, p\right)$
Mind energy	$\left(\frac{1}{c^2}\frac{\partial^2}{\partial t^2} - \nabla^2 + \frac{m^2c^2}{\hbar^2}\right)\psi(t,x) = 0$	$\psi(t,x) = \int \frac{d\omega}{2\pi\hbar} \int \frac{d^3k}{(2\pi\hbar)^3} e^{\mp i(\omega t - k.x)}\psi(\omega, k)$	$E^2 = p^2c^2 + m^2c^4$
Vector form	$(c + \mu^2)\psi(x) = 0,\ \mu = mc/\hbar$	$\psi(x) = \int \frac{d^4p}{(2\pi\hbar)^4} e^{-ip.\frac{x}{\hbar}}\psi(p)$	$p^2 = \pm m^2c^2$

Table 14.2 The Klein-Gordon equation in natural units with metric of $\eta_{\mu\nu} =$ diag $(\pm 1, \mp 1, \mp 1, \mp 1)$

	Position space	Fourier transformation $\omega = E,\ k = p$	Momentum space $p = (E, p)$
Mind energy	$\left(\partial_t^2 - \nabla^2 + m^2\right)\psi(t,x) = 0$	$\psi(t,x) = \int \frac{dw}{2\pi} \int \frac{d^3k}{(2\pi)^3} e^{\mp i(wt - k.x)} \psi(w,k)$	$E^2 = p^2 + m^2$
Vector form	$(c + m^2)\psi(x) = 0$	$\psi(x) = \int \frac{d^4p}{(2\pi)^4} e^{-ip.x} \psi(p)$	$p^2 = \pm m^2$

Here, $\pm\eta^{\mu\nu}\partial_\mu\partial_\nu$ is representing the d'Alembert function, ∇^2 is representing the Laplace variable, c is representing the light speed, and ħ is the Planck constant, where $c = \hbar = 1$ (Table 14.2).

Here, the Klein–Gordon equation permits two variables where one variable is positive and the other is negative, which can be obtained by describing a relativistic wave function of the energy state travel machine (T_m) as:

$$\left[\nabla^2 - \frac{m^2c^2}{\hbar^2}\right]\psi(r) = 0 \tag{14.22}$$

Since the Klein-Gordon function is a standard unit, $(c + m^2)\psi(x) = 0$, with the metric signature of $\eta_{\mu\nu} =$ diag $(\pm 1, -1, -1, -1)$, thus, it is solved by Fourier transformation as:

$$\psi(x) = \int \frac{d^4p}{(2\pi)^4} e^{-ip.x} \psi(p) \tag{14.23}$$

and using spherical dimension of the body, the complex exponentials are described here as:

$$p^2 = \left(p^0\right)^2 - p^2 = m^2 \tag{14.24}$$

This result will confirm the momentums of the energy state travel machine's energy dynamics as:

$$p^0 = \pm E(p) \text{ where } E(p) = \sqrt{p^2 + m^2}. \tag{14.25}$$

Using a new set of constant $C(p)$, the solution will become:

$$\psi(x) = \int \frac{d^4p}{(2\pi)^4} e^{ip.x} C(P) \delta\left(P^0\right)^2 - E(P)^2) \tag{14.26}$$

Subsequently, the electromagnetic fields of the energy state travel machine (T_m) can be further clarified by the Klein-Gordon equation of e identity, i.e., $p^2c^2 + m^2c^4 = E^2$, which will give:

$$\left((-i\hbar\nabla)^2c^2 + m^2c^4\right)\psi = \left(i\hbar\frac{\partial}{\partial t}\right)^2\psi \tag{14.27}$$

which simplifies to:

$$-\hbar^2c^2\nabla^2\psi + m^2c^4\psi = \hbar^2\frac{\partial^2}{\partial t^2}\psi \tag{14.28}$$

Rearranging terms yields:

$$\frac{1}{c^2}\frac{\partial^2}{\partial t^2}\psi - \nabla^2\psi + \frac{m^2c^2}{\hbar^2}\psi = 0 \tag{14.29}$$

Rewriting these functions considering the inverse metric ($-c^2$, 1, 1, 1) in relation to Einstein's summational variable, it can be written as:

$$-\eta^{\mu\nu}\partial_\mu\partial_\nu\psi = \sum_{\mu=0}^{\mu=3}\sum_{\nu=0}^{\nu=3} -\eta^{\mu\nu}\partial_\mu\partial_\nu\psi = \frac{1}{c^2}\partial_0^2\psi - \sum_{\nu=1}^{\nu=3}\partial_\nu\partial_\nu\psi = \frac{1}{c^2}\frac{\partial^2}{\partial t^2}\psi - \nabla^2\psi \tag{14.30}$$

Therefore, the Klein-Gordon equation can be expressed as a covariant function, which is the abbreviation of the form of $(c + \mu^2)\psi = 0$, where $\mu = \frac{mc}{\hbar}$ and $c = \frac{1}{c^2}\frac{\partial^2}{\partial t^2} - \nabla^2$, and, thus, the Klein-Gordon equation can be described as a field of energy state travel machine dynamics $V(\psi)$ as:

$$c\psi + \frac{\partial V}{\partial \psi} = 0 \tag{14.31}$$

Here, the gauge U(1) symmetry is a complex field $\varphi(x)\epsilon\mathbb{C}$ that can satisfy the Klein-Gordon equation as:

$$\partial_\mu j^\mu(x) = 0, j^\mu(x) = \frac{e}{2m}\left(\varphi^*(x)\partial^\mu\varphi(x) - \varphi(x)\partial^\mu\varphi^*(x)\right) \tag{14.32}$$

Subsequently, proof Klein-Gordon equation using algebraic manipulations from to form a complex field $\varphi(x)$ of mass m, can be expressed as a covariant notation as follows:

$$\left(c + \mu^2\right)\varphi(x) = 0 \tag{14.33}$$

and its complex conjugate:

$$\left(c+\mu^{2}\right)\varphi^{*}(x)=0 \tag{14.34}$$

Then, multiplying by the vectorial variables of $\varphi^{*}(x)$ and $\varphi(x)$, it can be expressed as:

$$\varphi^{*}\left(c+\mu^{2}\right)\varphi=0 \tag{14.35}$$

$$\varphi\left(c+\mu^{2}\right)\varphi^{*}=0 \tag{14.36}$$

Subtracting the former from the latter, the following can be obtained:

$$\varphi^{*}c\varphi-\varphi c\varphi^{*}=0 \tag{14.37}$$

$$\varphi^{*}\partial_{\mu}\partial^{\mu}\varphi-\varphi\partial_{\mu}\partial^{\mu}\varphi^{*}=0 \tag{14.38}$$

Then, it will confirm:

$$\partial_{\mu}(\varphi^{*}\partial^{\mu}\varphi)=\partial_{\mu}\varphi^{*}\partial^{\mu}\varphi+\varphi^{*}\partial_{\mu}\partial^{\mu}\varphi \tag{14.39}$$

from which it can be obtained the conservation law for the Klein-Gordon field:

$$\partial_{\mu}j^{\mu}(x)=0, j^{\mu}(x)\equiv\varphi^{*}(x)\partial^{\mu}\varphi(x)-\varphi(x)\partial^{\mu}\varphi^{*}(x) \tag{14.40}$$

Here, implementing the vectorial variable to the Klein-Gordon field, the scalar field-derived energy state travel machine can be calculated as:

$$T^{\mu\nu}=\frac{\hbar^{2}}{m}\left(\eta^{\mu\alpha}\eta^{\nu\beta}+\eta^{\mu\beta}\eta^{\nu\alpha}-\eta^{\mu\nu}\eta^{\alpha\beta}\right)\partial_{\alpha}\overline{\psi}\partial_{\beta}\psi-\eta^{\mu\nu}mc^{2}\overline{\psi}\psi \tag{14.41}$$

Here, the energy component T^{00} will confirm the positive frequency related to the particles of energy state travel machine as $\psi(x,t)=\varnothing(x,t)\mathrm{e}^{-\frac{i}{\hbar}mc^{2}t}$, where $\varnothing(x,t)=u_{E}(x)\mathrm{e}^{-\frac{i}{\hbar}Et}$. Thus, defining the energy state travel machine (T_{m}) of body $E'=E-mc^{2}=\sqrt{m^{2}c^{4}+c^{2}p^{2}}-mc^{2}\approx\frac{p^{2}}{2m}, E'\ll mc^{2}$ in the non-relativistic limit $v{\sim}p\ll c$ and hence $\left|i\hbar\frac{\partial\varnothing}{\partial t}\right|=E'\varnothing\ll mc^{2}\varnothing$. Thus, this yields the derivative ψ that can be written as:

$$\begin{aligned}\frac{\partial\psi}{\partial t}&=\left(-i\frac{mc^{2}}{\hbar}\varnothing+\frac{\partial\varnothing}{\partial t}\right)\mathrm{e}^{-\frac{i}{\hbar}mc^{2}t}\approx-i\frac{mc^{2}}{\hbar}\varnothing\mathrm{e}^{-\frac{i}{\hbar}mc^{2}t}\\\frac{\partial^{2}\psi}{\partial t^{2}}&\approx-\left(i\frac{2mc^{2}}{\hbar}\frac{\partial\varnothing}{\partial t}+\left(\frac{mc^{2}}{\hbar}\right)^{2}\varnothing\right)\mathrm{e}^{-\frac{i}{\hbar}mc^{2}t}\end{aligned} \tag{14.42}$$

Substituting it with the free Klein-Gordon equation, $c^{-2}\partial_t^2\psi = \nabla^2\psi - m^2\psi$, yields $-\frac{1}{c^2}\left(i\frac{2mc^2}{\hbar}\frac{\partial\varnothing}{\partial t} + \left(\frac{mc^2}{\hbar}\right)^2\varnothing\right)e^{-\frac{i}{\hbar}mc^2t} \approx \left(\nabla^2 - \left(\frac{mc^2}{\hbar}\right)^2\right)\varnothing e^{-\frac{i}{\hbar}mc^2t}$ that can be simplified as:

$$-i\hbar\frac{\partial\varnothing}{\partial t} = -\frac{\hbar^2}{2m}\nabla^2\varnothing \tag{14.43}$$

Since the free Klein-Gordon equation is a *classical* Schrödinger field, thus, it is further analyzed by the quantum field of Klein-Gordon equation of the body [3, 17, 40]. Subsequently, the quantum field of Klein-Gordon equation is being analyzed to confirm the symmetry of the energy state travel machine (T_m) function φ are in the local U(1) gauge of the body of $\varphi \to \varphi' = \exp(i\theta)\varphi$ of the body where $\theta(t, x)$ is a local vectorial angle, which transformation the functional phase as $\exp(i\theta) = \cos\theta + i\sin\theta$, where derivative ∂_μ can be replaced by gauge-covariant variable $D_\mu = \partial_\mu - ieA_\mu$, while the gauge fields transform as $eA_\mu \to eA'_\mu + \partial_\mu\theta$ which confirm the activation of energy state travel machine (T_m) of the human body as:

$$D_\mu D^\mu\varphi = -(\partial_t - ieA_0)^2\varphi + (\partial_i - ieA_i)^2\varphi = m^2\varphi \tag{14.44}$$

Results and Discussion

Activation of Imagination Power ($\mathbf{I}_p$)

To computationally confirm the formation of imagination power in the human brain, the vectorial momentum of its steady weak coupling energy field has been clarified by MATLAB software. Here, imagination power (I_p) is, thus, considered as the quantum field of the density of state (DOS) imagination proliferation of $J(\omega)$, which, in fact, will deform it into a mind energy magnitude of $V(\omega)$ [19, 21, 25]. Thus, imagination power (I_p) formation follows the Weisskopf-Wigner theory; therefore, all the generated imagination power is being passed through its all-dimensional modes (1D, 2D, and 3D) of the brain (Table 14.3).

Subsequently, all dimensions of 1D, 2D, and 3D of the brain are being further clarified by implementing frequency cut-off to avoid the bifurcation of the DOS to pave the path for imagination power (I_p) formation [18, 30, 34]. Thus, the DOS analyzed at various dimensions of the imagination has been calculated as $\varrho_{PC}(\omega)$, is being confirmed by the clarifying of imagination power (I_p) frequency formation of the Maxwell's theory of nanostructure correspond to the imagination power and, expressed by $\varrho_{PC}(\omega) \propto \frac{1}{\sqrt{\omega - \omega_e}}\Theta(\omega - \omega_e)$, where $\Theta(\omega - \omega_e)$ where ω_e is denoted as the frequency of the DOS [1, 31].

Table 14.3 Imagination power dynamics considering its densities of states (DOS) in the brain

Energy particle	Unit area $J(\omega)$ for different DOS*	Reservoir-induced self-energy induction $\Sigma(\omega)$*
1D	$\frac{C-1}{\pi}\frac{1}{\sqrt{\omega-\omega_e}}\Theta(\omega-\omega_e)$	$-\frac{C-1}{\sqrt{\omega_e-\omega}}$
2D	$C-\eta\left[\ln\left\|\frac{\omega-\omega_0}{\omega_0}\right\|-1\right]\Theta(\omega-\omega_e)\Theta(\Omega_d-\omega)$	$C-\eta\left[\mathrm{Li}_2\left(\frac{\Omega_d-\omega_0}{\omega-\omega_0}\right)-\mathrm{Li}_2\left(\frac{\omega_0-\omega_e}{\omega_0-\omega}\right)-\ln\frac{\omega_0-\omega_e}{\Omega_d-\omega_0}\ln\frac{\omega_e-\omega}{\omega_0-\omega}\right]$
3D	$\chi\sqrt{\frac{\omega-\omega_e}{\Omega_C}}\exp\left(-\frac{\omega-\omega_e}{\Omega_C}\right)\Theta(\omega-\omega_e)$	$\chi\left[\pi\sqrt{\frac{\omega_e-\omega}{\Omega_C}}\exp\left(-\frac{\omega-\omega_e}{\Omega_C}\right)\mathrm{erfc}\sqrt{\frac{\omega_e-\omega}{\Omega_C}}-\sqrt{\pi}\right]$

The unit areas $J(\omega)$ and the generation of self-energy inductions in the mind $\Sigma(\omega)$, demonstrated by the mind power corresponding to its variables of C, η, and χ as coupled forces among three-dimensional (1D, 2D, and 3D) point break of energy band (Lo et al. 2015: Sci. Rep)

Then, the density of state (DOS) has been implemented to confirm the qualitative form of the non-Weisskopf-Winger module that represents the imagination power (I_p), calculated by the dimensional clarification of the brains' sphere [4, 17, 28]. This DOS is thereafter determined considering the electromagnetic field of the brain which is determined by $\varrho_{PC}(\omega) \propto \frac{1}{\sqrt{\omega - \omega_e}}\Theta(\omega - \omega_e)$. Simply, here, the imagination power (I_p) of the DOS reveals a perfect algorithmic function, which is $\varrho_{PC}(\omega) \propto -[\ln|(\omega - \omega_0)/\omega_0| - 1]\Theta(\omega - \omega_e)$, where ω_e defines the prime tip point of the DOS distribution in the mind.

Subsequently, this distributed DOS imagination power will confirm the total deliberation of energy dynamics of mind corresponding to its quantum fields, and, thus, as defined above, $J(\omega)$ will confirm the quantum field of brain $V(\omega)$ corresponding to the imaginational power as:

$$J(\omega) = \varrho(\omega)|V(\omega)|^2 \tag{14.45}$$

Therefore, considering here the energy frequency of ω_c and the generated imagination power dynamic of $\langle a(t)\rangle = u(t, t_0)\langle a(t_0)\rangle$, the function $u(t, t_0)$ has been described as the imagination power structure as $u(t, t_0)$ by clarifying the integral-differential equation as:

$$u(t, t_0) = \frac{1}{1 - \Sigma'(\omega_b)} e^{-i\omega(t - t_0)} + \int_{\omega_e}^{\infty} d\omega \frac{J(\omega)e^{-i\omega(t - t_0)}}{[\omega - \omega_c - \Delta(\omega)]^2 + \pi^2 J^2(\omega)}, \tag{14.46}$$

where $\Sigma'(\omega_b) = [\partial\Sigma(\omega)/\partial\omega]_{\omega = \omega_b}$ and $\Sigma(\omega)$ confirms the imagination power induction into the reservoir:

$$\Sigma(\omega) = \int_{\omega_e}^{\infty} d\omega' \frac{J(\omega')}{\omega - \omega'}. \tag{14.47}$$

Here, the frequency ω_b represents the imagination power frequency mode ($0 < \omega_b < \omega_e$), calculated as $\omega_b - \omega_c - \Delta(\omega_b) = 0$, where $\lesssim\Delta(\omega) = \mathcal{P}\left[\int d\omega' \frac{J(\omega')}{\omega - \omega'}\right]$ is a primary-valued integral of the imagination [5, 11, 22]. In details, it can be explained that imagination energy dynamics (eV) in the brain is considered as the Fock imagination power state n_0, i.e., $\rho(t_0) = |n_0\rangle\langle n_0|$, which is obtained computationally from the real-time quantum field of its radial (r) dimension of the imagination at a time t and expressed as:

$$\rho(t) = \sum_{n=0}^{\infty} \mathcal{P}_n^{(n_0)}(t)|n_0\rangle\langle n_0| \tag{14.48}$$

$$\mathcal{P}_n^{(n_0)}(t) = \frac{[v(t,t)]^n}{[1+v(t,t)]^{n+1}}[1-\Omega(\mathrm{t})]^{n_0} \times \sum_{k=0}^{\min\{n_0,n\}} \binom{n_0}{k}\binom{n}{k}\left[\frac{1}{v(t,t)}\frac{\Omega(\mathrm{t})}{1-\Omega(\mathrm{t})}\right]^k \tag{14.49}$$

where $\Omega(\mathrm{t}) = \frac{|u(t,t_0)|^2}{1+v(t,t)}$. Thus, this result suggested that the Fock state imagination power is indeed generated into dynamic states of the imagination $\mathcal{P}_n^{(n_0)}(t)$ of $|n_0\rangle$ where the proliferation of energy deliberation $\mathcal{P}_n^{(n_0)}(t)$ is in the prime state $|n_0 = 5\rangle$ and in the steady state of mind thought $\mathcal{P}_n^{(n_0)}(t \to \infty)$, which will eventually reach at a non-equilibrium imagination power in order to transform it into mind energy.

Computation of Mind Energy (**M**$_e$)

To transform imagination power (I_p) into mind energy, the local symmetry of Higgs boson quantum field of the body has been simulated considering Abelian local symmetries by applying Higgs boson electromagnetic field of the mind [14, 16, 24]. Simply, the momentum of this energy dynamic will interrupt the gauge field symmetries of the mind, and thus, the Goldstone scalar field will act as the longitude modes of the Higgs boson field [15, 20, 32]. Therefore, the local symmetries of the quantum field of the body will be broken down as particle of T^α related to the gauge field of $A_\mu^\alpha\,(x)$, and thus, the Higgs boson quantum fields will then start to generate local U(1) phase symmetries to create mind energy [24, 26, 37]. So, this transformation process can be consisted of the scalar fields $\Phi(x)$ of energetically charged q paired with the electromagnetic field $A^\mu(x)$ of the mind energy that could be written as:

$$\mathcal{L} = -\frac{1}{4}\,F_{\mu\nu}F^{\mu\nu} + D_\mu\Phi^*\,D^\mu\Phi - V(\Phi^*\Phi), \tag{14.50}$$

where:

$$\begin{aligned} D_\mu\Phi(x) &= \partial_\mu\Phi(x) + iqA_\mu(x)\Phi(x) \\ D_\mu\Phi^*(x) &= \partial_\mu\Phi^*(x) - iqA_\mu(x)\Phi^*(x), \end{aligned} \tag{14.51}$$

and

$$V(\Phi^*\Phi) = \frac{\lambda}{2}\,(\Phi^*\Phi)^2 + m^2(\Phi^*\Phi). \tag{14.52}$$

Here, $\lambda > 0$ but $m^2 < 0$; therefore, $\Phi = 0$ is a local peak factor of the scalar field, and the precise form of quantum dynamics is $\Phi = \frac{v}{\sqrt{2}} * e^{i\theta}$ that can be expressed as:

$$v = \sqrt{\frac{-2m^2}{\lambda}} \text{ for any real } \theta. \tag{14.53}$$

Consequently, here, the scalar quantum dynamics $\Phi(x)$ will form a non-zero value of $\langle \Phi \rangle \neq 0$, which will form the local U(1) symmetries into electromagnetic field of mind energy. Therefore, in these local U(1) symmetries, the quantum dynamics of $\Phi(x)$ will be an activated scalar field of the mind energy that can be expressed as:

$$\Phi(x) = \frac{1}{\sqrt{2}} \Phi_r(x) * e^{i\Theta(x)}, \text{real } \Phi_r(x) > 0, \text{real } \Phi(x). \tag{14.54}$$

Since the scalar field of mind energy is a dynamic momentum of the quantum $\Phi(x) = 0$, thus, it will meet the principle of $\langle \Phi \rangle \neq 0$ in order to be considered as the instinctual function of the mind energy. Thus, considering the momentum field of $\phi_r(x)$ and $\Theta(x)$, the scalar field related to the radial field of ϕ_r of the mind energy can be simplified as:

$$V(\phi) = \frac{\lambda}{8} \left(\phi_r^2 - v^2\right)^2 + \text{const.} \tag{14.55}$$

Subsequently, the momentum field can be shifted by applying the variable scalars $\Phi_r(x) = v + \sigma(x)$ and expressed as:

$$\phi_r^2 - v^2 = (v + \sigma)^2 - v^2 = 2v\sigma + \sigma^2 \tag{14.56}$$

$$V = \frac{\lambda}{8} \left(2v\sigma - \sigma^2\right)^2 = \frac{\lambda v^2}{2} * \sigma^2 + \frac{\lambda v}{2} * \sigma^3 + \frac{\lambda}{8} * \sigma^4. \tag{14.57}$$

Thus, the functional derivative $D_\mu \phi$ will become:

$$D_\mu \phi = \frac{1}{\sqrt{2}} \left(\partial_\mu \left(\phi_r e^{i\Theta}\right) + iqA_\mu * \phi_r e^{i\Theta}\right) = \frac{e^{i\Theta}}{\sqrt{2}} \times \left(\partial_\mu \phi_r + \phi_r * i\partial_\mu \Theta + \phi_r * iqA_\mu\right) \tag{14.58}$$

$$\begin{aligned}|D_\mu\phi|^2 &= \frac{1}{2}\left|\partial_\mu\phi_\mathrm{r} + \phi_\mathrm{r} * i\partial_\mu\Theta + \phi_\mathrm{r} * iqA_\mu\right|^2 \\ &= \frac{1}{2}\left(\partial_\mu\phi_r\right) + \frac{\phi_r^2}{2} * \left(\partial_\mu\Theta qA_\mu\right)^2 \\ &= \frac{1}{2}\left(\partial_\mu\,\sigma\right)^2 + \frac{(v+\sigma)^2}{2} * \left(\partial_\mu\Theta + qA_\mu\right)^2.\end{aligned} \tag{14.59}$$

The Lagrangian is then given by:

$$\mathcal{L} = \frac{1}{2}\left(\partial_\mu\sigma\right)^2 - v(\sigma) - \frac{1}{4}\,F_{\mu\nu}F^{\mu\nu} + \frac{(v+\sigma)^2}{2} * \left(\partial_\mu\Theta + qA_\mu\right)^2. \tag{14.60}$$

Subsequently, it will initiate to generate energy state travel machine (T_m) into the quantum field of the mind considering the scalar particles of λv^2 corresponding to its energy dynamics of the mind energy [3, 27, 33].

Computation of Energy State Travel Machine (T$_m$)

Finally, the electromagnetic field of the energy state travel machine (T_m) is being modeled by Higgs boson quantum dynamics considering its local symmetry of U(1) analysis in order to from gauge-variable QED, in terms of gauge particles $\varnothing\,' \rightarrow e^{i\alpha(x)}\varnothing$ to transform mind energy into energy state travel machine. This mechanism is being confirmed by the following variable derivatives considering the specific transformational laws of the scalar field:

$$\begin{aligned}&\partial_\mu \rightarrow D_\mu = \partial_\mu = ieA_\mu && [\text{covariant derivatives}] \\ &A'_\mu = A_\mu + \frac{1}{e}\partial_\mu\alpha && [A_\mu \text{ derivatives}]\end{aligned} \tag{14.61}$$

Here, the local U(1) gauge invariant Lagrangian is being considered as the perplex scalar field that is expressed by:

$$\mathcal{L} = (D^\mu)^\dagger\left(D_\mu\varnothing\right) - \frac{1}{4}F_{\mu\nu}F^{\mu\nu} - V(\varnothing). \tag{14.62}$$

Thus, the equation of the Lagrangian $\mathcal{L}$ is being considered as the quantum field of scalar particles ϕ_1 and ϕ_2 and a mass μ of the energy state travel machine; thus, here, $\mu^2 < 0$ will confirm an infinite number of quanta to satisfy the equation $\phi_1^2 + \phi_2^2 = -\mu^2/\lambda = v^2$. So, the quantum field can be clarified as $\phi_0 = \frac{1}{\sqrt{2}}[(\upsilon + \eta) + i\xi]$, and then the derivative of the Lagrangian can be expressed as kinetic mode of:

$$\begin{aligned}\mathcal{L}_{\text{kin}}(\eta,\xi) &= (D^{\mu}\phi)^{\dagger}(D^{\mu}\phi) \\ &= (\partial^{\mu}+ieA^{\mu})\phi^{*}\left(\partial_{\mu}-ieA_{\mu}\right)\phi\end{aligned} \tag{14.63}$$

Here, $V(\eta,\xi) = \lambda\ \upsilon^2\eta^2$ is the final term to the second order, and thus, the full Lagrangian can be expressed as:

$$\mathcal{L}_{\text{kin}}(\eta,\xi) = \frac{1}{2}\left(\partial_{\mu}\eta\right)^2 - \lambda\upsilon^2\eta^2 + \frac{1}{2}\left(\partial_{\mu}\xi\right)^2 - \frac{1}{4}\ F_{\mu\nu}F^{\mu\nu} + \frac{1}{2}\ e^2\upsilon^2A_{\mu}^2 - e\upsilon A_{\mu}(\partial^{\mu}\xi) \tag{14.64}$$

Here, η represents the mass, ξ represents the massless, μ represents the mass for the quanta, and thus A_{μ} is defined as the term $\partial_{\mu}\alpha$, as is the function of the quantum field [23, 27, 36]. Naturally, A_{μ} and ϕ can be changed spontaneously, so the equation can be rewritten as a Lagrangian scalar to confirm the formation of the energy state travel machine (T_{m}) particles within the quantum field of the body:

$$\begin{aligned}\mathcal{L}_{\text{scalar}} &= (D^{\mu}\phi)^{\dagger}(D^{\mu}\phi) - V\left(\phi^{\dagger}\phi\right) \\ &= (\partial^{\mu}+ieA^{\mu})\frac{1}{\sqrt{2}}\ (v+h)\left(\partial_{\mu}-ieA_{\mu}\right)\frac{1}{\sqrt{2}}\ (v+h) - V\ \left(\phi^{\dagger}\phi\right)\end{aligned} \tag{14.65}$$

$$= \frac{1}{2}\left(\partial_{\mu}h\right)^2 + \frac{1}{2}\ e^2A_{\mu}^2\ (v+h)^2 - \lambda v^2h^2 - \lambda vh^3 - \frac{1}{4}\ \lambda h^4 + \frac{1}{4}\ \lambda h^4 \tag{14.66}$$

Here, the Lagrangian scalar thus revealed that the Higgs boson quantum field can certainly be initiated to from energy state travel machine (T_{m}). To confirm and determine this energy state travel machine (T_{m}) generation, a further calculation considering its isotropic distributed kinetic energy has been conducted with respect to the angle θ from the dimensional axis of the body and the differential density of energy $\in$ and written as:

$$\mathrm{d}n = \frac{1}{2}n(\in)\ \sin\theta\mathrm{d}\in\mathrm{d}\theta. \tag{14.67}$$

Here, the speed of high-energy frequency is being implemented as c (1–cosθ) considering the release of energy by expressing the following equation:

$$\frac{\mathrm{d}\tau_{\text{abs}}}{\mathrm{d}x} = \int\int\frac{1}{2}\sigma n\ (\in)(1-\cos\theta)\ \sin\theta\mathrm{d}\in\mathrm{d}\theta. \tag{14.68}$$

Rewriting these variables as integral over s instead of θ, by (79) and (80), it has been determined as:

$$\frac{d\tau_{abs}}{dx} = \pi r_0^2 \left(\frac{m^2c^4}{E}\right)^2 \int_{\frac{m^2c^4}{E}}^{\infty} \in^{-2} n(\in)\,\overline{\phi}\,[s_0\,(\in)]\;de, \tag{14.69}$$

where:

$$\overline{\phi}[s_0(\in)] = \int_1^{s_0(\in)} s\overline{\sigma}\,(s)ds, \overline{\sigma}(s) = \frac{2\sigma(s)}{\pi r_0^2}. \tag{14.70}$$

Simply, this result confirms the representation of the energy state travel machine (T_m) formation that readily allows the calculation of $\overline{\phi}\,[s_0]$ to determine the net travel state energy formation of s_0 [21, 23, 35]. Therefore, the net functional asymptotic formula is expressed as follows to confirm the energy state travel machine formation:

$$\overline{\phi}[s_0] = \begin{array}{l} 2s_0(\ln 4s_0 – 2) + \ln 4s_0(\ln 4s_0 – 2) – \frac{(\pi^2-9)}{3} + s_0^{-1}\left(\ln 4s_0 + \frac{9}{8}\right) \\ + \ldots (s_0 \gg 1); \end{array} \tag{14.71}$$

$$\overline{\phi}[s_0] = \left(\frac{2}{3}\right)\,(S_0-1)^{\frac{3}{2}} + \left(\frac{5}{3}\right)(S_0-1)^{\frac{5}{2}} – \left(\frac{1507}{420}\right)\,(S_0-1)^{\frac{7}{2}} + \ldots (s_0 – 1 \ll 1). \tag{14.72}$$

The function $\frac{\overline{\phi}[s_0]}{(s_0-1)}$ is revealed as $1 < s_0 < 10$; at larger s_0, it confirms a standard algorithm function of s_0. Thus, the energy spectra of the energy state travel machine can be written as $n(\in) \propto \in^m$, and thus, the calculation of the energy state travel machine in terms of energy spectra can be expressed as:

$$\begin{aligned} n(\in) &= 0,\ \in\ < \in_0 \\ &= C_{\in}^{-\alpha} v D_{\in}^{\beta}, \in_0 <\ \in\ < \in_m \\ &= 0,\ \in\ > \in_m \end{aligned} \tag{14.73}$$

Then, it can be transformed as:

$$\left(\frac{d\tau_{abs}}{dx}\right)_{\alpha} = \pi r_0^2 C \left(\frac{m^2c^4}{E}\right)^{1-\alpha} \times \begin{cases} 0, & E < E_m \\ [F_\alpha(1) – F_\alpha(\sigma_m)], & E_m < E < E_0 \\ [F_\alpha(\sigma_0) – F_\alpha(\sigma_m)], & E > E_0; \end{cases} \tag{14.74}$$

$$\left(\frac{d\tau_{abs}}{dx}\right)_{\beta} = \pi r_0^2 D\left(\frac{m^2c^4}{E}\right)^{1+\beta} \times \begin{cases} 0, & E < E_m \\ \left[F_\beta(\sigma_m), \right. & E_m < E < E_0 \\ \left[F_\beta(\sigma_m) - F_\beta(\sigma_0)\right], & E > E_0. \end{cases} \tag{14.75}$$

In these functional variables, the energy state travel machine can be properly defined by asymptotic formula. Thus, the term $\Gamma_\gamma^{\mathrm{LPM}}$ defines the energy generation in relation to the energy dynamics as:

$$\Gamma_\gamma \equiv \frac{dn_\gamma}{dVdt}. \tag{14.76}$$

Here, the contributions $\Gamma_\gamma^{\mathrm{LPM}}$, the rate of energy state travel machine formation are being confirmed as $O(\alpha_{EM}\ \alpha_s)$. Thus, it has been confirmed by implementing the contributed energy state travel machine formation $\Gamma_\gamma^{\mathrm{LPM}}$ to energy-physical reaction μ of the body by expressing the following equation:

$$\frac{d\Gamma_\gamma^{\mathrm{LPM}}}{d^3k} = \frac{d_F q_s^2 \alpha_{EM}}{4\pi^2 k} \int_{-\infty}^{\infty} \frac{dp_\parallel}{2\pi} \int \frac{d^2 p_\perp}{(2\pi)^2} A\left(p_\parallel, k\right) \mathrm{Re}\left\{2P_\perp \cdot f\left(p_\perp; p_\parallel, k\right)\right\} \tag{14.77}$$

where d_F is the variable state of the energy state travel machine particles [N_c in SU (N_c)], q_s is the Abelian charge of the energy state travel machine, and $k \equiv |k|$, and thus, the kinetic functional mode $A(p\|, k)$ is being expressed by

$$A\left(p_\parallel, k\right) \equiv \begin{cases} \dfrac{n_b\left(k + p_\parallel\right)\left[1 + n_b\left(p_\parallel\right)\right]}{2p_\parallel\left(p_\parallel + k\right)} & \text{scalars} \\ \dfrac{n_f\left(k + p_\parallel\right)\left[1 - n_f\left(p_\parallel\right)\right]}{2\left[p_\parallel\left(p_\parallel + k\right)\right]^2}\left[p_\parallel{}^2 + \left(p_\parallel + k\right)^2\right], & \text{fermions} \end{cases} \tag{14.78}$$

with

$$n_b(p) \equiv \frac{1}{\exp[\beta(p - \mu)] - 1}, \quad n_f(p) \equiv \frac{1}{\exp[\beta(p - \mu)] + 1} \tag{14.79}$$

The function $f(p\perp; p\|, k)$ is then integrated to resolve the below equation which suggested that the energy state travel machine proliferation in the body are very much possible which can be expressed as following derivative of energy generation:

$$2p_{\perp} = i\delta E f\left(p_{\perp}; p_{\parallel}, k\right) + \frac{\pi}{2} C_{F} g_{s}^{2} m_{D}^{2} \int \frac{d^{2}q_{\perp}}{(2\pi)^{2}} \frac{dq_{\parallel}}{2\pi} \frac{dq^{0}}{2\pi} 2\pi\delta\left(q^{0} - q_{\parallel}\right)$$
$$\times \frac{T}{|q|} \left[\frac{2}{\left|q^{2} - \Pi_{L}(Q)\right|^{2}} + \frac{\left[1 - \left(q^{0}/\left|q_{\parallel}\right|\right)^{2}\right]^{2}}{\left|\left(q^{0}\right)^{2} - q^{2} - \Pi_{T}(Q)\right|^{2}} \right] \left[f\left(p_{\perp}; p_{\parallel}, k\right) - f\left(q + p_{\perp}; p_{\parallel}, k\right) \right] \tag{14.80}$$

Simply, this energy state travel machine formation is derived from the explicit forms of mind energy corresponding to its given energy function of $f(p\perp; p\|, k)$ to transform mind energy into energy state travel machine (T_m) of the human body to travel anywhere in the universe and multiverse based on imagination.

Conclusions

The formation of imagination power, mind energy, and energy state travel machine are being modeled through a series of transformation process to confirm that imagination has the extreme power that can be transformed into mind energy and then can be further transformed into energy state travel machine to travel anywhere based on imagination. It is because the results of this computational modeling suggested that the availability of imagination power is very much doable in human brain, which can be transformed into mind energy by implementing the quantum electrodynamics (QED) of the imagination power. Subsequently, this mind energy can also be reformed into the energy state travel machine by integrating Higgs boson [BR ($H \rightarrow \gamma\gamma^{-}$)] quantum dynamics of the human body in order to travel physically anywhere in the multiverse. Here, the transformation mechanism of imagination power into mind energy and then into energy state travel machine shall indeed be the most innovative scientific discovery to develop ultimate transportation system that has ever been thought before.

Acknowledgments The author, Md. Faruque Hossain, declares that any findings, predictions, and conclusions described in this article are solely performed by the author, and it is confirmed that there is no conflict of interest for publishing this research paper in a suitable journal and/or publisher.

References

1. Boettcher, I., Pawlowski, J. M., & Diehl, S. (2012). Ultracold atoms and the functional renormalization group. *Nuclear Physics B-Proceedings Supplements, 228*, 63–135.
2. Broz, M., Contreras, J. G., & Takaki, J. T. (2020). A generator of forward neutrons for ultra-peripheral collisions: nOOn. *Computer Physics Communications, 253*, 107181.

3. Chang, D. E., Sørensen, A. S., Demler, E. A., & Lukin, M. D. (2007). A single-photon transistor using nanoscale surface plasmons. *Nature Physics, 3*(11), 807–812.
4. Chen, G., Chen, S., Li, C., & Chen, Y. (2013). Examining non-locality and quantum coherent dynamics induced by a common reservoir. *Scientific Reports, 3*(1), 1–6.
5. Chen, J., Wang, C., Zhang, R., & Xiao, J. (2012). Multiple plasmon-induced transparencies in coupled-resonator systems. *Optics Letters, 37*(24), 5133–5135.
6. Cheng, M., & Song, Y. (2012). Fano resonance analysis in a pair of semiconductor quantum dots coupling to a metal nanowire. *Optics Letters, 37*(5), 978–980.
7. Douglas, J. S., Habibian, H., Hung, C., Gorshkov, A. V., Kimble, H. J., & Chang, D. E. (2015). Quantum many-body models with cold atoms coupled to photonic crystals. *Nature Photonics, 9*(5), 326–331.
8. Dupuis, N., Canet, L., Eichhorn, A., Metzner, W., Pawlowski, J. M., Tissier, M., & Wschebor, N. (2021). The nonperturbative functional renormalization group and its applications. *Physics Reports, 910*, 1–114.
9. Eichler, J., & Stöhlker, T. (2007). Radiative electron capture in relativistic ion–atom collisions and the photoelectric effect in hydrogen-like high-Z systems. *Physics Reports, 439*(1–2), 1–99.
10. Englund, D., Majumdar, A., Faraon, A., Toishi, M., Stoltz, N., Petroff, P., & Vučković, J. (2010). Resonant excitation of a quantum dot strongly coupled to a photonic crystal nanocavity. *Physical Review Letters, 104*(7), 073904.
11. Fernández, J., & Martín, F. (2009). Electron and ion angular distributions in resonant dissociative photoionization of H2 and D2 using linearly polarized light. *New Journal of Physics, 11*(4), 043020.
12. Gazi, V., & Passino, K. M. (2005). Stability of a one-dimensional discrete-time asynchronous swarm. *IEEE Transactions on Systems, Man, and Cybernetics, Part B (Cybernetics), 35*(4), 834–841.
13. Hencken, K., Baur, G., & Trautmann, D. (2006). Transverse momentum distribution of vector mesons produced in ultraperipheral relativistic heavy ion collisions. *Physical Review Letters, 96*(1), 012303.
14. Jahnke, V., Luna, A., Patino, L., & Trancanelli, D. (2014). More on thermal probes of a strongly coupled anisotropic plasma. *Journal of High Energy Physics, 2014*(1), 1–40.
15. Jiaqi, L., Yi, Z., Chengkai, T., & Xingxing, Z. (2019). *INS aided high dynamic single-satellite position algorithm*. Paper presented at the 2019 IEEE international conference on signal processing, communications and computing (ICSPCC), pp. 1–5.
16. Johnson, B. R., Reed, M. D., Houck, A. A., Schuster, D. I., Bishop, L. S., Ginossar, E., & Girvin, S. M. (2010). Quantum non-demolition detection of single microwave photons in a circuit. *Nature Physics, 6*(9), 663–667.
17. Md. Faruque Hossain. (2022). Ultraviolet germicidal irradiation (UVGI) application in building design to terminate pathogens naturally. *Material Today Sustainability*. https://doi.org/10.1016/j.mtsust.2022.100161. (Springer).
18. Md. Faruque Hossain. (2021). Sustainable building technology: Thermal control of solar energy to cool and heat the building naturally. *Environment, Development and Sustainability*. https://doi.org/10.1007/s10668-020-01212-z. (Springer).
19. Md. Faruque Hossain. (2020). Modeling of global temperature control. *Environment, Development and Sustainability*. https://doi.org/10.1007/s10668-020-00924-6. (Springer).
20. Md. Faruque Hossain. (2018). Green science: Advanced building design technology to mitigate energy and environment. *Renewable and Sustainable Energy Reviews, 81*(2), 3051–3060. (Elsevier).
21. Md. Faruque Hossain. (2018). Photon energy amplification for the design of a micro-PV panel. *International Journal of Energy Research*. https://doi.org/10.1002/er.4118. (Wiley).
22. Md. Faruque Hossain. (2018). Design and construction of ultra-relativistic collision PV panel and its application into building sector to mitigate total energy demand. *Journal of Building Engineering, 9*, 147–154. (Elsevier).

23. Md. Faruque Hossain. (2017). Green science: Independent building technology to mitigate energy, environment, and climate change. *Renewable and Sustainable Energy Reviews, 73*, 695–705. (Elsevier).
24. Md. Faruque Hossain. (2016). Solar energy integration into advanced building design for meeting energy demand. *International Journal of Energy Research, 40*, 1293–1300. (Wiley).
25. Md. Faruque Hossain. (2016). Theory of global cooling. *Energy, Sustainability, and Society, 6*, 24. (Springer).
26. Naghiloo, M., Foroozani, N., Tan, D., Jadbabaie, A., & Murch, K. W. (2016). Mapping quantum state dynamics in spontaneous emission. *Nature Communications, 7*(1), 1–7.
27. Reinhard, P., & Nazarewicz, W. (2021). Nuclear charge densities in spherical and deformed nuclei: Toward precise calculations of charge radii. *Physical Review C, 103*(5), 054310.
28. Shalm, L. K., Hamel, D. R., Yan, Z., Simon, C., Resch, K. J., & Jennewein, T. (2013). Three-photon energy–time entanglement. *Nature Physics, 9*(1), 19–22.
29. Stehle, C., Zimmermann, C., & Slama, S. (2014). Cooperative coupling of ultracold atoms and surface plasmons. *Nature Physics, 10*(12), 937–942.
30. Szafron, R., & Czarnecki, A. (2016). High-energy electrons from the muon decay in orbit: Radiative corrections. *Physics Letters B, 753*, 61–64.
31. Tame, M. S., McEnery, K. R., Özdemir, Ş. K., Lee, J., Maier, S. A., & Kim, M. S. (2013). Quantum plasmonics. *Nature Physics, 9*(6), 329–340.
32. Md. Faruque Hossain. (2022). Implementation of hybrid wind and solar energy in the transportation sector to mitigate global energy and environmental vulnerability. *Clean Technologies and Environmental Policy*. https://doi.org/10.1007/s10098-022-02437-4
33. Md. Faruque Hossain. (2020). Application of wind energy into the transportation sector. *International Journal of Precision Engineering and Manufacturing-Green Technology*. https://doi.org/10.1007/s40684-020-00235-1
34. Md. Faruque Hossain. (2019). Green technology: Transformation of transpiration vapor to mitigate global water crisis. *Polytechnica*. https://doi.org/10.1007/s41050-019-00009-y
35. Md. Faruque Hossain. (2018). Global environmental vulnerability and the survival period of all living beings on earth. *International Journal of Environmental Science and Technology*. https://doi.org/10.1007/s13762-018-1722-y
36. Md. Faruque Hossain. (2018). Transformation of dark photon into sustainable energy. *International Journal of Energy and Environmental Engineering, 9*, 99–110.
37. Md. Faruque Hossain. (2017). Invisible transportation infrastructure technology to mitigate energy and environment. *Energy, Sustainability, and Society, 7*, –27. https://doi.org/10.1186/s13705-017-0128-x
38. Md. Faruque Hossain. (2017). Application of advanced technology to build a vibrant environment on planet Mars. *International Journal of Environmental Science and Technology, 14*(12), 2709–2720.
39. Md. Faruque Hossain. (2016). Integration of wind into running vehicles to meet its total energy demand. *Energy, Ecology, and Environment, 2*(1), 35–48.
40. Md. Faruque Hossain. (2016). Production of clean energy from cyanobacterial biochemical products. *Strategic Planning for Energy and the Environment, 3*, 6–23.

Part V
Sustainable Society

Chapter 15
Application of Photophysical Reaction Technology to Eliminate Pathogens Naturally from Earth

Introduction

A pathogen enters the human body by absorption, adsorption, inhalation, and ingestion, producing toxins in human body tissues, suppressing the human immune system, and causing serious illness or death [1–3]. The present danger, COVID-19, a deadly mutant of SARS virus that attacks the respiratory system of the human body and causes deadly respiratory syndrome (SARS-CoV-2), leads a person to death within a few weeks [4–6]. SARS-CoV-2 has reportedly been found in almost all nations in the world since its outbreak was first detected in November 2019; thus, the World Health Organization (WHO) declared this COVID-19 outbreak a pandemic [7–9]. With the way the pandemic has had a major effect on global health since more than four million people have died worldwide, it is obvious that the emergence of a more advanced technology is required in building design technology to mitigate these deadly pathogens naturally. This emergence of technology creates a major concern, especially since these pathogens have mutation characteristics that allow them to adapt to vaccines, making them extremely hard to control and posing a threat to human survival on Earth in the near future [9–11]. Simply, strategies such as prevention technology rather than treatment technology are urgently needed to eliminate these deadly pathogens before they enter the human body. In recent years, ultraviolet germicidal irradiance (UVGI) for in-duct airborne bioaerosol disinfection research has been conducted by Luo and Zhong [3], and the spatial analysis of the impact of UVGI technology in occupied rooms using ray-tracing simulation has been conducted by Hou et al. [8]. Although there are interesting findings, the knowledge gap in utilizing UVGI to eliminate pathogens inside buildings remains unresolved. Thus, in this research, UVGI application was modeled by using an exterior curtain wall skin of a building to eliminate pathogens dramatically since the photonic short-range wavelengths at 254–280 nm of UVGI are an excellent powerful tool with the ability to quickly kill pathogens regardless of their mutation capability [13–15]. Thus, this research is being conducted to study the photonic

M. F. Hossain, *Global Sustainability*, https://doi.org/10.1007/978-3-031-34575-3_15

irradiance on the exterior glazing wall surface, which originates from solar energy, to deliver UVGI to disrupt the nucleic acid and twist the DNA and/or RNA bonding of the pathogens to eventually cause them to die. This research hence conducted the application of photophysical reactions in advanced building design technology that has the ability to form UVGI from solar irradiance to kill every pathogen naturally inside the building before it penetrates the human body.

Materials and Methods

UVGI Production by the Exterior Glazing Wall

A quantum field of sunlight has been clarified to determine the emission of UVGI light at 254–280 nm wavelength on the surface of the outer glazing wall of a building in a predefined period before people enter the building. Naturally, photonic wavelength generation is being analyzed considering the penetration of solar irradiance by a perfect semiconductor connected with MATLAB software to clarify the photonic quantum V, charges I, and reflection of light R on the surface of the exterior glazing wall (Fig. 15.1).

Then, the photon dynamics have been analyzed through the clarification of the wavelength spectrum into the exterior glazing wall surface for the determination of the photonic UVGI release rate from sunlight, which is expressed by the Hamiltonian equation [15, 16]:

$$H = \sum \omega_{ci} a_i^\dagger a_i + \sum_K \omega_k b_k^\dagger b_k + \sum_{ik} \left(V_{ik} a^\dagger b_k + V_{ik}^* b_k^\dagger a_i \right) \tag{15.1}$$

where $a_i\left(a_i^\dagger\right)$ represents the kinetics of the photon energy mode, $b_k\left(b_k^\dagger\right)$ represents the kinematics of the photodynamic mode, and the cofactors V_{ik} denote the mode of photon energy in the exterior wall surface at various azimuthal angles (Fig. 15.2).

Considering the first solar irradiance dynamic proliferation, here, the whole electromagnetic photonic spectrum is being structured considering excited energy state photons in the outer glazing wall surface and expressed by the following equation [17–19]:

$$\begin{aligned} \rho(t) = & -i\left[H_c'(t)\rho(t)\right] + \sum_{ij} \left\{ k_{ij}(t) \left[2a_j\rho(t)a_i^\dagger - a_i^\dagger a_j \rho(t) - \rho(t) a_i^\dagger a_j \right] \right. \\ & \left. + k_{ij}(t) \left[a_i^\dagger \rho(t) a_j + a_j \rho(t) a_i^\dagger - a_i^\dagger a_j \rho(t) - \rho(t) a_j a_i^\dagger \right] \right\} \end{aligned} \tag{15.2}$$

In this equation, $\rho(t)$ represents the attenuated density of solar irradiance states, and $H_c'(t) = \sum_{ij} \omega_{cij}'(t) a_i^\dagger a_j$ represents restandardized solar irradiance frequencies

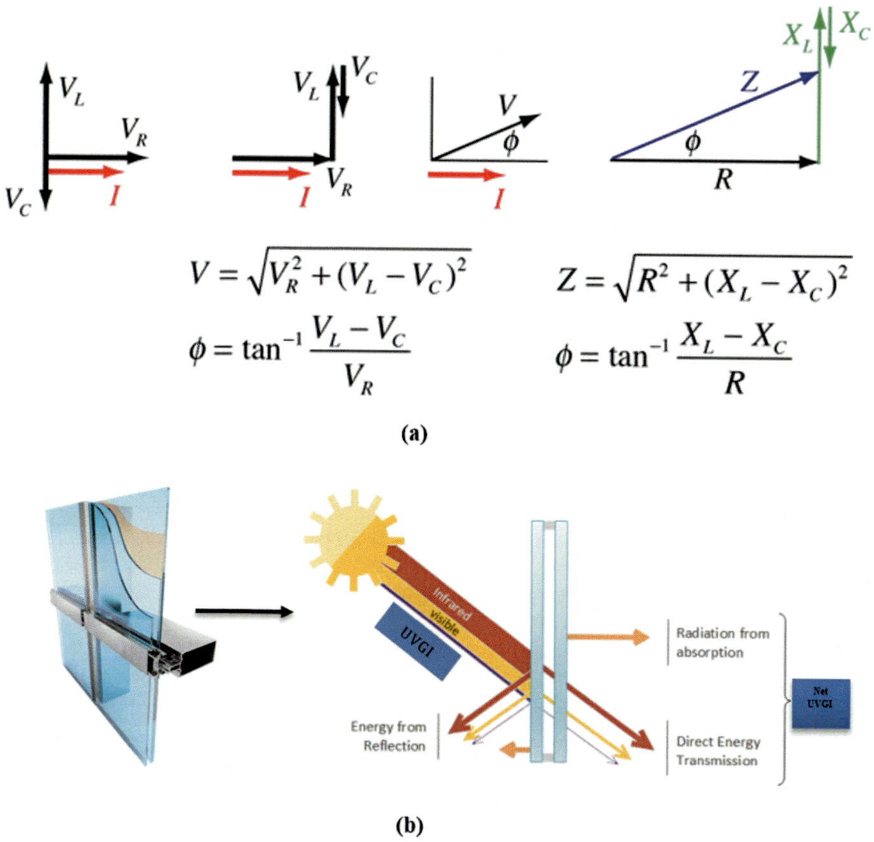

Fig. 15.1 The modeling of light irradiance showing (**a**) the clarification of the photonic quantum V, charges I, and reflection of light R on the surface of exterior glazing wall and (**b**) the penetration of the sun's radiation in the exterior glazing wall surface

$\omega'_{cii}(t) = \omega'_{ci}(t)$, which is factored-induced solar irradiance $\omega'_{cij}(t)$. The factors $\kappa_{ij}(t)$ and $\tilde{\kappa}_{ij}(t)$ are clarified here for irradiance photon proliferation in the glazing wall connected to a semiconductor. Consequently, the wavelength frequencies, ω'_{cij}, and time-related function, $\kappa_{ij}(t)$ and $\tilde{\kappa}_{ij}(t)$, are clarified by the integration vector-perturbative principle, and thus, the solar irradiance dynamic is denoted as $H_{\mathrm{I}} = \sum_k \lambda_k x q_k$, where x and q_k represent the area of the solar irradiance holder on top of the glazing wall. Considering the quantum dynamics of the solar irradiance, the net area of the reservoir of the photon dynamics of the glazing wall skin is modified as $H_{\mathrm{I}} = \sum_k V_k \left(a^\dagger b_k + b_k^\dagger a + a^\dagger b_k^\dagger + a b_k \right)$ to determine the kind of insolation entry within the glazing wall surface. Thus, solar irradiance dynamics are clarified by the dissipation of the solar irradiance functions $\kappa\,(t)$ and $\tilde{\kappa}(t)$ (subindices (i, j) in Eq. (15.2)) and, therefore, its modification as a variable equation [20, 21]:

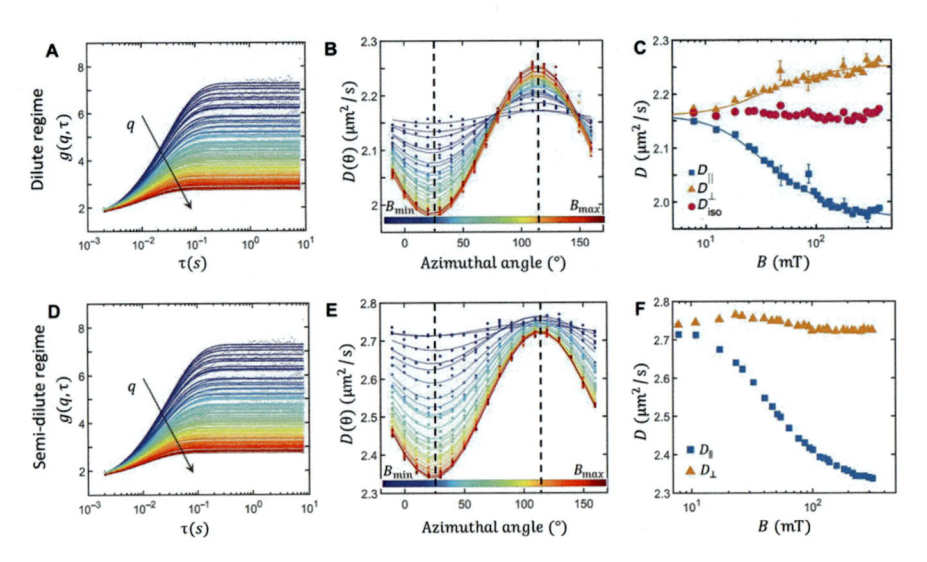

Fig. 15.2 (**a, d**) Solar energy dynamics (q,c) at various wavelengths of the outer wall surface at a period of (s). (**b, e**) Photon energy deliberation mode ($\mu m^2/s$) at various azimuthal angles of irradiance penetration on the surface of outer glazing wall. (**c, f**) The photon energy density ($\mu m^2/s$) at the surface unit (mT) on the outer glass considering the kinetics of the photon energy mode of $a_i\left(a_i^\dagger\right)$, kinematics of the photodynamic mode of $b_k\left(b_k^\dagger\right)$, and their cofactors of V_{ik}

$$\omega_c'(t) = -\,\mathrm{Im}[u\,(t,t_0)/u\,(t,t_0)] \qquad (15.3)$$

$$k(t) = -\,\mathrm{Re}[u\,(t,t_0)/u\,(t,t_0)] \qquad (15.4)$$

$$\widehat{k}(t) = \dot{v}\,(t,t) + 2v\,(t,t)\,k(t). \qquad (15.5)$$

where $u(t,t_0)$ represents the photonic irradiance area and $v(t,t)$ denotes the irradiation proliferation on top of the exterior glazing wall. Further modification of the equation using the following equilibrium photonic dynamic mechanism on the outer glazing wall surface is being calculated as follows [22, 23]:

$$\dot{u}\,(t,t_0) = -\,i\omega_c u\,(t,t_0) - \int_{t_0}^{t} dt'g(t-t')u(t',t_0) \qquad (15.6)$$

$$v(t,t) = \int_{t_0}^{t} dt \int_0^t dt_2\ u^*(t_1,t_0)\widehat{g}\,(t_1 - t_2)u(t_2,t_0) \qquad (15.7)$$

where v_c denotes the primary frequency of solar irradiance and scalar functions in Eqs. (15.6) and (15.7) are determined by the backup function within the solar irradiance and the amount of energy from the photons formed on top of the outer glazing wall per *unit* area of $J(\varepsilon)$ considering the relationship of $g(t-t') = \int d\omega J(\omega)e^{-i\omega(t-t')}$ and $\tilde{g}(t-t') = \int d\omega J(\omega)\overline{n}(\omega,T)e^{-i\omega(t-t')}$. Here, $\overline{n}(\omega,T) = 1/\left[e^{\hbar\omega/k_B T} - 1\right]$ denotes the photon proliferation in the outer glazing wall surface at temperature T. The *unit* area $J(\omega)$ is clarified in conjunction with the density of state (DOS) $\varrho(\omega)$ energy of photon generation inside the glazing wall surface at the magnitude of V_k between the photon dynamics and glazing wall skin:

$$J(\omega) = \sum_k |V_k|^2 \, \delta(\omega - \omega_k) = \varrho(\omega)|V(\omega)|^2 = [n * e(1 + 2n)]^4 \qquad (15.8)$$

Finally, the solar irradiance dynamic rate has been simplified into the glazing wall skin as $V_k \rightarrow V(\omega)$ and i of V_{ik} at the nonequilibrium photon capture $J(\omega)$ and is expressed as the following equation:

$$J(\omega) = [n * e(1 + 2n)]^4 \qquad (15.9)$$

where $n = E = hf$ represents the amount of irradiation generation on top of the glazing wall, e denotes the photonic constant, and n is the electromagnetic radiation of photons, which has been further clarified to calculate the estimated electromagnetic (EM) radiation through the sunlight needed to kill the pathogens inside the building naturally.

Electromagnetic Radiation

Clarification of the EM radiation emitted through the surface of an outer glazing wall of the building considering nodal plane (B) and anti-nodal plane (E) has been analyzed using a computerized connected conductor. Since the EM is a linearly polarized sinusoidal electromagnetic wave, it has been propagating in the direction through an isotropic, dissipation-less medium of the glazing wall surface [24, 25]. Therefore, the electric fields denoted with blue arrows are oscillating by the following $\pm x$-directional factor, and the orthogonal magnetic fields denoted with arrows of the red color are oscillating by the following $\pm y$-directional factors in various phases considering the electric fields on the glazing wall surface (Fig. 15.3).

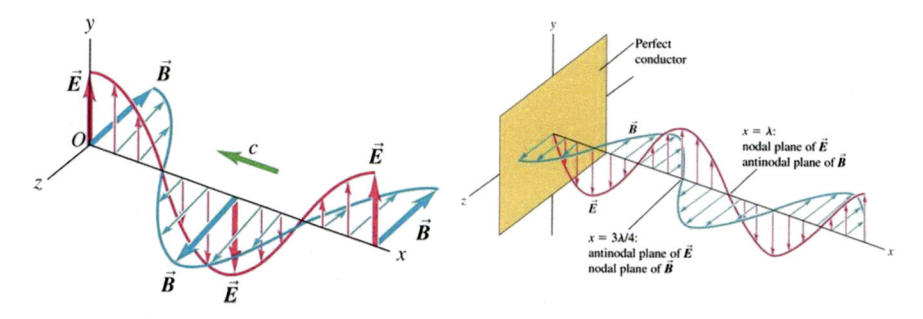

Fig. 15.3 Electromagnetic spectrum and sinusoidal EM plane waves of the solar irradiance transmitted on the surface of the glazing wall considering both nodal (B) and anti-nodal (E) surfaces

Since photon particles are primarily passed by electromagnetic waves, these waves will certainly react with other photon charges [15, 26]. The energy of the photons is hence carried out by the EM to interact with the photonic momentum of its solar irradiance [27, 28]. Consequently, the electromagnetic radiation associated with EM waves has been clarified by using MATLAB software to release the radiation continuously without any perturbation of the photonic flow [5, 29, 30]. Since the EM of visibility has lower frequencies of *nonionizing radiation*, its photonic wavelength will have enough energy needed to break the biochemical bond of the pathogens. Simply, the effects of these shortwave radiations on biochemical systems on pathogenic cells are solely the sunlight heating mechanism of the energy transfer from low-frequency photon energy, which will have enough energy to break chemical bonds and cause severe damage to pathogenic cell structures.

Killing Pathogens

Since the function of UVGI on pathogens depends on the frequency of the solar radiation and its wavelength, the formation of UVGI from solar irradiance is being controlled by a computerized *photometer* connected with a semiconductor. Thereafter, the function of EM on pathogens, including coronaviruses (COVID-19), is being analyzed using polymerase chain reaction (PCR) considering the wavelengths in the 254–280 nm spectrum to break the biomolecular binding of microorganismal DNA and/or RNA-producing thymine dimers ("T" of ATGC) to confirm their inability to function biochemically and reproduce in the host cell (Fig. 15.4).

Simply, the usage of UVGI light to form the outer glazing wall will deactivate pathogens by disrupting their DNA and/or RNA, because when DNA and/or RNA of the pathogens absorbs UVGI light, the covalent bonds among the same nucleic acids of the pathogens will be broken, and their failure to form covalent bond dimers will suppress the transcription of DNA and/or RNA, which in turn will result in the malfunction of pathogen biochemical reactions and replication and will cause the pathogens to die immediately.

Results and Discussion

UVGI Production by the Exterior Glazing Wall Skin

To mathematically calculate the generation of UVGI by the exterior glazing wall, the photon dynamics have been determined considering the functional unit area $J(\omega)$ of the outer glazing wall surface [31–33]. Since the unit area $J(\omega)$ has a consistent solar irradiance emission rate, UVGI formation from solar irradiance has a precise dynamic mode of 1D, 2D, and 3D into the outer glazing wall surface, which is calculated as the transmission of UVGI through the window as 50% penetration

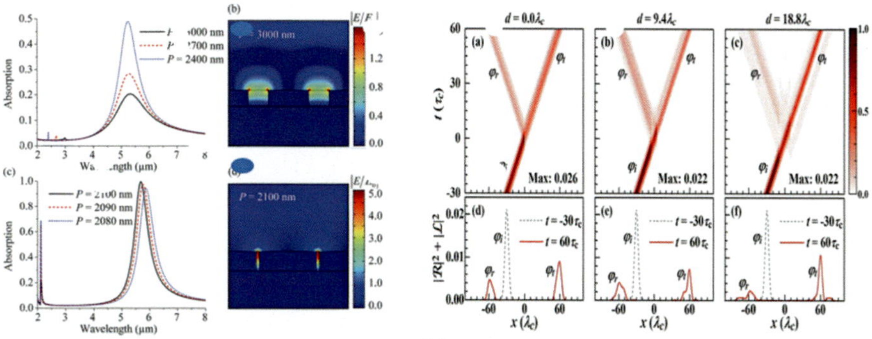

(a)

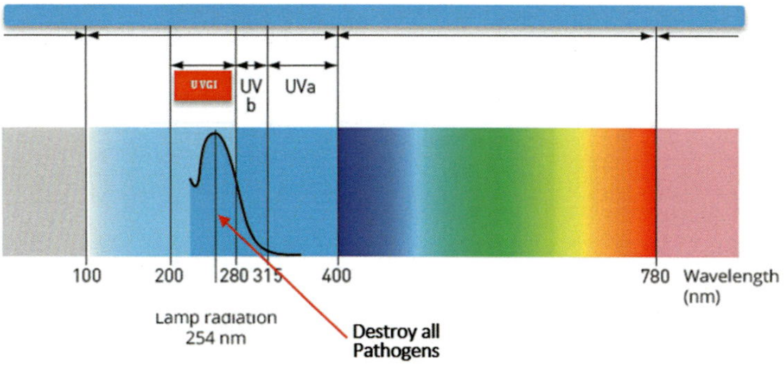

(b)

Fig. 15.4 (**a**) The photophysical radiation absorption in considering the wavelength of the photon irradiance of UVGI and (**b**) the application of UVGI radiation at 254–280 nm to eliminate the pathogens inside the building naturally

through the glazing wall. Simply, the generation of transmitted UVGI ultimately evolves from the light spectrum equilibrium of the solar irradiance penetration per unit area in relation to the photon reservoir-induced energy factor, as described in Table 15.1.

Consequently, a high-level solar irradiance frequency cut-off Ω_C is being determined from the bifurcation of the DOS in a 3D outer glazing wall surface. Subsequently, a sharpened solar irradiance high-level frequency cut-off was employed at Ω_d, maintaining the total DOS in the 2D and 1D outer glazing wall surfaces. Since the function $\mathrm{Li}_2(x)$ acts as an algorithm and $e_{rfc}(x)$ acts as an additional function, the DOS at the outer glazing wall surface, $\varrho_{PC}(\omega)$, is confirmed by the calculation of solar irradiance cut-off frequencies on the outer glazing wall surface [4, 9, 34]. The determination of the combination of the total DOS on the outer glazing wall surface is confirmed as $\varrho_{PC}(\omega) \propto \frac{1}{\sqrt{\omega - \omega_e}}\Theta(\omega - \omega_e)$, where $\Theta(\omega - \omega_e)$ denotes the variable function and ω_e denotes the solar irradiance frequency considering DOS on the outer glazing wall surface [35–37].

Table 15.1 The formation of UVGI from the solar irradiance at variable C, η, and χ functional energy emissions on the surface glazing wall in the 1D, 2D, and 3D directional modes corresponding to various DOSs per unit area $J(\omega)$ and solar irradiance induction $\Sigma(\omega)$ on the glazing wall surface

UVGI	Unit area $J(\omega)$ at various DOS solar energy	Solar irradiance induction $\Sigma(\omega)$
1D	$\frac{C-1}{\pi}\frac{1}{\sqrt{\omega-\omega_e}}\Theta(\omega-\omega_e)$	$-\frac{C-1}{\sqrt{\omega_e-\omega}}$
2D	$-\eta\left[\ln\left\|\frac{\omega-\omega_0}{\omega_0}\right\|-1\right]\Theta(\omega-\omega_e)\Theta(\Omega_d-\omega)$	$\eta\left[\mathrm{Li}_2\left(\frac{\Omega_d-\omega_0}{\omega-\omega_0}\right)-\mathrm{Li}_2\left(\frac{\omega_0-\omega_e}{\omega_0-\omega}\right)-\ln\frac{\omega_0-\omega_e}{\Omega_d-\omega_0}\ln\frac{\omega_e-\omega}{\omega_0-\omega}\right]$
3D	$\chi\sqrt{\frac{\omega-\omega_e}{\Omega_C}}\exp\left(-\frac{\omega-\omega_e}{\Omega_C}\right)\Theta(\omega-\omega_e)$	$\chi\left[\pi\sqrt{\frac{\omega_e-\omega}{\Omega_C}}\exp\left(-\frac{\omega-\omega_e}{\Omega_C}\right)\mathrm{erfc}\sqrt{\frac{\omega_e-\omega}{\Omega_C}}-\sqrt{\pi}\right]$

This DOS is, therefore, calculated from the glazing wall surface by conducting 3D analysis of the electromagnetic field (EMF) to confirm the accurate qualitative state of solar irradiance on the glazing wall surface as DOS: $\varrho_{PC}(\omega) \propto \frac{1}{\sqrt{\omega - \omega_e}}\Theta(\omega - \omega_e)$, [2, 34, 38]. To conduct the 2D and 1D analysis of the glazing wall surface, the photonic DOS was calculated from the pure algorithm of PBE, as $\varrho_{PC}(\omega) \propto -[\ln|(\omega - \omega_0)/\omega_0| - 1]\Theta(\omega - \omega_e)$, where ω_e denotes the peak algorithm, and thus, the unit area $J(\omega)$ was determined from the calculation of DOS in the glazing wall surface by the fine photonic magnitude $V(\omega)$ and expressed as [6, 39]:

$$J(\omega) = \varrho(\omega)|V(\omega)|^2 \tag{15.10}$$

Here, the PB frequency ω_c has been considered, and thus, the photon dynamics have been used as the function $u(t, t_0)$ of the UVGI proliferation structure in the relationship $\langle a(t)\rangle = u(t, t_0)\langle a(t_0)\rangle$, which was thus confirmed using the integral differential calculation given in Eq. (15.10):

$$u(t, t_0) = \frac{1}{1 - \Sigma'(\omega_b)} e^{-i\omega(t - t_0)} + \int_{\omega_e}^{\infty} d\omega \frac{J(\omega)e^{-i\omega(t - t_0)}}{[\omega - \omega_c - \Delta(\omega)]^2 + \pi^2 J^2(\omega)} \tag{15.11}$$

where $\Sigma'(\omega_b) = [\partial\Sigma(\omega)/\partial\omega]_{\omega = \omega_b}$ and $\Sigma(\omega)$ denotes the solar irradiance induction:

$$\Sigma(\omega) = \int_{\omega_e}^{\infty} d\omega' \frac{J(\omega')}{\omega - \omega'} \tag{15.12}$$

Here, the frequency ω_b in Eq. (15.2) represents the UVGI frequency mode in the PBG ($0 < \omega_b < \omega_e$) and is calculated using the pole condition $\omega_b - \omega_c - \Delta(\omega_b) = 0$, where $\lesssim\Delta(\omega) = \mathcal{P}\left[\int d\omega' \frac{J(\omega')}{\omega - \omega'}\right]$ is a principal-value integral.

Subsequently, the detailed UVGI spectrum, considering the emission magnitude $|u(t, t_0)|$, on the glazing wall surface was calculated and is shown in Table 15.1 for 1D, 2D, and 3D considering the variable detuning integration of solar irradiance [11, 38, 39]. The photonic spectrum deliberation rate is shown in Fig. 15.5, which reveals that photonic UVGIs are released at a modest rate once they cross the surface of the glazing wall, corresponding to the observation that the UVGI release rates within the glazing wall surface and in the vicinity of the glazing wall surface are almost exponential [4, 18, 40]. Therefore, the curb line of photon intensity sharply increases with respect to the time on the surface area, which confirms that this UVGI proliferation of dynamics confirms that the DOS rate in the vicinity of the glazing wall surface is in an equilibrium modest rate (Fig. 15.5).

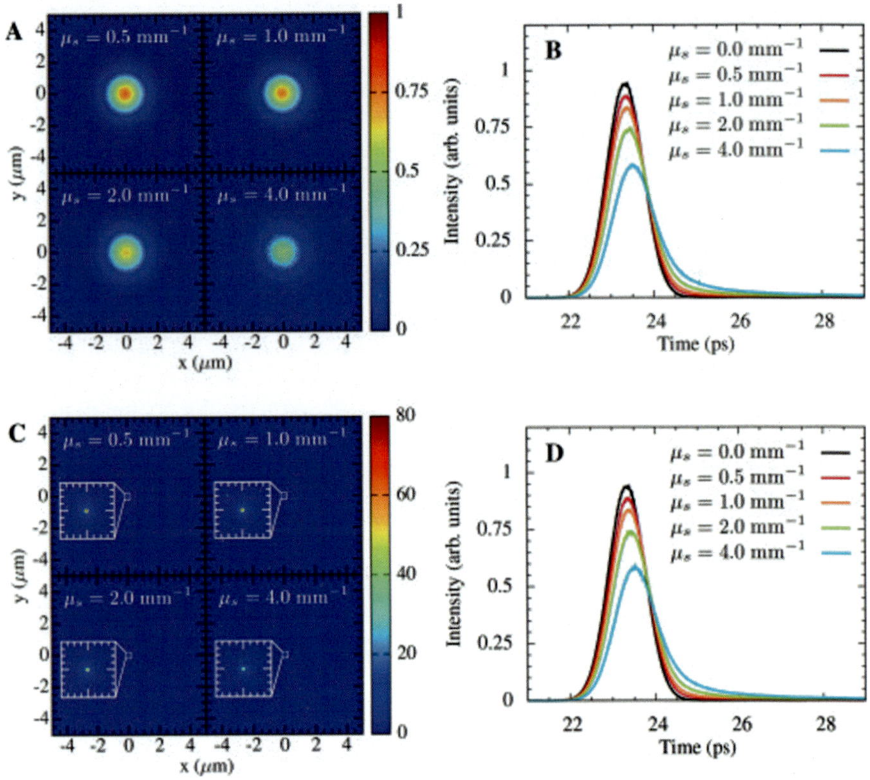

Fig. 15.5 The UVGI release rate shown (**a, c**) from the photon irradiance generation rate on the surface area (μm) and (**b, d**) the DOS of the photon intensity (*arb .units*) at the time (ps) on the surface of the outer glazing wall

Then, the UVGI spectrum dynamics on the glazing wall surface are determined considering thermal variation with respect to the photon correlation function $v(t, t)$ by determining the nonequilibrium photon scattering theorem [14, 41]:

$$v(t,t) = \int_{t_0}^{t} dt_1 \int_{t_0}^{t} dt_2 u^*(t_1, t_0)\tilde{g}(t_1, t_2)u(t_2, t_0) \tag{15.13}$$

Here, the time-related function $\tilde{g}(t_1, t_2) = \int d\omega J(\omega)\overline{n}(\omega, T)e^{-i\omega(t-t')}$ reveals the photonic UVGI release initiated by the surface relativistic condition of the glazing wall, where $\overline{n}(\omega, T) = 1/\left[e^{\hbar\omega/k_B T} - 1\right]$ is the proliferation of the UVGI emission in the glazing wall surface at temperature T and expressed as:

$$v(t,t\rightarrow\infty)=\int_{\omega_e}^{\infty}d\omega\mathcal{V}(\omega)\quad\text{with }\mathcal{V}(\omega)=\bar{n}(\omega,T)[\mathcal{D}_l(\omega)+\mathcal{D}_d(\omega)]\tag{15.14}$$

Here, Eq. (21) is being simplified to determine the nonequilibrium condition: $\mathcal{V}(\omega)=\bar{n}(\omega,T)\mathcal{D}_d(\omega)$. Under low-temperature conditions, Einstein's photon fluctuation dissipation is not functionally viable at the photonic band (PB) but connects the within its band gap to release UVGI, which is measurable as $n(t)=\langle a^{\dagger}(t)a(t)\rangle=|u(t,t_0)|^2 n(t_0)v(t,t)$, where $n(t_0)$ represents the primary PB [3, 8, 14]. Therefore, in Fig. 15.5b, the rate of release of UVGI versus temperature is plotted to confirm nonequilibrium proliferated photon energy, as shown by the solid blue curve. Therefore, PB has been considered as the Fock state as the UVGI release rate n_0, i. e. $\rho(t_0)=|n_0\rangle\langle n_0|$ at time t and is expressed as:

$$\rho(t)=\sum_{n=0}^{\infty}\mathcal{P}_n^{(n_0)}(t)|n_0\rangle\langle n_0|\tag{15.15}$$

$$\mathcal{P}_n^{(n_0)}(t)=\frac{[v(t,t)]^n}{[1+v(t,t)]^{n+1}}[1-\Omega(\text{t})]^{n_0}\times\sum_{k=0}^{\min\{n_0,n\}}\binom{n_0}{k}\binom{n}{k}\left[\frac{1}{v(t,t)}\frac{\Omega(\text{t})}{1-\Omega(\text{t})}\right]^k\tag{15.16}$$

where $\Omega(\text{t})=\dfrac{|u(t,t_0)|^2}{1+v(t,t)}$.

Simply, this result revealed that a Fock state photon acts as the catalyst to release photon energy at various $|n_0\rangle is\ \mathcal{P}_n^{(n_0)}(t)$ values, and thus, the UVGI release has been determined as $\mathcal{P}_n^{(n_0)}(t)$ as the primary state $|n_0=5\rangle$ and steady-state limit $\mathcal{P}_n^{(n_0)}(t\rightarrow\infty)$ and expressed as:

$$\mathcal{P}_n^{(n_0)}(t\rightarrow\infty)=\frac{[\bar{n}(\omega_c,T)]^n}{[1+\bar{n}(\omega_c,T)]^{n+1}}\tag{15.17}$$

To probe this UVGI production in the outer glazing wall surface, it has been further broken down through extreme coherent states considering the proliferation state of UVGI and expressed as:

$$\rho(t)=\mathcal{D}[\alpha(t)]\rho_T[v(t,t)]\mathcal{D}^{-1}[\alpha(t)]\tag{15.18}$$

where $\mathcal{D}[\alpha(t)]=\exp\{\alpha(t)a^{\dagger}-\alpha^*(t)a\}$ denotes the displacement function of $\alpha(t)=u(t,t_0)\alpha_0$ and:

$$\rho_T[v(t,t)]=\sum_{n=0}^{\infty}\frac{[v(t,t)^n]}{[1+v(t,t)]^{n+1}}\,|\,n\rangle\langle n|\tag{15.19}$$

Here, ρ_T denotes a standard state photon particle $v(t,t)$, where Eq. (15.19) confirms the net UVGI generation from the glazing wall surface EM radiation.

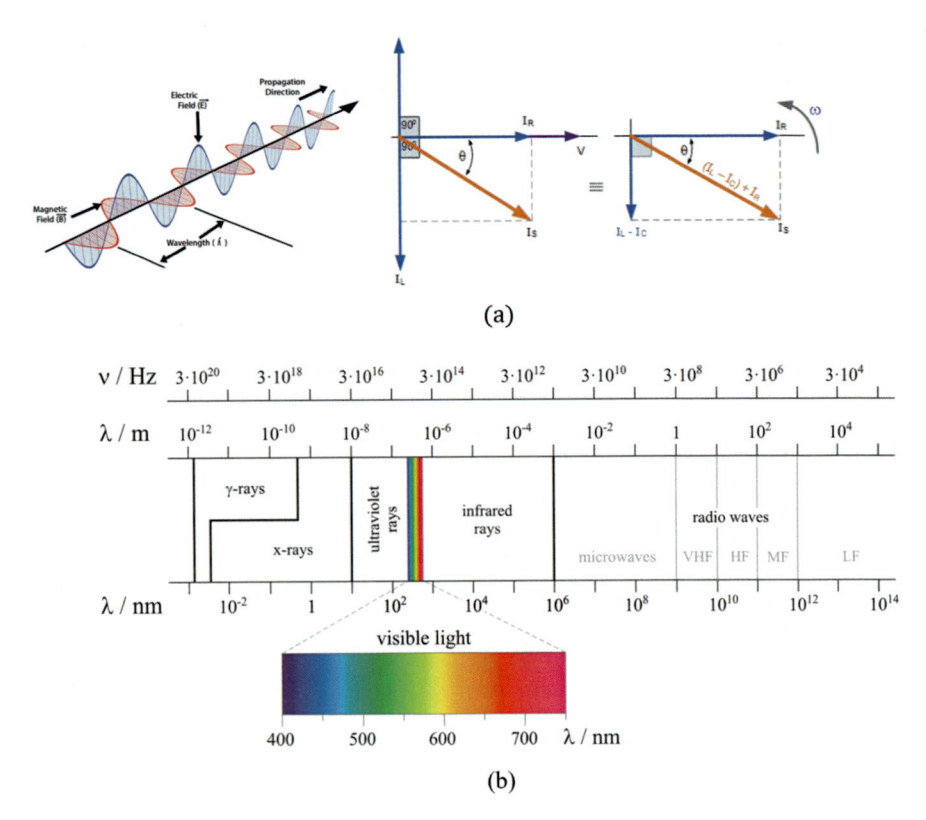

(a)

(b)

Fig. 15.6 (**a**) The formation of photonic electromagnetic wave considering the wavelength and magnetic and electric field of the photon irradiance on the glazing wall surface. (**b**) The EM radiation of UVGI where the photons electron get excited to damage the DNA and/or RNA of all pathogens by the ultraviolet ray of 254–280 nm

Electromagnetic Radiation

Then, this EM radiation is clarified by the wavelength of UVGI sinusoidal mono-chromatic waves, which in turn are classified from the EM spectrum of the photon particles. Thus, the EM waveform is calculated using the spectral analysis of photonic frequency considering their *power* content, which is the spectral density of the solar irradiance. Thus, the random electromagnetic radiation of UVGI is determined from the wide band forms of solar radiation from the unique sinusoidal wave field of the solar irradiance (Fig. 15.6).

Killing Pathogens

Since the EM field of the solar irradiance on the exterior glazing wall surface emitted a light spectrum between 254 and 280 nm, the released electromagnetic field from the glazing wall allowed the addition of a mass-term solar irradiance transformation into UVGI to target the chemical bond of the pathogens. Hence, the DNA and/or RNA of the pathogens are the components of the chemical bond of adenine (A), thymine (T), guanine (G), and cytosine (C), and these bonds are very sensitive to UVGI; thus, UVGI interacts directly with these bonds of DNA and/or RNA and subsequently damages their cyclobutane pyrimidine dimers (CPD) and 6–4 pyrimidine-pyrimidone photoproducts (6-4PPs) and its Dewar isomers (Fig. 15.7), leading to the development of an unstable oxetane or azetidine intermediate end base of the thymine or cytosine dimers [1, 34, 39]. Consequently, spontaneous rearrangement of these intermediate ends gives rise to 6-4PP, and eventually, the pyrimidine dimers caused a kink into the DNA strands, stopping transcription and protein synthesis and causing pathogen damage immediately, which was estimated

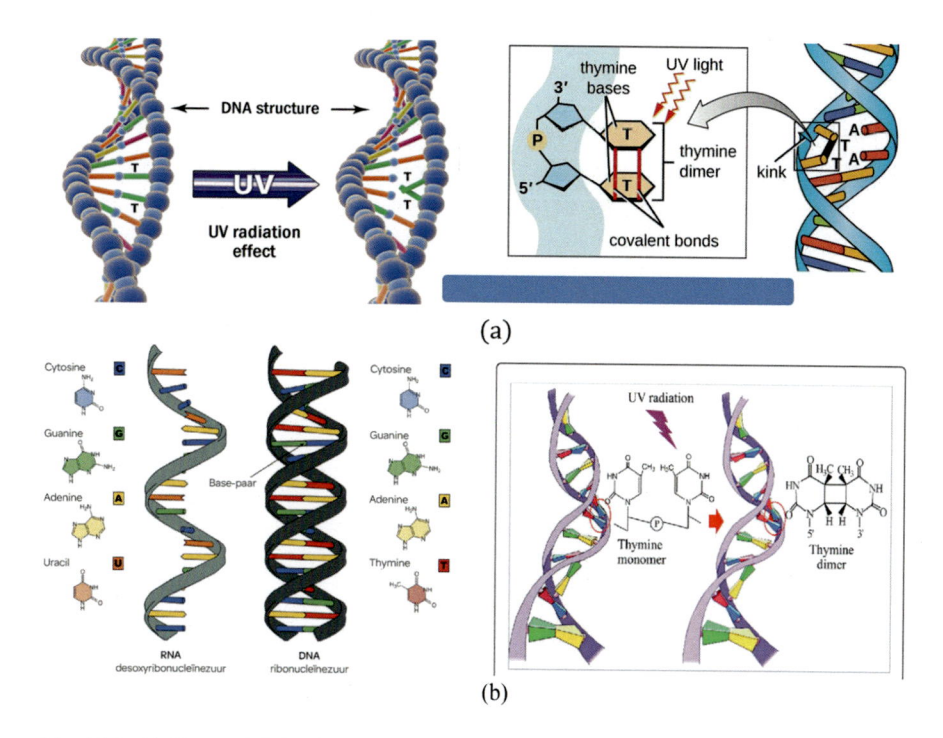

Fig. 15.7 (**a**) How UV light affects on DNA structure, (**b**) detailed RNA and DNA structure of pathogens, and (**c**) UV radiation for disinfection involvement to generate desired germicidal mechanism to destroy the thymine dimer of the pathogens to ultimately disable their biochemical function and reproduction

Table 15.2 The logarithmic study of the degree of pathogen damage in response to UVGI irradiance was conducted by using MATLAB software, which shows that various factors, including COVID-19, are more effective than bacteria in responding to UVGI

Name of pathogens	Irradiance time (s)	Degree of pathogen damage				
		$1\ m^2$ area	$2\ m^2$ area	$3\ m^2$ area	$4\ m^2$ area	$5\ m^2$ area
Bacteria (average)	1	2.7×10^1	5.4×10^1	2.7×10^2	5.4×10^2	2.7×10^3
	10	2.7×10^2	5.4×10^2	2.7×10^3	5.4×10^3	2.7×10^4
	20	2.7×10^3	5.4×10^3	2.7×10^4	5.4×10^4	2.7×10^5
	30	2.7×10^4	5.4×10^4	2.7×10^5	5.4×10^5	2.7×10^6
	60	2.7×10^8	5.4×10^8	2.7×10^{10}	5.4×10^{10}	2.7×10^{12}
Viruses (average)	1	3.1×10^1	3.1×10^1	3.1×10^2	5.4×10^2	2.7×10^4
	10	3.1×10^2	3.1×10^2	3.1×10^3	5.4×10^3	2.7×10^5
	20	3.1×10^3	3.4×10^3	3.1×10^4	5.4×10^4	2.7×10^7
	30	3.1×10^4	3.1×10^4	3.1×10^5	5.4×10^5	2.7×10^7
	60	3.1×10^8	3.1×10^8	3.1×10^{10}	5.4×10^{10}	2.7×10^{15}
COVID-19	1	2.1×10^1	3.1×10^1	3.1×10^2	5.4×10^2	2.7×10^4
	10	2.1×10^2	3.1×10^2	3.1×10^3	5.4×10^3	2.7×10^5
	20	2.1×10^3	3.4×10^3	3.1×10^4	5.4×10^4	2.7×10^7
	30	2.1×10^4	3.1×10^4	3.1×10^5	5.4×10^5	2.7×10^7
	60	2.1×10^8	3.1×10^8	3.1×10^{10}	5.4×10^{10}	2.7×10^{16}

considering the logarithmic value in response to irradiance time and degree of pathogen damage calculated by using MATLAB software (Table 15.2).

Simply, the application of UVGI radiation to the pathogen caused severe damage to DNA and/or RNA via the absorption of photons by DNA chromophores, which eventually generated the reactive oxygen charge that oxidized the DNA bases, prohibiting the complete replication and mutation process (Fig. 15.7). Thus, the application of UVGI to pathogens suggests that UVGI is very functional in disrupting the DNA and/or RNA bonding of pathogens, including COVID-19, to eventually kill them.

Conclusions

The research has incorporated the use of irradiation in an advanced building design technology that involves the application of photophysical reaction through an outer glazing wall surface forming short-range wavelengths of 254–280 nm of *ultraviolet germicidal irradiation* (UVGI) released by the sun to kill pathogens inside the building. The technology explored the use of short ultraviolet wavelengths of UVGI as being successful in the denaturation of pathogen nucleic acid functions. Since some of these pathogens, such as COVID-19, have mutation characteristics, making the effectiveness of vaccines very difficult, this research identified the use of a prevention technology instead of treatment technology that is much more effective. Hence, the study of solar irradiance on the outer glazing wall of the building is being controlled by using a semiconductor to form a 254–280 nm wavelength of UVGI to kill the pathogen naturally by destroying its nucleic acid bonds. Simply, denaturing the nucleic acids (thymine dimers) of the pathogen using short wavelengths of 254–280 nm of the UVGI formed from the outer glazing wall to damage the DNA and/or RNA structure of the pathogens will indeed be an interesting field of building design science to eradicate these dangerous pathogens before they attack the human body.

Acknowledgments The author assures that the research has no conflict of interest and can be published in any suitable journal. Any findings, predictions, and conclusions described in this article are solely those of the author.

References

1. Grinin, A., Matveev, A., Yost, D. C., Maisenbacher, L., Wirthl, V., Pohl, R., Hänsch, T. W., & Udem, T. (2020). Two-photon frequency comb spectroscopy of atomic hydrogen. *Science, 370*, 1061–1066. https://doi.org/10.1126/science.abc7776
2. Reinhard, A., Volz, T., Winger, M., Badolato, A., Hennessy, K. J., Hu, E. L., & Imamoğlu, A. (2012). Strongly correlated photons on a chip. *J. Healthc. Eng.Nat. Photonics, 6*, 93–96. https://doi.org/10.1038/nphoton.2011.321
3. Sayrin, C., Dotsenko, I., Zhou, X., Peaudecerf, B., Rybarczyk, T., Gleyzes, S., Rouchon, P., Mirrahimi, M., Amini, H., Brune, M., Raimond, J.-M., & Haroche, S. (2011). Real-time quantum feedback prepares and stabilizes photon number states. *Nature, 477*, 73–77. https://doi.org/10.1038/nature10376
4. Lei, C. U., & Zhang, W.-M. (2012). A quantum photonic dissipative transport theory. *Ann. Phys., 327*, 1408–1433. https://doi.org/10.1016/j.aop.2012.02.005
5. O'Shea, D., Junge, C., Volz, J., & Rauschenbeutel, A. (2013). Fiber-optical switch controlled by a single atom. *Physical Review Letters, 111*, 193601. https://doi.org/10.1103/PhysRevLett.111.193601
6. Armani, D. K., Kippenberg, T. J., Spillane, S. M., & Vahala, K. J. (2003). Ultra-high-Q toroid microcavity on a chip. *Nature, 421*, 925–928. https://doi.org/10.1038/nature01371
7. Englund, D., Majumdar, A., Faraon, A., Toishi, M., Stoltz, N., Petroff, P., & Vucković, J. (2010). Resonant excitation of a quantum dot strongly coupled to a photonic crystal nanocavity. *Physical Review Letters, 104*, 073904. https://doi.org/10.1103/PhysRevLett.104.073904
8. Roy, D. (2013). Two-photon scattering of a tightly focused weak light beam from a small atomic ensemble: An optical probe to detect atomic level structures. *Physical Review A, 87*, 063819. https://doi.org/10.1103/PhysRevA.87.063819. Hou, M., Pantelic, J., Aviv, D. (2021). Spatial analysis of the impact of UVGI technology in occupied rooms using ray-tracing simulation. *Indoor Air 31*, 1625–1638. https://doi.org/10.1111/ina.12827
9. Saloux, E., Teyssedou, A., & Sorin, M. (2011). Explicit model of photovoltaic panels to determine voltages and currents at the maximum power point. *Solar Energy, 85*, 713–722. https://doi.org/10.1016/j.solener.2010.12.022
10. Wright, J. B., Li, Q., Brener, I., Luk, T. S., Wang, G. T., Chow, W. W., Lester, L. F. (2012). *Gallium nitride single-mode nanowire lasers*. In: Conference on Lasers and Electro-Optics Optical Society of America, Sydney, Australia, pp. CTh4M–5.

11. Yan, G. A., Lu, H., & Chen, A.-X. (2016). Single-photon router: Implementation of information-holding of quantum states. *International Journal of Theoretical Physics, 55*, 3366–3374. https://doi.org/10.1007/s10773-016-2965-3
12. Baur, G., Hencken, K., Trautmann, D., Sadovsky, S., & Kharlov, Y. (2002). Dense laser-driven electron sheets as relativistic mirrors for coherent production of brilliant X-ray and γ-ray beams. *Physics Reports, 364*, 359–450. https://doi.org/10.1016/S0370-1573(01)00101-6
13. Zhang, G., & Pan, Y. (2020). On the dynamics of two photons interacting with a two-qubit coherent feedback network. *Automatica, 117*, 108978. https://doi.org/10.1016/j.automatica.2020.108978
14. Luo, H., & Zhong, L. (2021). Ultraviolet germicidal irradiation (UVGI) for in-duct airborne bioaerosol disinfection: Review and analysis of design factors. *Build Environment, 197*, 107852. https://doi.org/10.1016/j.buildenv.2021.107852
15. Chen, J., Wang, C., Zhang, R., & Xiao, J. (2012). Multiple plasmon-induced transparencies in coupled-resonator systems. *Optics Letters, 37*, 5133–5135. https://doi.org/10.1364/OL.37.005133
16. Huang, J. F., Shi, T., Sun, C. P., & Nori, F. (2013). Controlling single-photon transport in waveguides with finite cross section. *Physical Review A, 88*, 013836. https://doi.org/10.1103/PhysRevA.88.013836
17. Hou, J., Wang, H., & Liu, P. (2018). Applying the blockchain technology to promote the development of distributed photovoltaic in China. *International Journal of Energy Research, 42*, 2050–2069. https://doi.org/10.1002/er.3984
18. Douglas, J. S., Habibian, H., Hung, C. L., Gorshkov, A. V., Kimble, H. J., & Chang, D. E. (2015). Quantum many-body models with cold atoms coupled to photonic crystals. *Nature Photonics, 9*, 326–331. https://doi.org/10.1038/nphoton.2015.57
19. Liao, J. Q., & Law, C. K. (2010). Correlated two-photon transport in a one-dimensional waveguide side-coupled to a nonlinear cavity. *Physical Review A, 82*, 053836. https://doi.org/10.1103/PhysRevA.82.053836
20. Birnbaum, K. M., Boca, A., Miller, R., Boozer, A. D., Northup, T. E., & Kimble, H. J. (2005). Photon blockade in an optical cavity with one trapped atom. *Nature, 436*, 87–90. https://doi.org/10.1038/nature03804
21. Hossain, M. F. (2018). Photon energy amplification for the design of a micro PV panel. *International Journal of Energy Research, 42*, 3861–3876. https://doi.org/10.1002/er.4118
22. Hossain, M. F. (2019). Sustainable technology for energy and environmental benign building design. *Journal of Building Engineering, 22*, 130–139. https://doi.org/10.1016/j.jobe.2018.12.001
23. Hossain, M. F. (2018). Transforming dark photons into sustainable energy. *International Journal of Energy and Environmental Engineering, 9*, 99–110. https://doi.org/10.1007/s40095-017-0257-1
24. Hossain, M. F. (2018). Green science: Advanced building design technology to mitigate energy and environment. *Renewable and Sustainable Energy Reviews, 81*, 3051–3060. https://doi.org/10.1016/j.rser.2017.08.064
25. Hossain, M. F. (2017). Design and construction of ultra-relativistic collision PV panel and its application into building sector to mitigate total energy demand. *Journal of Building Engineering, 9*, 147–154. https://doi.org/10.1016/j.jobe.2016.12.005
26. Hossain, M. F. (2017). Green science: Independent building technology to mitigate energy, environment, and climate change. *Renewable and Sustainable Energy Reviews, 73*, 695–705. https://doi.org/10.1016/j.rser.2017.01.136
27. Hossain, M. F. (2016). Solar energy integration into advanced building design for meeting energy demand and environment problem. *International Journal of Energy Research, 40*, 1293–1300. https://doi.org/10.1002/er.3525
28. Hou, M. M., Pantelic, J., & Aviv, D. (2021). Spatial analysis of the impact of UVGI technology in occupied rooms using ray-tracing simulation. *Indoor Air, 31*, 1625–1638.

29. Cheng, M. T., & Song, Y. Y. (2012). Fano resonance analysis in a pair of semiconductor quantum dots coupling to a metal nanowire. *Optics Letters, 37*, 978–980. https://doi.org/10.1364/ol.37.000978
30. Artemyev, N., Jentschura, U. D., Serbo, V. G., & Surzhykov, A. (2012). Strong electromagnetic field effects in ultra-relativistic heavy-ion collisions. *European Physical Journal C: Particles and Fields, 72*, 1935. https://doi.org/10.1140/epjc/s10052-012-1935-z
31. Zhong, N., Dai, Q., Liang, R., Li, X., Tan, X., Zhang, X., Wei, Z., Wang, F., Liu, H., & Meng, H. (2017). Analogue of electromagnetically induced absorption with double absorption windows in a plasmonic system. *PLoS One, 12*, e0179609. https://doi.org/10.1371/journal.pone.0179609
32. Longo, P., Schmitteckert, P., & Busch, K. (2011). Few-photon transport in low-dimensional systems. *Physical Review A, 83*, 063828. https://doi.org/10.1103/PhysRevA.83.063828
33. Kolchin, P., Oulton, R. F., & Zhang, X. (2011). Nonlinear quantum optics in a waveguide: Distinct single photons strongly interacting at the single atom level. *Physical Review Letters, 106*, 113601. https://doi.org/10.1103/PhysRevLett.106.113601
34. Doumane, R., Balistrou, M., Logerais, P. O., Riou, O., Durastanti, J. F., & Charki, A. (2015). A circuit-based approach to simulate the characteristics of a silicon photovoltaic module with aging. *Journal of Solar Energy Engineering, 137*, 021020. https://doi.org/10.1115/1.4029541
35. Maxwell, R. M., & Condon, L. E. (2016). Connections between groundwater flow and transpiration partitioning. *Science, 353*, 377–380. https://doi.org/10.1126/science.aaf7891
36. Huang, Y., Min, C., & Veronis, G. (2011). Subwavelength slow-light waveguides based on a plasmonic analogue of electromagnetically induced transparency. *Applied Physics Letters, 99*, 143117. https://doi.org/10.1063/1.3647951
37. S. Xue, Petersen, I. R. (2015). *A root locus approach to constructing Markovian coupled oscillators to capture the dynamics of a non-Markovian quantum system*. In: 2015 10th Asian Control Conference (ASCC), ASCC, Kota Kinabalu, Malaysia, pp. 1–6.
38. Pregnolato, T., & Song, E. (2015). Single-photon non-linear optics with a quantum dot in a waveguide. *Nature Communications, 6*, 86–95.
39. Shi, T., Fan, S., & Sun, C. P. (2011). Two-photon transport in a waveguide coupled to a cavity in a two-level system. *Physical Review A, 84*, 063803. https://doi.org/10.1103/PhysRevA.84.063803
40. Wang, Y., Zhang, Y., Zhang, Q., Zou, B., & Schwingenschlogl, U. (2016). Dynamics of single photon transport in a one-dimensional waveguide two-point coupled with a Jaynes-Cummings system. *Scientific Reports, 6*, 33867. https://doi.org/10.1038/srep33867
41. Xiao, Y., Meng, C., Wang, P., Ye, Y., Yu, H., Wang, S., Gu, F., Dai, L., & Tong, L. (2011). Single-nanowire single-mode laser. *Nano Letters, 11*, 1122–1126. https://doi.org/10.1021/nl1040308

Chapter 16
Theory of Sustainable Design of Time and Space

Introduction

There is a misconception about the perception of the climate condition of the time and space in the multiverse, which is vulnerable to sustain any life [4, 12, 23]. In fact, this knowledge gap is nothing but mystery since no one had done any research about the climate condition of the time and space considering the multiverse to confirm that it is having vulnerable climate condition to sustain lives and, thus, the problem remains unresolved; therefore, the myth remains a controversy. In this research, a novel mathematical modeling is being proposed primary on probing the existence of the light energy (L_e), dark energy (D_e), and antimatter (A_m) in the time and then, secondarily, how energies of the time and space can be decoded from its energy momentum to cool and heat the multiverse naturally at a comfort level to sustain lives. Simply, the energy momentum of the time and space is being modeled to absorb its energy band edges (PBE) by its nano-point break waveguides by using its QED in order to form cool-state energy to cool the time and space [19, 30]. Subsequently, this cool-state energy is also being modeled to convert it into heat-state energy via the energy radiations (PR) irradiated by its quantum of Higgs-Bosons ($H \rightarrow \gamma\gamma^{-}$) by the induction of its electromagnetic fields to heat the time and space, respectively [11, 13, 15]. This cooling and heating conversion mechanism of the time and space shall indeed be a noble scientific discovery ever to control the climate condition of any time and space in the multiverse that will confirm to sustain lives up there.

M. F. Hossain, *Global Sustainability*, https://doi.org/10.1007/978-3-031-34575-3_16

Methods and Simulation

Computation of Light Energy (L_e)

The computation of light energy considering the sphericity and the orbital parameters of the time and space, a mathematical modeling is being conducted to initiate the presence of light energy in the time and space by expressing law of cosines as:

$$\rho \cos(c) = \rho \cos(a) \cos(b) + \rho \sin(a) \sin(b) \cos(C) \tag{16.1}$$

Here *a*, *b*, and *c* represent as the arc lengths in radians corresponding to the sides of a spherical triangle where *C* represents the angle of the vertex corresponding to the other side of *arc* with the it length *c* [8, 24, 46]. Naturally, considering the implementation of solar zenith angle Θ, the following terms can be written as the laws of cosines: $C = h = r$; $c = \Theta$; $a = \frac{1}{2}\pi - \phi$; $b = \frac{1}{2}\pi - \delta$, and thus, the equation can be expressed as:

$$\rho \cos(\Theta) = \sin(\phi) \sin(\delta) + \cos(\phi) \cos(\delta) \cos(r) \tag{16.2}$$

In order to modify this calculation, a furthered clarification has been conducted considering a general derivative as below:

$$\begin{aligned} \rho \cos(\theta) = {} & \rho \sin(\phi) \sin(\delta) \cos(\beta) \\ & + \sin(\delta) \cos(\phi) \sin(\beta) \cos(\gamma) + \cos(\phi) \cos(\delta) \cos(\beta) \cos(h) \\ & - \cos(\delta) \sin(\phi) \sin(\beta) \cos(\gamma) \cos(h) - \cos(\delta) \sin(\beta) \sin(\gamma) \sin(h) \end{aligned} \tag{16.3}$$

Here, β represents the angle corresponding to the horizon and γ corresponding to the azimuth angle and, thus, the sphere of any time and space in the multiverse, which is originated from its energy dynamics, denoted here R_E and the mean distance is denoted as R_O represents here as the average distant corresponding of astronomical unit (*a.u.*) and the energy constant is named here S_O [7, 14, 36]. Thus, the energy flux onto the plane tangent of the sphere of the time and space is calculated as:

$$Q = \begin{cases} S_o \dfrac{R_0^2}{R_E^2} \cos(\theta) & \cos(\theta) > 0 \\ 0 & \cos(\theta) \le 0 \end{cases} \tag{16.4}$$

Here, the mean Q over a time is the mean value of Q in relation to its rotation within the time angle of $h = \pi$ to $h = -\pi$, and thus, calculation can be expressed as:

$$Q^{-\text{time}} = -\frac{1}{2\pi}\int_{\pi}^{-\pi} Q dh \tag{16.5}$$

Since h_0 is the time-related angle once Q is the positive, it will be feasible during the light energy presence duration in the time and space when $\Theta = 1/2\,\pi$ or for h_O, and thus the equation can be solved by:

$$\rho \sin(\phi)\sin(\delta) + \cos(\phi)\cos(\delta)\cos\left(h_{0)}\right) = 0 \tag{16.6}$$

or

$$\rho \cos(h_0) = -\tan(\phi)\tan(\delta) \tag{16.7}$$

Since $\tan(\varphi)\tan(\delta) > 1$, the light energy will confirm its existence at $h = \pi$, so $h_o = \pi$. Subsequently, in a Q^{-} time $= 0\frac{R_0^2}{R_E^2}$ would remain mostly constant during the period of light energy presence in a specific space in the multiverse, which can be expressed by considering the following integral:

$$\begin{aligned}\int_{\pi}^{-\pi} Q dh = \int_{h_0}^{-h_0} Q dh = S_0\frac{R_0^2}{R_E^2}\int_{h_0}^{-h_0} \cos(\theta) dh \\ = S_0\frac{R_0^2}{R_E^2}\left[h \sin(\phi)\sin(\delta) + \cos(\phi)\cos(\delta)\sin(h)\right]_{h=h_0}^{h=-h_0} \\ = -2S_0\frac{R_0^2}{R_E^2}\left[h_0 \sin(\phi)\sin(\delta) + \cos(\phi)\cos(\delta)\sin\left(h_{0)}\right)\right]\end{aligned} \tag{16.8}$$

Therefore:

$$Q^{-\text{time}} = \frac{S_0}{\pi}\frac{R_0^2}{R_E^2}\left[h_0 \sin(\phi)\sin(\delta) + \cos(\phi)\cos(\delta)\sin\left(h_{0)}\right)\right] \tag{16.9}$$

Since the θ is representing here as mainstream angle that describes the orbit of the time and space, $\theta = 0$ corresponding to its vernal equinox of δ at its orbital position that can be written as:

$$\delta = \varepsilon \sin(\theta) \tag{16.10}$$

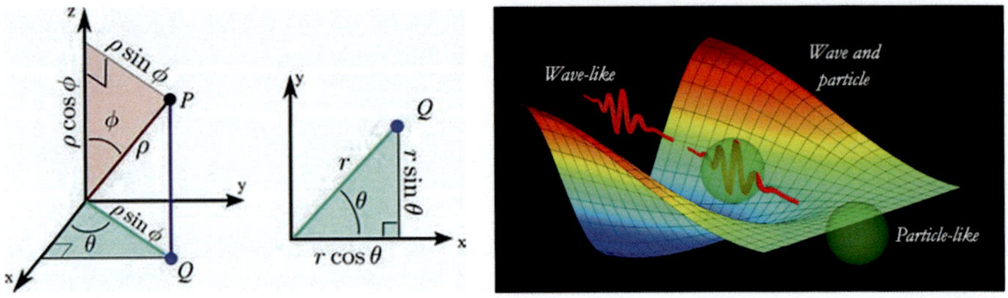

(a) (b)

Fig. 16.1 The distribution of light energy in the time and space, (**a**) clarify the presence of light photon with respect law of cosines, (**b**) schematic diagram of the light energy dynamic on the time and space's sphericity and orbital parameter

Here ε represents the mainstream longitude of sphere ϖ that is related to the vernal equinox, and thus, the equation can be written as:

$$R_E = \frac{R_0}{1 + e\cos(\theta - \omega)} \tag{16.11}$$

or

$$\frac{R_0}{R_E} = 1 + e\cos(\theta - \omega) \tag{16.12}$$

Given the above clarification of ϖ, ε, s_o, and e from astrodynamical computation, light energy L_e can be computed for any latitude φ and θ of the time and space (Fig. 16.1). Here, $\theta = 0°$ is representing the precise time of the vernal equinox, $\theta = \infty$ is representing the precise space of the equinox, and thus, the equation can be simplified for light energy irradiance presence on a given time in the space that can be expressed as:

$$L_p = S_0\left(1 + \cos\left(2\pi\frac{n}{x}\right)\right) \tag{16.13}$$

Here S_o is the energy constant, n is the unit of time, x is the duration of time, and π is the sphere of the space that optimized the generation of light energy (L_e) in the time and space, respectively.

Computation of Dark Energy ($\mathbf{D_e}$)

Since the dark energy is a hidden particle, which is a force resemble to the light energy, the particle of the dark energy is being clarified considering their quantum electrodynamics (QED) by analyzing its minimal way of new U (1) gauge field considering its vectorial (V) dynamics of energy of its tangential (T) field of unknown Z dimension of the time and space [1, 9, 10]. Here, the new U (1) gauge symmetry and its kinetic energy proliferation corresponding to the dark energy field, the *Standard Model Hypercharge Field*, are being computed in order to confirm the determination of striking electron flow from the dark energy, which will initial to probe the existence of this energy (Fig. 16.2). After striking the electron, the dark energy ($\overrightarrow{De}$) will be converted to a regular light energy and scatters at an angle θ with speed of c, which is being expressed as the real fermion of the energy momentum of four vectors of each constituent particle of the time and space:

$$\overrightarrow{De} = \left| \frac{Edp}{c}, \ \overrightarrow{Pd} \right| + \overrightarrow{Xef} \left[\frac{Eef}{c}, \overrightarrow{Pef} \right] \tag{16.14}$$

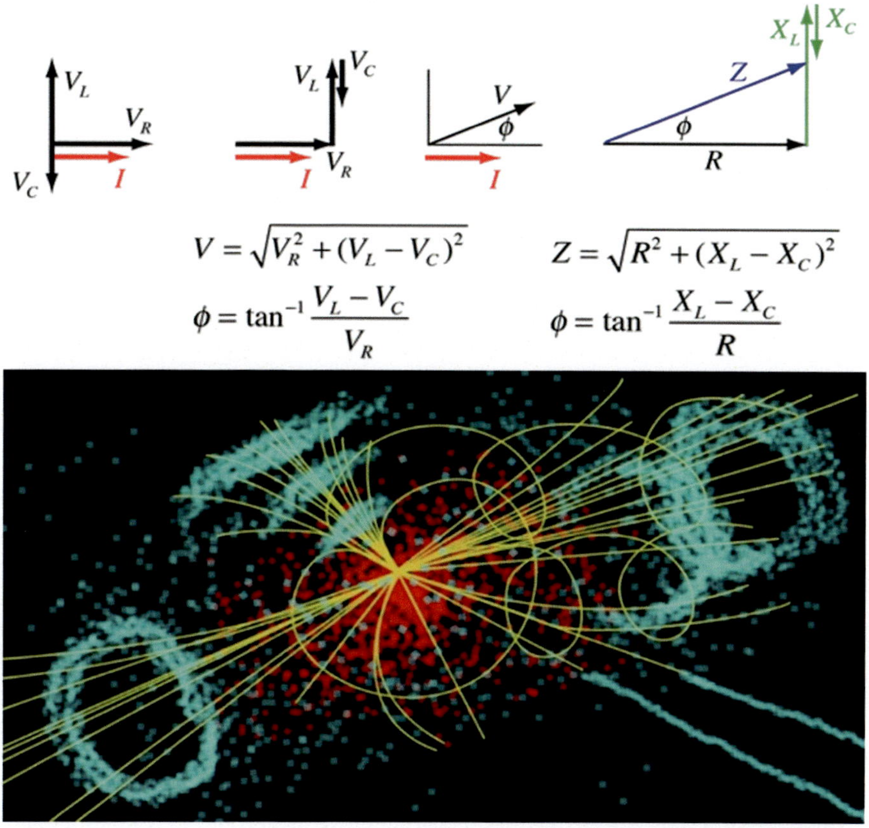

Fig. 16.2 The basic model of dark energy U (1) gauge symmetry in the time and space corresponding to its vectorial dynamics of tangential striking electron within the kinetic energy field parameters

Since energy momentum is conserved and thus, $\Delta P = 0$, therefore, the sum of the initial momentums must be equal to the sum of the final momentum [44], thus, it can be rewritten the sum of the four vectors as:

$$\overrightarrow{De} + \overrightarrow{Xei} = \overrightarrow{Xp} + \overrightarrow{Xef} \tag{16.15}$$

Since the isolated momentums of the outgoing electron are being calculated and thus the square of both sides of the equation can be rewritten as:

$$\left(\overrightarrow{De} + \overrightarrow{Xei} - \overrightarrow{Xp}\right)^2 = \overrightarrow{Xef}^2 \tag{16.16}$$

Multiplying the energy momentum of four vectors together, the cross term here can be rewritten as:

$$2\,\overrightarrow{Xd}.\overrightarrow{Xei} = 2 * \left[\frac{Ed}{c}, \overrightarrow{Pd}\right].\left[m_e c, \vec{0}\right] = 2Edm_e \tag{16.17}$$

Subsequently, multiplying together the four momentums of the initially stationary electron considering the generation of dark energy, the equation can be rewritten as follows:

$$m_d^2c^2 - m_e^2c^2 + -2m_eE_p - 2\left(\frac{E_dE_p}{c^2} - \frac{P_dE_p\cos(\theta)}{c}\right) + 2E_dm_e = m_e^2c^2 \tag{16.18}$$

which can be simplified as:

$$m_d^2c^2 - 2\left(m_eE_p + \left(\frac{E_dE_p}{c^2} - \frac{P_dE_p}{c}\cos(\theta)\right) - E_dm_e\right) = 0 \tag{16.19}$$

Here, to confirm the energy generation from dark energy of the momentums, the vector energy momentum has been calculated by conducting the following mathematical probe as:

$$2\left(m_eE_p + \left(\frac{E_dE_p}{c^2}\right) - \frac{P_dE_p}{c}\cos(\theta)\,)\right) = m_d^2c^2 - 2E_dm_e \tag{16.20}$$

Here, the expression for the energy generation is related to the vectorial momentum of dark energy, thus, the limit that $m_d \to 0$ that can be expressed the case of scattering. Since the $m_d \to 0$, the $P_d \to E_c$ and thus, the energy generation can be expressed by adding the differential scattering cross section of the configuration of the dark energy which will confirm the presence of dark energy at any time and space in the multiverse by representing the following equation:

$$De = \frac{\partial \sigma_{kn}}{\partial \Omega} = \frac{r_0^2}{2} \frac{\left(P_d^2 c^2 + m_d^2 c^4\right)}{\left(\frac{-\left(\frac{m_d^2 c^2}{2m_e} - E_d\right)}{\left(1+\left(\frac{E_d}{m_e c^2} - \frac{P_d}{m_e^c}\cos(\theta)\right)\right)}\right)^2} \left[\left(\left(P_d^2 c^2 + m_d^2 c^4\right)^{\frac{\frac{1}{2}}{\frac{-\left(\frac{m_d^2 c^2}{2m_e} - E_d\right)}{\left(1+\left(\frac{E_d}{m_e c^2} - \frac{P_d}{m_e c}\cos(\theta)\right)\right)}}}\right)\right.$$

$$\left. + \frac{-\left(\frac{m_d^2}{2m_e} - E_d\right)}{\frac{\left(1+\left(\frac{E_d}{m_e c^2} - \frac{P_d}{m_e c}\cos(\theta)\right)\right)}{\left(P^2 C^2 + m_d^2 C^4\right)^{\frac{1}{2}}} - \sin(\theta)^2}\right] \quad (16.21)$$

Computation of Antimatter (A_m)

Since the antimatter is the matter consisting of the anti-particles that represents particles in standard regular matter, nanoscopic numbers of anti-particles are being calculated here in relation to its particle's accelerators [2, 6]. Since a standard particle and its anti-particle (proton-antiproton) have the same mass, its electric charge will have opposite quantum numbers in order to confirm that proton having a positive charge and anti-proton having a negative charge [16, 34, 47]. Simply, the energy content within the antimatter will be absorbed from its surrounding time and space and then will be released as a mass energy of $E = mc^2$. Therefore, it is confirmed evidence that any time and space in the multiverse is consisting entirely of standard matter, which is equal to antimatter (H^*) once its free quantum fields (n) are in both activated and excited states (Fig. 16.3).

Here, the quantum field of the antimatter is the components of all free quantum momentum of time and space; thus, its components (t, x) are being represented into a vector $x = (ct, x)$ considering superposition of the energy momentum in a common metric signature convention of $\eta_{\mu\nu} =$ diag $(\pm 1, \mp 1, \mp 1, \mp 1)$ of Klein–Gordon equations that can be expressed by the following table (Tables 16.1 and 16.2):

Here, the Klein–Gordon equation permits two variables where one variable is positive and another is negative, which can be obtained by describing a relativistic wavefunction of the antimatter as:

$$\left[\nabla^2 - \frac{m^2 c^2}{\hbar^2}\right]\psi(r) = 0 \quad (16.22)$$

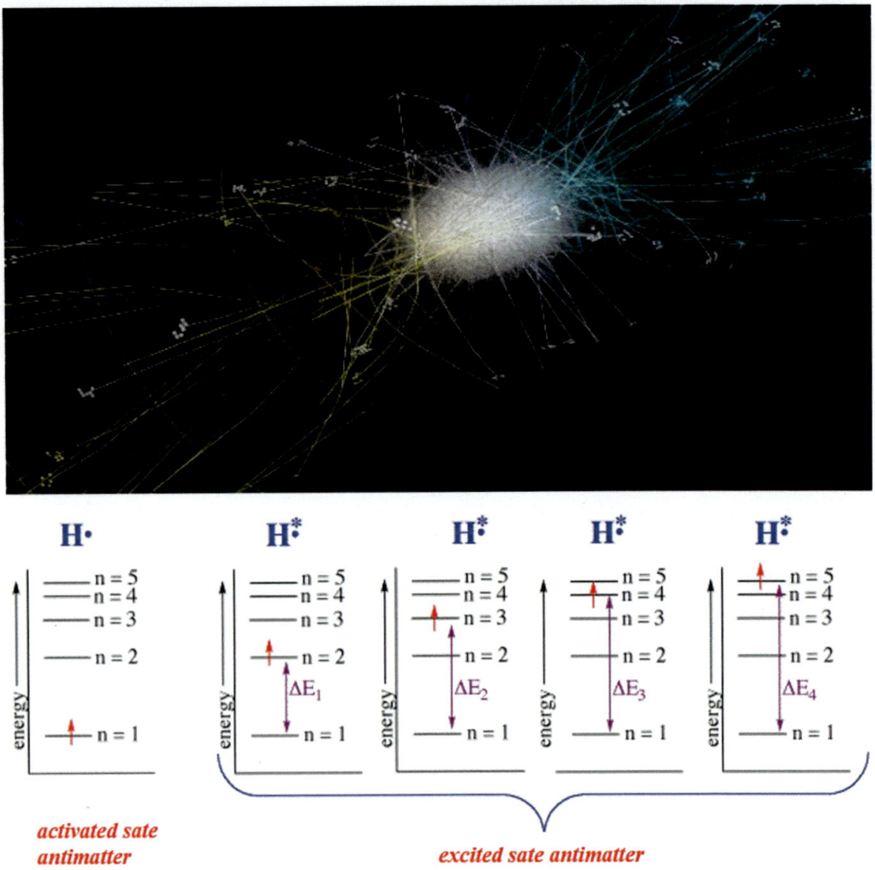

Fig. 16.3 The concept of antimatter distribution energy (H^*) in the time and space due to the quantum mechanical functions of its free quantum fields (n) corresponding to its vectorial superposition of activated state and excited state

Since the Klein–Gordon function is a standard unit, $(c + m^2)\psi(x) = 0$,with the metric signature of $\eta_{\mu\nu} = \text{diag}\ (\pm 1, -1, -1, -1)$, it is solved by Fourier transformation as:

$$\psi(x) = \int \frac{d^4 p}{(2\pi)^4} e^{-ip.x} \psi(p) \tag{16.23}$$

and using spherical dimension of time and space, the complex exponentials are described here as:

Table 16.1 Klein–Gordon equation in normal units with metric of $\eta_{\mu v} =$ diag $(\pm 1, \mp 1, \mp 1, \mp 1)$

	Position space $x = (ct, x)$	Fourier transformation $\omega = \frac{E}{\hbar}$, $k = p/\hbar$	Momentum space $p = \left(\frac{E}{c}, p\right)$
Time and space	$\left(\frac{1}{c^2}\frac{\partial^2}{\partial t^2} - \nabla^2 + \frac{m^2c^2}{\hbar^2}\right)\psi(t,x) = 0$	$\psi(t,x) = \int \frac{d\omega}{2\pi\hbar} \int \frac{d^3k}{(2\pi\hbar)^3} e^{\mp i(\omega t - k.x)} \psi(\omega, k)$	$E^2 = p^2c^2 + m^2c^4$
Four-vector form	$(c + \mu^2)\psi(x) = 0,\ \mu = mc/\hbar$	$\psi(x) = \int \frac{d^4p}{(2\pi\hbar)^4} e^{-ip.\frac{x}{\hbar}} \psi(p)$	$p^2 = \pm m^2c^2$

Here, $\pm\eta^{\mu v}\partial_\mu \partial_v$ is representing the d'Alembert function and ∇^2 is representing the Laplace variable, while c is representing the light speed, and ħ is the Planck constant, while $c = \hbar = 1$.

Table 16.2 Klein–Gordon equation in natural unit metric of $\eta_{\mu\nu} =$ diag $(\pm 1, \mp 1, \mp 1, \mp 1)$

	Position space	Fourier transformation $\omega = E,\ k = p$	Momentum space $p = (E, p)$
Time and space	$\left(\partial_t^2 - \nabla^2 + m^2\right)\psi(t, x) = 0$	$\psi(t, x) = \int \frac{dw}{2\pi} \int \frac{d^3k}{(2\pi)^3} e^{\mp i(wt - k.x)} \psi(w, k)$	$E^2 = p^2 + m^2$
Four-vector form	$(c+ m^2)\psi(x) = 0$	$\psi(x) = \int \frac{d^4p}{(2\pi)^4} e^{-ip.x} \psi(p)$	$p^2 = \pm m^2$

$$p^2 = \left(p^0\right)^2 - p^2 = m^2 \tag{16.24}$$

This result will confirm the momentums of the antimatter's positive and negative energy dynamics as:

$$p^0 = \pm E(p) \text{ where } E(p) = \sqrt{p^2 + m^2}. \tag{16.25}$$

Using a new set of constant $C_{(p)}$,the solution will become

$$\psi(x) = \int \frac{d^4p}{(2\pi)^4} e^{ip.x} C(P) \delta\left(P^0\right)^2 - E(P)^2) \tag{16.26}$$

It is a standard practice to segregate the positive and negative energy dynamics while working with the positive energy only as:

$$\psi(x) = \int \frac{d^4p}{(2\pi)^4} \delta\left(\left(p^0\right)^2 - E(p)^2\right) \left(A(p) e^{-ip^0x^0 + ip^ix^i} + B(p) e^{+ip^0x^0 + ip^ix^i}\right) \theta\left(p^0\right) \tag{16.27}$$

$$= \int \frac{d^4p}{(2\pi)^4} \delta(\left(p^0\right)^2 - E(p)^2) \left(A(P) e^{-ip^0x^0 + ip^ix^i} + B(-P) e^{+ip^0x^0 - ip^ix^i}\right) \theta\left(p^0\right)$$
$$\to \int \frac{d^4p}{(2\pi)^4} \delta(\left(p^0\right)^2 - E(p)^2) (A(p) e^{-ip.x} + B(P) e^{+ip.x}) \theta\left(p^0\right) \tag{16.28}$$

Here, $B(p) \to B(-p)$ can be considered p^0- in order to pick up the positive frequency from its delta as:

$$\psi(x) = \int \frac{d^4p}{(2\pi)^4} \frac{\delta(p^0 - E(p))}{2E(p)} (A(p) e^{-ip.x} + B(p) e^{+ip.x}) \theta\left(p^0\right) \tag{16.29}$$

$$= \int \frac{d^3p}{(2\pi)^3} \frac{1}{2E(p)} (A(p) e^{-ip.x} + B(p) e^{+ip.x}) |_{p^0 = +E(p)} \tag{16.30}$$

Since it is a general solution to the Klein–Gordon equation, Lorentz invariant quantities of $p.\ x = p_\mu x^\mu$ is the final variable to solve the Klein–Gordon equation [24, 37, 39]. Subsequently, the Lorentz invariance will absorb the ½$E(p)$ factor into the coefficient $A(p)$and $B(p)$, and thus, the antimatter energy of a free particle can be written as:

$$\frac{p^2}{2m} = E \tag{16.31}$$

By quantizing this, it can confirm the non-relativistic Schrödinger equation for a free particle energy as:

$$\frac{\widehat{p}^2}{2m}\psi = \widehat{E}\psi \tag{16.32}$$

where $\widehat{p} = -i\hbar\nabla$ is representing the momentum function (∇ is the delta variable) and $\widehat{E} = i\hbar\frac{\partial}{\partial t}$ is the energy dynamic.

Since the Schrödinger equation is a non-relativistic invariant, it is standard to use its free particle as the momentum energy as:

$$\sqrt{p^2c^2 + m^2c^4} = E \tag{16.33}$$

Here, implementing the quantum momentum of the free particle, the energy yields can be written as:

$$\sqrt{(-i\hbar\nabla)^2c^2 + m^2c^4\psi} = i\hbar\frac{\partial}{\partial t}\psi \tag{16.34}$$

Subsequently, the electromagnetic fields of the energy yield can be further clarified by the Klein and Gordon equation of e identity, i.e., $p^2c^2 + m^2c^4 = E^2$, which will give:

$$\left((-i\hbar\nabla)^2c^2 + m^2c^4\right)\psi = \left(i\hbar\frac{\partial}{\partial t}\right)^2\psi \tag{16.35}$$

which simplifies to

$$-\hbar^2c^2\nabla^2\psi + m^2c^4\psi = \hbar^2\frac{\partial^2}{\partial t^2}\psi \tag{16.36}$$

Rearranging terms yields

$$\frac{1}{c^2}\frac{\partial^2}{\partial t^2}\psi - \nabla^2\psi + \frac{m^2c^2}{\hbar^2}\psi = 0 \tag{16.37}$$

Rewriting these functions considering the inverse metric $(-c^2, 1, 1, 1)$ in relation to Einstein's summational variable, it can be written as:

$$-\eta^{\mu v}\partial_\mu\partial_v\psi = \sum_{\mu=0}^{\mu=3}\sum_{v=0}^{v=3} -\eta^{\mu v}\partial_\mu\partial_v\psi = \frac{1}{c^2}\partial_0^2\psi - \sum_{v=1}^{v=3}\partial_v\partial_v\psi = \frac{1}{c^2}\frac{\partial^2}{\partial t^2}\psi - \nabla^2\psi \tag{16.38}$$

Therefore, the Klein–Gordon equation can be expressed as a covariant function, which is the abbreviation of the form of $(c + \mu^2)\psi = 0$, where $\mu = \frac{mc}{\text{h}}$ and $c = \frac{1}{c^2}\frac{\partial^2}{\partial t^2} - \nabla^2$, and thus, the Klein–Gordon equation can be described as a field of energy dynamics $V(\psi)$ as:

$$c\psi + \frac{\partial V}{\partial \psi} = 0 \tag{16.39}$$

Here, the gauge U(1) symmetry is a complex field $\varphi(x)\epsilon\mathbb{C}$ that can satisfy the Klein–Gordon as:

$$\partial_\mu j^\mu(x) = 0, j^\mu(x) = \frac{e}{2m}(\varphi^*(x)\partial^\mu\varphi(x) - \varphi(x)\partial^\mu\varphi^*(x)) \tag{16.40}$$

Subsequently, proof Klein–Gordon equation using algebraic manipulations to form a complex field $\varphi(x)$ of mass m can be expressed as a covariant notation as follows:

$$(c + \mu^2)\varphi(x) = 0 \tag{16.41}$$

And its complex conjugate

$$(c + \mu^2)\varphi^*(x) = 0 \tag{16.42}$$

Then, multiplying by the vectorial variables of $\varphi^*(x)$ and $\varphi(x)$, it can be expressed as:

$$\varphi^*(c + \mu^2)\varphi = 0 \tag{16.43}$$

$$\varphi(c + \mu^2)\varphi^* = 0 \tag{16.44}$$

Subtracting the former from the latter, it can be obtained

$$\varphi^* c\varphi - \varphi c\varphi^* = 0 \tag{16.45}$$

$$\varphi^*\partial_\mu\partial^\mu\varphi - \varphi\partial_\mu\partial^\mu\varphi^* = 0 \tag{16.46}$$

Then it will confirm

$$\partial_\mu(\varphi^* \partial^\mu \varphi) = \partial_\mu \varphi^* \partial^\mu \varphi + \varphi^* \partial_\mu \partial^\mu \varphi \tag{16.47}$$

From which it can be obtained the conservation law for the Klein–Gordon field:

$$\partial_\mu j^\mu(x) = 0, \qquad j^\mu(x) \equiv \varphi^*(x)\partial^\mu \varphi(x) - \varphi(x)\partial^\mu \varphi^*(x) \tag{16.48}$$

Here, implementing of the vectorial variable to the Klein–Gordon field, the scalar field-derived stress–energy tensor can be calculated as:

$$T^{\mu\nu} = \frac{\hbar^2}{m}\left(\eta^{\mu\alpha}\eta^{\nu\beta} + \eta^{\mu\beta}\eta^{\nu\alpha} - \eta^{\mu\nu}\eta^{\alpha\beta}\right)\partial_\alpha \overline{\psi}\partial_\beta \psi - \eta^{\mu\nu} mc^2 \overline{\psi}\psi \tag{16.49}$$

Here, the time–time component T^{00} will confirm the positive frequency related to the particles with free energy of antimatter as $\psi(x,t) = \varnothing(x,t)e^{-\frac{i}{\hbar}mc^2 t}$ where $\varnothing(x,t) = u_E\ (x)e^{-\frac{i}{\hbar}Et}$. Thus, defining the kinetic energy $E' = E - mc^2 = \sqrt{m^2c^4 + c^2p^2} - mc^2 \approx \frac{p^2}{2m}, E' \ll mc^2$ in the non-relativistic limit $v{\sim}p < < c$, and hence $\left|i\hbar \frac{\partial \varnothing}{\partial t}\right| = E'\varnothing \ll mc^2\varnothing$, this yield of the derivative ψ can be written as:

$$\begin{aligned} \frac{\partial \psi}{\partial t} &= \left(-i\frac{mc^2}{\hbar}\varnothing + \frac{\partial \varnothing}{\partial t}\right)e^{-\frac{i}{\hbar}mc^2 t} \approx -i\frac{mc^2}{\hbar}\varnothing e^{-\frac{i}{\hbar}mc^2 t} \\ \frac{\partial^2 \psi}{\partial t^2} &\approx -\left(i\frac{2mc^2}{\hbar}\frac{\partial \varnothing}{\partial t} + \left(\frac{mc^2}{\hbar}\right)^2 \varnothing\right)e^{-\frac{i}{\hbar}mc^2 t} \end{aligned} \tag{16.50}$$

Substituting it with the free Klein–Gordon equation, $c^{-2}\partial_t^2 \psi = \nabla^2 \psi - m^2 \psi$, yield $-\frac{1}{c^2}\left(i\frac{2mc^2}{\hbar}\frac{\partial \varnothing}{\partial t} + \left(\frac{mc^2}{\hbar}\right)^2 \varnothing\right)e^{-\frac{i}{\hbar}mc^2 t} \approx \left(\nabla^2 - \left(\frac{mc^2}{\hbar}\right)^2\right)\varnothing e^{-\frac{i}{\hbar}mc^2 t}$ can be simplified as:

$$-i\hbar \frac{\partial \varnothing}{\partial t} = -\frac{\hbar^2}{2m}\nabla^2 \varnothing \tag{16.51}$$

Since free Klein–Gordon equation is a *classical* Schrödinger field, it is further analyzed by quantum field of Klein–Gordon of the time and space [3, 43]. Subsequently, the quantum field of Klein–Gordon equation is being analyzed to confirm the symmetry of the antimatter energy wavefunction φ is in the local U (1) gauge of the sphere of $\varphi \to \varphi' =$ exp $(i\theta)\varphi$ of the time and space. Where $\theta(t, x)$ is a local vectorial angle, which transformation the wavefunction into a complex phase as exp $(i\theta) =$ cos θ + i sin θ, where derivative ∂_μ can be replaced by gauge-covariant variable $D_\mu = \partial_\mu - ieA_\mu$, while the gauge fields transform as $eA_\mu \to eA'_\mu + \partial_\mu\theta$ wchic confirm that presence of (−,+,+,+) metric signature of the antimatter in the time and space as:

$$D_\mu D^\mu \varphi = -(\partial_t - ieA_0)^2 \varphi + (\partial_i - ieA_i)^2 \varphi = m^2 \varphi \tag{16.52}$$

where A is the scalar potential and which can be simplified as functional term of $D_\mu D^\mu \varphi + AF^{\mu\nu} D_\mu \varphi D_\nu (D_\alpha D^\alpha \varphi) = 0$, and thus, the scalar field can be multiplied by i to confirm the for a field of energy dynamics of A_m is sufficiently presence in the time and space by expressing as:

$$A_m = \int_x \eta^{\mu\nu} (\partial_\mu \varphi^* + ieA_\mu \varphi^*)(\partial_\nu \varphi - ieA_\nu \varphi) = \int_x |D\varphi|^2 \tag{16.53}$$

Since the existence of light energy L_e, dark energy D_e, and antimatter A_p in the time and space are being proved by the above mathematical modeling, these proved energies are being further remodeled computationally to pave the time and space cool and heat by controlling its energy dynamics to secure a comfortable climate condition.

Cooling Mechanism of the Time and Space

Since light energy, dark energy, and antimatter are the primary elements for the time and space, the $HhE^-(L_p; D_p; A_p)$ energies for time and space are being modeled independently to confirm its existence within the multiverse. Here, the computed HhE^- has been clarified mathematically for any time and space in the multiverse to deform its cooling-state energy by analyzing its nano-point break waveguide energy momentum [5, 17, 40]. Then, the cluster of its energy momentum is being calculated considering its nano-point deflect waveguide to decode the quantum dynamic into cool-state energy (HcE^-) by expressing the following *Hossain* equation as:

$$H = C + \sum \omega_{ci} a_i^\dagger a_i + \sum_K \omega_k b_k^\dagger b_k + \sum_{ik} \left(V_{ik} a^\dagger b_k + V_{ik}^* b_k^\dagger a_i \right), \tag{16.54}$$

where $a_i \left(a_i^\dagger \right)$ and $b_k \left(b_k^\dagger \right)$ present the function of the nano-point break mode and its energy-dynamic modules of its cofactor, V_{ik}, denotes the magnitude of the energy mode within its nanostructures.

Simply, this cool-state energy modules will form HcE^- from its energy dynamics to cool the time and space considering its the internuclear distances of 1D-H ($n = 1$) + H($n = 2, \ldots, \infty$), 2D-H($n = 2$) + H($n = 2, \ldots, \infty$), and 3D-H ($n = \infty$) + H(n = unlimited,...,∞), where n and l are the primary and vectorial quanta of the cool-state energy. Thus, the energy state quanta considering time-dependent energy intensity (k) and space-dependent energy wavelength (λ) confirm the mode of energy dynamics within the time and space to activate this energy into cool-state energy (Fig. 16.4).

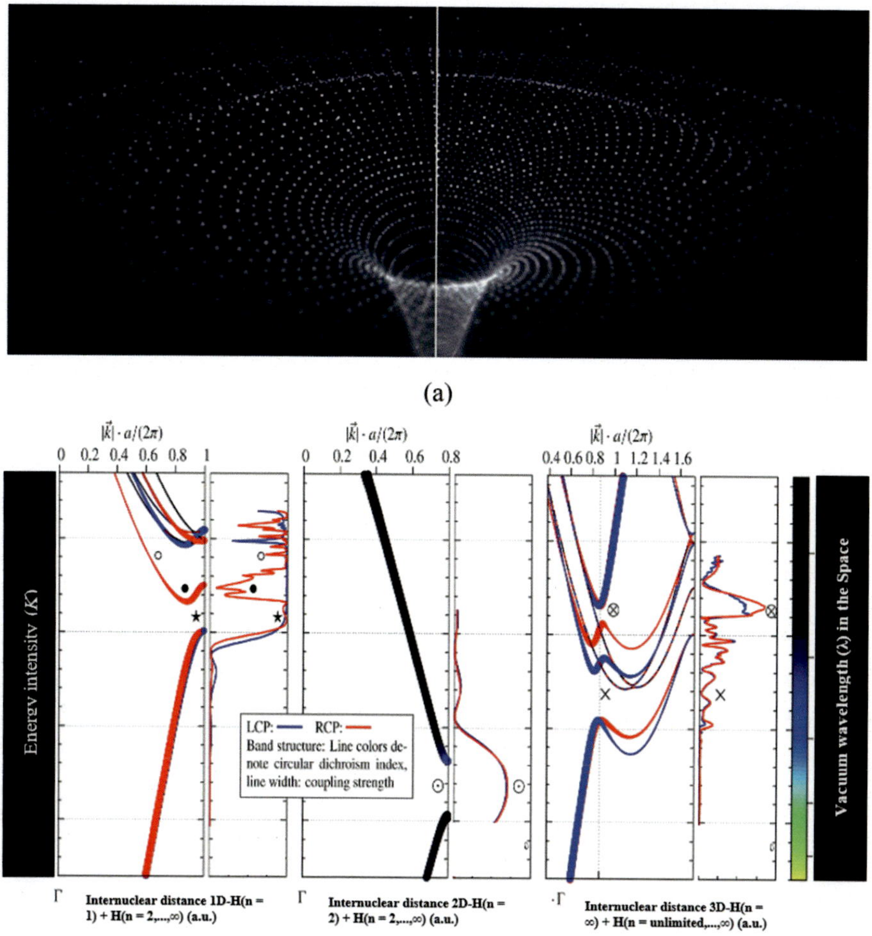

Fig. 16.4 (**a**) The conceptual array of the cooling energy distributions in the time and space, (**b**) semiclassical pathways of energy band structure (LCP and RCP) that generate cool-state energy within the internuclear distance (a.u.) where red and blue representing the excited and activated states of cool-energy corresponding to its wavelength in the space and energy intensity of (k) to the time (Lo et al. 2015: Sci. Rep)

This cool-state energy can be defined by **hidden energy** transformation characteristics of the cool energy quanta that can be expressed by:

$$E = \quad EL - EO \left\{ \exp\left[\frac{q(V + I_{\mathrm{RS}})}{AkTc}\right] - 1 \right\} - \frac{(V + I_{\mathrm{RS}})}{R_{\mathrm{S}}}, \tag{16.57}$$

where E_L presents the cool energy generation, EO is the energy dynamics, and R_s is the variable function for the adverse condition. Here A, representation of the activated module energy in related to Boltzmann constant, k (= 1.38 × 10^{-23} W/m^{2}K) where, q (= 1.6 × 10^{-19}C) is the electrons of activated energy, and T_C is the dynamic proliferation of the energy and thus, the energy dynamics can be expressed by:

$$E_O = E_{RS}\left(\frac{T_C}{T_{ref}}\right)^3 \exp\left[\frac{qEG\left(\frac{1}{T_{ref}} - \frac{1}{T_C}\right)}{KA}\right]. \tag{16.58}$$

Here, E_{RS} denotes the functional energy that relies on the energy speed and qEG denotes the energy flow per unit area of the cool-state energy [18, 38, 41].

Simply, in this cooling energy mode, the energy dynamics is related to T–R module of the time and space that can be expressed by:

$$T = -IR_s + K \log\left[\frac{I_L - I + I_O}{I_O}\right], \tag{16.59}$$

where K represents the constant $\left(= \frac{AkT}{q}\right)$ and I_{mo} and V_{mo} are the energy dynamics. Therefore, the correlation between I_{mo} and V_{mo} is similar as the T–R module of the time and space, which can be expressed precisely as:

$$T_{mo} = -I_{mo}R_{Smo} + K_{mo} \log\left(\frac{I_{Lmo} - I_{mo} + I_{Omo}}{I_{Omo}}\right) \tag{16.60}$$

where I_{Lmo} denotes the net energy generation, I_{Omo} denotes the activated energy dynamics, R_{smo} is the inactivated energy dynamics, and K_{mo} is the constant cofactor related to $K_{mo} = N_S \times K$, and thus, net cool energy dynamic in the time and space can be written as:

$$T_{mo} = -I_{mo}N_S R_S + N_S K \log\left(\frac{I_L - I_{mo} + I_o}{I_o}\right). \tag{16.61}$$

Here, energy-generated cooling thermal conductivity relies primarily on energy flux and relativistic thermal state of the time and space; thus, the thermal conductivity could be expressed as follows:

$$I_L = G\left[I_{SC} + K_I\left(T_{cool}\right)\right]V_{mo} \tag{16.63}$$

$$T_{cool} = T\left(\frac{I_L}{(G * V_{mo})x(I_{sc} + K1)}\right) \tag{16.64}$$

where I_{sc} is the energy dynamics per unit area, K_I is the relativistic energy coefficient, T_{cool} is the cooling thermal of the energy, and G is the cool energy thermal conductivity of the time and space.

Heating Mechanism of the Time and Space

To transform cool-state energy into heat-state energy, the local symmetry of Higgs-Boson quantum field of the time and space has been used. Simply, the time and space has been simulated considering Abelian local symmetries by applying Higgs-Boson electromagnetic field of the time and space to form congenial heat on it [27, 28, 42]. Simply the momentum of energy dynamic will interrupt the gauge-field symmetries of the time and space, and thus, the Goldstone scalar field will act as the longitude modes of the Higgs-Boson field. Therefore, the local symmetries of quantum field of the time and space will be broken down as particle of T^{α} related to the gauge field of $A_{\mu}^{\alpha}\,(x)$, and thus, the Higgs-Boson quantum fields will then start to generate local U (1) phase symmetries in order to create heat [31, 32, 45]. So, this transformation process can be consisting of the precise scalar fields $\Phi\,(x)$ of energically charged q paired with the electromagnetic field $A^{\mu}(x)$ of the time and space that could be written as:

$$\mathcal{L} = -\frac{1}{4}\,F_{\mu\nu}F^{\mu\nu} + D_{\mu}\Phi^{*}\,D^{\mu}\Phi - V(\Phi^{*}\Phi), \tag{16.65}$$

where

$$\begin{aligned} D_{\mu}\Phi(x) &= \partial_{\mu}\Phi(\mathrm{x}) + \mathrm{iqA_{\mu}(x)}\Phi\,(\mathrm{x}) \\ D_{\mu}\Phi^{*}(x) &= \partial_{\mu}\Phi^{*}(\mathrm{x}) - \mathrm{iqA_{\mu}(x)}\Phi^{*}(\mathrm{x}) \end{aligned} \tag{16.66}$$

and

$$V(\Phi^{*}\Phi) = \frac{\lambda}{2}\,(\Phi^{*}\Phi)^{2} + m^{2}(\Phi^{*}\Phi). \tag{16.67}$$

Here $\lambda > 0$ but $m^2 < 0$; therefore, $\Phi = 0$ is a local peak factor of the scalar field, and the precise form of quantum dynamics is $\Phi = \frac{v}{\sqrt{2}} * e^{i\theta}$ that can be expressed as:

$$v = \sqrt{\frac{-2m^{2}}{\lambda}}\ \text{for any real}\ \theta. \tag{16.68}$$

Consequently, here, the scalar quantum dynamics $\Phi\,(x)$ will form a nonzero value of $\langle\Phi\rangle \neq 0$, which will form the local U (1) symmetries into electromagnetic field of time and space. Therefore, in this local U (1) symmetries, the quantum dynamics of Φ (x) will be an activated scalar field of the time and space that can be expressed as:

$$\Phi\ (\mathrm{x}) = \frac{1}{\sqrt{2}}\ \Phi_{\mathrm{r}}(\mathrm{x}) * \mathrm{e}^{\mathrm{i}\Theta(\mathrm{x})}, \mathrm{real}\ \Phi_{\mathrm{r}}(\mathrm{x}) > 0, \mathrm{real}\ \Phi(\mathrm{x})\ . \tag{16.69}$$

Since the scalar field of time and space is a dynamic momentum of the quantum Φ $(x) = 0$, it will meet the principle of $\langle \Phi \rangle \neq 0$ in order to be considered as the instinctual function of the time and space. Thus, considering the momentum field of $\phi_r(x)$ and $\Theta(x)$, the scalar filed related to the radial field of ϕ_r of the time and space can be simplified:

$$V(\phi) = \frac{\lambda}{8}\ \left(\phi_r^2 - v^2\right)^2 + \mathrm{const.} \tag{16.70}$$

Subsequently, the momentum field can be shifted by applying the variable scalars, $\Phi_r(x) = v + \sigma(x)$,and expressed as:

$$\phi_r^2 - v^2 = (v + \sigma)^2 - v^2 = 2v\sigma + \sigma^2 \tag{16.71}$$

$$V = \frac{\lambda}{8}\ \left(2v\sigma - \sigma^2\right)^2 = \frac{\lambda v^2}{2} * \sigma^2 + \frac{\lambda v}{2} * \sigma^3 + \frac{\lambda}{8} * \sigma^4\ . \tag{16.72}$$

Thus, the functional derivative $D_\mu \phi$ will become

$$\begin{aligned} D_\mu \phi &= \frac{1}{\sqrt{2}}\ \left(\partial_\mu \left(\phi_r e^{i\Theta}\right) + iqA_\mu * \phi_r e^{i\Theta}\ \right) \\ &= \frac{e^{i\Theta}}{\sqrt{2}}\ \left(\partial_\mu \phi_r + \phi_r * i\partial_\mu \Theta + \phi_r * iqA_\mu\right) \end{aligned} \tag{16.73}$$

$$\begin{aligned} \left|D_\mu \phi\right|^2 &= \frac{1}{2} \left|\partial_\mu \phi_r + \phi_r{}^* i\partial_\mu \Theta + \phi_r{}^* iqA_\mu\right|^2 \\ &= \frac{1}{2}\left(\partial_\mu \phi_r\right) + \frac{\phi_r^{2*}}{2}\ \left(\partial_\mu \Theta qA_\mu\right)^2 \\ &= \frac{1}{2}\left(\partial_\mu\ \sigma\right)^2 + \frac{(v+\sigma)^{2*}}{2}\ \left(\partial_\mu \Theta + qA_\mu\right)^2. \end{aligned} \tag{16.74}$$

The Lagrangian is then given by

$$\mathcal{L} = \frac{1}{2}\ \left(\partial_\mu\ \sigma\right)^2 - v\ (\sigma) - \frac{1}{4}\ F_{\mu\nu} F^{\mu\nu} + \frac{(v+\sigma)^2}{2} * \left(\partial_\mu \Theta + qA_\mu\right)^2. \tag{16.75}$$

Simply, to conduct the heat $(\mathcal{L}_{\mathrm{heat}})$ generation through electromagnetic field of the time and space considering this *Lagrangian*, it has been further expanded $\mathcal{L}_{\mathrm{heat}}$ as the energy dynamics of the electromagnetic field of the time and space that can be expressed as free particles of heat:

(a)

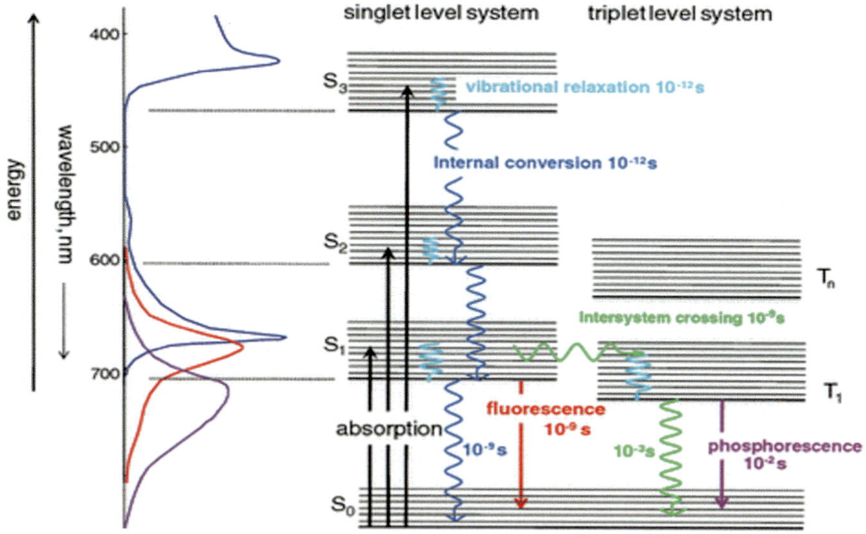

(b)

Fig. 16.5 (**a**) The concept of heating mechanism of time and space by the theoretical function of energy dynamics of optimum energy formation and (**b**) the mode of energy formation corresponding to its wavelength of energy spectrum in relation to its absorbance of singlet and triplet level system. (Ulrich E. Steiner; 2014. F. Phot. Photochem. and Photobio)

$$\mathcal{L}_{\text{heat}} = \frac{1}{2}\left(\partial_\mu \sigma\right)^2 - \frac{\lambda v^2}{2} * \sigma^2 - \frac{1}{4} F_{\mu\nu}F^{\mu\nu} + \frac{v^2}{2} * \left(qA_\mu + \partial_\mu \Theta\right)^2. \tag{16.76}$$

Subsequently, it will initiate to generate heating energy into the quantum field of the time and space considering the scalar particles of λv^2 corresponding to its energy spectrum of singlet and triplet system of the time and space (Fig. 16.5).

Results and Discussion

Cooling Mechanism of Time and Space

To computationally confirm the formation of cool-state energy (HcE^-) in the time and space, the vectorial momentum of its steady weak coupling energy field has been clarified in order to determine its absorbed energy rate per unit area of $J(\omega)$. Here, $J(\omega)$ is thus considered as the quantum field of the density of states (DOS) energy proliferation, which, in fact, is the deformed cooling energy magnitudes $V(\omega)$, which is formed from its nano-point break waveguide energy momentum [20, 22, 26]. Thus, cooling energy formation follows the Weisskopf–Wigner theory is therefore all the generated HcE^- shall pass through all dimensional modes (1D, 2D, and 3D) within the time and space (Table 16.3).

Subsequently, all dimensions of 1D, 2D, and 3D of the time and space are being further clarified by implanting frequency cutoff to avoid the bifurcation of the DOS to pave the path for cooling energy formation [33, 35]. Thus, the DOS analyzed at various dimensions of the time and space has been calculated, here as $\varrho_{\text{PC}}(\omega)$, is being confirmed by the clarifying of cooling energy frequency formation of Maxwell's theory of nanostructure correspond to the time and space (Fig. 16.6). The DOS is thus here expressed by $\varrho_{\text{PC}}(\omega) \propto \frac{1}{\sqrt{\omega - \omega_e}}\Theta(\omega - \omega_e)$, where $\Theta(\omega - \omega_e)$ where ω_e is denoted as the frequency of the PBE at the used DOS.

Then, the density of state (DOS) has been implemented to confirm the qualitative form of the non-Weisskopf–Wigner module that represents the precise cooling-state energy, calculated by the dimensional clarification of the time and space [20, 22, 29]. This DOS is thereafter determined considering the electromagnetic field of time and space, which is determined by $\varrho_{\text{PC}}(\omega) \propto \frac{1}{\sqrt{\omega - \omega_e}}\Theta(\omega - \omega_e)$. Simply, here, the cooling energy of DOS reveals a perfect algorithmic function close to the PBE, which is $\varrho_{\text{PC}}(\omega) \propto -[\ln|(\omega - \omega_0)/\omega_0| - 1]\Theta(\omega - \omega_e)$, where ω_e defines the prime tip point of the DOS distribution in the time and space.

Subsequently, this distributed DOS energy will transform it into the cool state, which will confirm the total deliberation of energy dynamics of the time and space corresponding to its quantum fields, and, thus, defined above, $J(\omega)$ will confirm the quantum field of cooling energy magnitudes $V(\omega)$ corresponding to its photonic band (PB), photonic band edge (PBE), and photonic band gap (PBG) in the time and space as:

$$J(\omega) = \varrho(\omega)|V(\omega)|^2. \tag{16.77}$$

Therefore, considering here, the energy frequency of ω_c and the generated cooling energy dynamic of $\langle a(t)\rangle = u(t, t_0)\langle a(t_0)\rangle$, the function $u(t, t_0)$ has been described as the photon structure as $u(t, t_0)$ by clarifying the integral–differential equation as:

Table 16.3 Cooling energy dynamics considering its densities of states (DOS) in the time and space. The unit areas $J(\omega)$ and the generation of self-energy inductions in the time and space reservoir $\Sigma(\omega)$, demonstrated by the cooling energy corresponding to its variables of C, η, and χ as coupled forces among three-dimensional (1D, 2D, and 3D) point break of energy band

Energy particle	Unit area $J(\omega)$ for different DOS*	Reservoir-induced self-energy induction $\Sigma(\omega)$*
1D	$\frac{C-1}{\pi}\frac{1}{\sqrt{\omega-\omega_e}}\Theta(\omega-\omega_e)$	$-\frac{C-1}{\sqrt{\omega_e-\omega}}$
2D	$C-\eta\left[ln\left\|\frac{\omega-\omega_0}{\omega_0}\right\|-1\right]\Theta(\omega-\omega_e)\Theta(\Omega_d-\omega)$	$C-\eta\left[Li_2\left(\frac{\Omega_d-\omega_0}{\omega-\omega_0}\right)-Li_2\left(\frac{\omega_0-\omega_e}{\omega_0-\omega}\right)-ln\,\frac{\omega_0-\omega_e}{\Omega_d-\omega_0}\,ln\,\frac{\omega_e-\omega}{\omega_0-\omega}\right]$
3D	$\chi\sqrt{\frac{\omega-\omega_e}{\Omega_C}}exp\left(-\frac{\omega-\omega_e}{\Omega_C}\right)\Theta(\omega-\omega_e)$	$\chi\left[\pi\sqrt{\frac{\omega_e-\omega}{\Omega_C}}exp\left(-\frac{\omega-\omega_e}{\Omega_C}\right)erfc\sqrt{\frac{\omega_e-\omega}{\Omega_C}}-\sqrt{\pi}\right]$

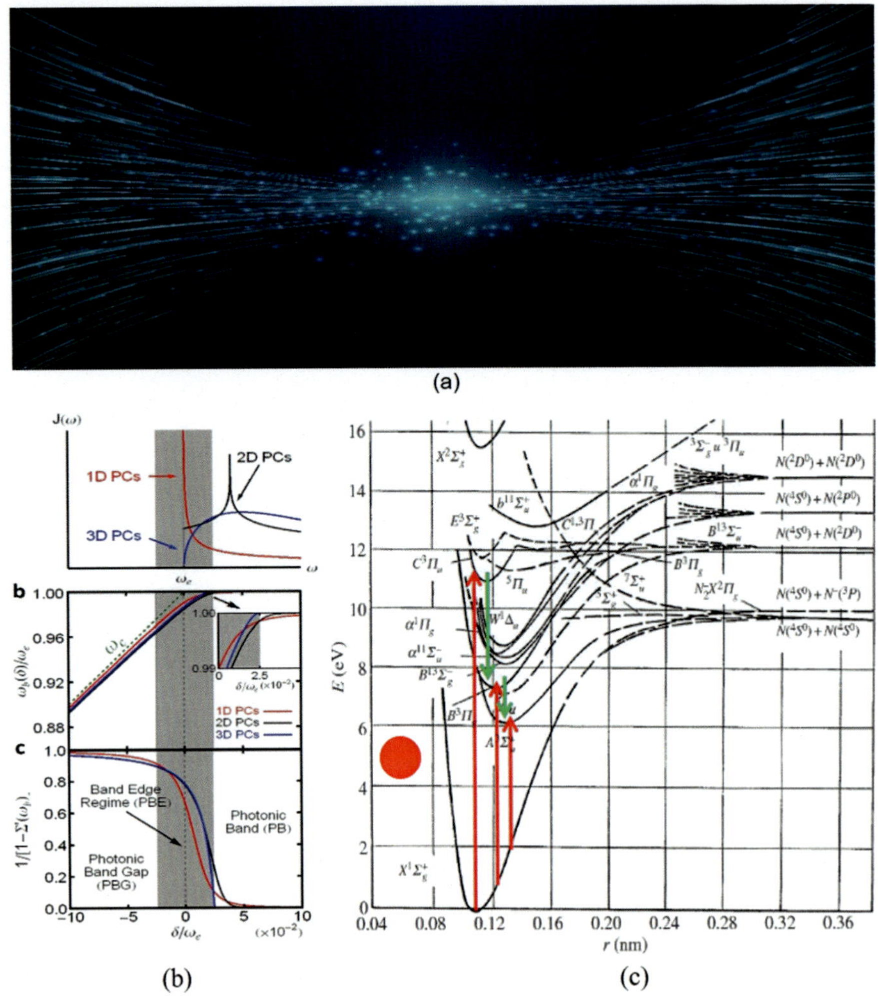

Fig. 16.6 (**a**) The concept of cooling-state energy dynamics in the time and space, (**b**) the cooling energy band structures and energy transformation mode at various DOSs of 1D, 2D, and 3D in the time and space (Lo et al. 2015: Sci. Rep.), and (**c**) the cooling energy formation magnitude (*eV*) in relation to the radial (*r*) dimension of time and space. (Starikovskiy et al. 2015; P. T. Royal Society A: Math. Phys. and Eng. Sci)

$$u(t,t_0)=\frac{1}{1-\Sigma'(\omega_b)}e^{-i\omega(t-t_0)}+\int_{\omega_e}^{\infty}d\omega\frac{J(\omega)e^{-i\omega(t-t_0)}}{[\omega-\omega_c-\Delta(\omega)]^2+\pi^2J^2(\omega)},\qquad(16.78)$$

where $\Sigma'(\omega_b)=[\partial\Sigma(\omega)/\partial\omega]_{\omega=\omega_b}$ and $\Sigma(\omega)$ confirm the self-energy induction into the reservoir,

$$\Sigma(\omega) = \int_{\omega_e}^{\infty} d\omega' \frac{J(\omega')}{\omega - \omega'} \quad . \tag{16.79}$$

Here, the frequency ω_b represents the cooling energy frequency mode $(0 < \omega_b < \omega_e)$,calculated as $\omega_b - \omega_c - \Delta(\omega_b) = 0$, where $\lesssim\Delta(\omega) = \mathcal{P}\left[\int d\omega' \frac{J(\omega')}{\omega - \omega'}\right]$ is a primary-valued integral of the time and space. In details, it can be explained that energy dynamics (eV) is in the time and space is considered as the Fock cooling state n_0, *i. e.* , $\rho(t_0) = |n_0\rangle\langle n_0|$, which is obtained computationally from the real-time quantum field of its radial (r) dimension of the space at a time t and expressed as:

$$\rho(t) = \sum_{n=0}^{\infty} \mathcal{P}_n^{(n_0)}(t)|n_0\rangle\langle n_0| \tag{16.80}$$

$$\mathcal{P}_n^{(n_0)}(t) = \frac{[v(t,t)]^n}{[1 + v(t,t)]^{n+1}}[1 - \Omega(\mathrm{t})]^{n_0} \text{ x } \sum_{k=0}^{\min\{n_0,n\}} \binom{n_0}{k}\binom{n}{k}\left[\frac{1}{v(t,t)}\frac{\Omega(\mathrm{t})}{1 - \Omega(\mathrm{t})}\right]^k \tag{16.81}$$

where $\Omega(t) = \frac{|u(t,t_0)|^2}{1+v(t,t)}$. Thus, this result suggested that the Fock state cooling energy is indeed generated into dynamic states of the time and space $\mathcal{P}_n^{(n_0)}(t)$ of $|n_0\rangle$ where the proliferation of energy deliberation $\mathcal{P}_n^{(n_0)}(t)$ is in the prime state $|n_0 = 5\rangle$ and in the steady-state limits $\mathcal{P}_n^{(n_0)}(t \to \infty)$, which will eventually reach at a non-equilibrium cool-state energy in order to cool the time and space finally.

Heating Mechanism of Time and Space

Since the electromagnetic field of the time and space is being modeled by Higgs-Boson quantum dynamics, the local symmetry of U (1) here will allow to form gauge-variable QED, in terms of gauge particles $\varnothing' \to e^{i\alpha(x)}\varnothing$, which is the cooling-state energy that can transform into thermal-state energy in order to heat the time and space at a comfort level. This mechanism is being confirmed by the following variable derivatives considering the specific transformational laws of the scalar field, written as:

$$\begin{aligned} &\partial_\mu \to D_\mu = \partial_\mu = ieA_\mu && [\text{covariant derivatives}] \\ &A'_\mu = A_\mu + \frac{1}{e}\,\partial_\mu\alpha && [A\mu \quad \text{derivatives}] \end{aligned} \tag{16.82}$$

Here, the local U (1) gauge-invariant Lagrangian is being considered as the perplex scalar field that is expressed by:

$$\mathcal{L} = (D^{\mu})^{\dagger}\left(D_{\mu}\varnothing\right) - \frac{1}{4}F_{\mu\nu}F^{\mu\nu} - V(\varnothing). \tag{16.83}$$

The term $\frac{1}{4}F_{\mu\nu}F^{\mu\nu}$ is the kinetic motion of the gauge field for considering thermal energy and $V(\varnothing)$ is denoted the kinetic term written as $V(\varnothing^{*}\varnothing) = \mu^{2}(\varnothing^{*}\varnothing) + \lambda\,(\varnothing^{*}\varnothing)^{2}$.

Thus, the equation of the Lagrangian $\mathcal{L}$ is being considered as the quantum field of scalar particle ϕ_1 and ϕ_2 and a mass μ; thus, here, $\mu^2 < 0$ will confirm an infinite number of quanta to satisfy the equation $\phi_1^2 + \phi_2^2 = -\mu^2/\lambda = v^2$. So, the quantum field can be clarified as $\phi_0 = \frac{1}{\sqrt{2}}\left[(\upsilon + \eta) + i\xi\right]$, and then the derivative of the Lagrangian can be expressed as kinetic mode of:

$$\begin{aligned}\mathcal{L}_{\text{kin}}(\eta,\xi) &= (D^{\mu}\phi)^{\dagger}(D^{\mu}\phi)\\ &= (\partial^{\mu} + ieA^{\mu})\phi^{*}\left(\partial_{\mu} - ieA_{\mu}\right)\phi\end{aligned} \tag{16.84}$$

Here, $V(\eta,\xi) = \lambda\,\upsilon^2\eta^2$ is the final term to the second order, and thus, the full Lagrangian can be expressed as:

$$\mathcal{L}_{\text{kin}}(\eta,\xi) = \frac{1}{2}\left(\partial_{\mu}\eta\right)^2 - \lambda\upsilon^2\eta^2 + \frac{1}{2}\left(\partial_{\mu}\xi\right)^2 - \frac{1}{4}F_{\mu\nu}F^{\mu\nu} + \frac{1}{2}e^2\upsilon^2A_{\mu}^2 - e\upsilon A_{\mu}(\partial^{\mu}\xi) \tag{16.85}$$

Here, η represents the mass, ξ represents the massless, μ represents the mass for the quanta, and thus A_{μ} is defined as the term $\partial_{\mu}\alpha$, as is the function of the quantum field. Naturally, A_{μ} and ϕ can be changed spontaneously, so the equation can be rewritten as a Lagrangian scalar to confirm the formation of the thermal energy particles within the quantum field of the time and space:

$$\begin{aligned}\mathcal{L}_{\text{scalar}} &= (D^{\mu}\phi)^{\dagger}(D^{\mu}\phi) - V\left(\phi^{\dagger}\phi\right)\\ &= (\partial^{\mu} + ieA^{\mu})\frac{1}{\sqrt{2}}(v+h)\left(\partial_{\mu} - ieA_{\mu}\right)\frac{1}{\sqrt{2}}(v+h) - V\left(\phi^{\dagger}\phi\right)\end{aligned} \tag{16.86}$$

$$= \frac{1}{2}\left(\partial_{\mu}h\right)^2 + \frac{1}{2}e^2A_{\mu}^2(v+h)^2 - \lambda v^2h^2 - \lambda vh^3 - \frac{1}{4}\lambda h^4 + \frac{1}{4}\lambda h^4 \tag{16.87}$$

Here, the Lagrangian scalar thus revealed that the Higgs-Boson quantum field can certainly be initiated to form thermal energy in the time and space. In order to confirm to determine this heat-energy generation in the time and space, a further calculation considering its isotropic distributed kinetic energy has been conducted with respect to the angle θ from the dimensional axis of the time and space and the differential density of energy $\in$ and written as:

$$dn = \frac{1}{2}n\,(\in)\,\sin\theta d\in d\theta. \tag{16.88}$$

Here, the speed of high-energy frequency is being implemented as c (1–$cos\theta$) considering the absorption of energy per unit area by expressing the following equation:

$$\frac{d\tau_{\text{abs}}}{dx} = \int\int \frac{1}{2}\sigma n\ (\in)\ (1-\cos\theta)\ \sin\theta d \in d\theta. \tag{16.89}$$

Rewriting these variables as integral over s instead of θ, by (16.88) and (16.89), it has been determined as:

$$\frac{d\tau_{\text{abs}}}{dx} = \pi r_0^2 \left(\frac{m^2c^4}{E}\right)^2 \int_{\frac{m^2c^4}{E}}^{\infty} \in^{-2} n(\in)\ \overline{\phi}\ [s_0\ (\in)]\ de, \tag{16.90}$$

where

$$\overline{\phi}\ [s_0\ (\in)] = \int_1^{s_0(\in)} s\overline{\sigma}\ (s)ds, \overline{\sigma}(s) = \frac{2\sigma(s)}{\pi r_0^2}. \tag{16.91}$$

Simply, this result confirms the representation of the heating energy generation in that is readily allow the calculation of $\overline{\phi}$ [s_0] to determine the net heating energy formation of s_0 [21, 25, 29] in the time and space. Therefore, the net functional asymptotic formula is expressed as follows to confirm the heating energy formation:

$$\overline{\phi}[s_0] = 2\ s_0(\ln 4s_0\text{–}2) + \ln 4s_0(\ln 4s_0\text{–}2)\text{–}\frac{(\pi^2-9)}{3} + s_0^{-1}\left(\ln 4s_0 + \frac{9}{8}\right) + \ldots (s_0 > > 1); \tag{16.95}$$

$$\overline{\phi}\ [s_0] = \left(\frac{2}{3}\right)\ (S_0-1)^{\frac{3}{2}} + \left(\frac{5}{3}\right)\ (S_0-1)^{\frac{5}{2}}\text{–}\left(\frac{1507}{420}\right)\ (S_0-1)^{\frac{7}{2}} + \ldots (s_0\text{–}1 < < 1). \tag{16.96}$$

The function $\frac{\overline{\phi}[s_0]}{(s_0-1)}$ is revealed as $1 < s_0 < 10$; at larger s_0, it confirms a standard algorithm function of s_0. Thus, the energy spectra of the thermal energy can be written as $n(\in) \propto \in^m$, and thus, the calculation of the thermal energy in terms of energy spectra in the time and space can be expressed as:

$$\begin{aligned} n(\in) &= 0,\quad \in\ <\ \in_0 \\ &= C_{\in}^{-\alpha}\ \text{or}\ D_{\in}^{\beta},\ \in_0\ <\ \in\ < \in_m \\ &= 0,\quad \in\ >\ \in_m \end{aligned} \tag{16.97}$$

Then, it can be transformed as:

$$\left(\frac{d\tau_{\text{abs}}}{dx}\right)_{\alpha} = \pi r_0^2 C\left(\frac{m^2c^4}{E}\right)^{1-\alpha}$$
$$\times\left\{\begin{array}{cc} 0, & E < E_m \\ [F_\alpha(1)\text{–}\mathrm{F}_\alpha(\sigma_m)], & E_m < E < E_0\,. \\ [F_\alpha(\sigma_0)\text{–}\mathrm{F}_\alpha(\sigma_m)], & E > E_0 \end{array}\right\} \tag{16.98}$$

$$\left(\frac{d\tau_{\text{abs}}}{dx}\right)_{\beta} = \pi r_0^2 D\left(\frac{m^2c^4}{E}\right)^{1+\beta}$$
$$\times\left\{\begin{array}{ll} 0, & E < E_m \\ \left[F_\beta(\sigma_m),\right. & E_m < E < E_0\,. \\ \left[F_\beta(\sigma_m)\text{–}\mathrm{F}_\beta(\sigma_0)\right], & E > E_0 \end{array}\right\} \tag{16.99}$$

In these functional variables, the heating energy spectra on the time and space can be properly defined by asymptotic formula. Thus, the term Γ_{γ}^{LPM} defines the energy generation in relation to the irradiated energy dynamics of per unit area as:

$$\Gamma_{\gamma} \equiv \frac{dn_{\gamma}}{dVdt}. \tag{16.100}$$

Here, the contributions Γ_{γ}^{LPM}, the rate of thermal energy generation are being confirmed as $O(\alpha_{EM}\ \alpha_s)$. Thus, it has been confirmed by implementing the contributed thermal energy rate Γ_{γ}^{LPM} to thermodynamically control the temperature T and energy-physical reaction μ of the time and space by expressing the following equation:

$$\frac{d\Gamma_{\gamma}^{LPM}}{d^3k} = \frac{d_F q_s^2 \alpha_{EM}}{4\pi^2 k}\int_{-\infty}^{\infty}\frac{dp_{\parallel}}{2\pi}\int\frac{d^2\mathrm{p}_{\perp}}{(2\pi)^2}\mathrm{A}\left(p_{\parallel},k\right)\ \mathrm{Re}\left\{2\mathrm{P}_{\perp}\cdot f\left(\mathrm{p}_{\perp};p_{\parallel},k\right)\right\} \tag{16.101}$$

where d_{F} is the variable state of the energy particles [N_c in SU(N_c)] and q_s is the Abelian charge of the energy quark, $k \equiv |k|$; thus, the kinetic functional mode A(p||, k) is being expressed by:

$$A\left(p_{\|}, k\right) \equiv \begin{cases} \dfrac{n_b\left(k+p_{\|}\right)\left[1+n_b\left(p_{\|}\right)\right]}{2p_{\|}\left(p_{\|}+k\right)} & \text{scalars} \\ \dfrac{n_f\left(k+p_{\|}\right)\left[1-n_f\left(p_{\|}\right)\right]}{2\left[p_{\|}\left(p_{\|}+k\right)\right]^2}\left[p_{\|}{}^2+\left(p_{\|}+k\right)^2\right], & \text{fermions} \end{cases} \tag{16.102}$$

with

$$n_b(p) \equiv \frac{1}{\exp[\beta(p-\mu)]-1}, \quad n_f(p) \equiv \frac{1}{\exp[\beta(p-\mu)]+1} \tag{16.103}$$

The function $f(p\perp; p\|, k)$ is then integrated to resolve the below equation, which suggested that the heating energy proliferation in the time and space is very much possible, which can be expressed as following derivative of energy generation:

$$\begin{aligned} 2p_{\perp} = i\delta E f\left(p_{\perp}; p_{\|}, k\right) + \frac{\pi}{2} C_F g_s^2 m_D^2 \int \frac{d^2 q_{\perp}}{(2\pi)^2} \frac{dq_{\|}}{2\pi} \frac{dq^0}{2\pi} 2\pi\delta\left(q^0 - q_{\|}\right) \text{x} \\ \times \frac{T}{|\mathrm{q}|}\left[\frac{2}{|\mathrm{q}^2 - \Pi_{\mathrm{L}}(\mathrm{Q})|^2} + \frac{\left[1-\left(q^0/|\mathrm{q}_{\|}|\right)^2\right]^2}{\left|(q^0)^2 - \mathrm{q}^2 - \Pi_{\mathrm{T}}(\mathrm{Q})\right|^2}\right] \\ \times \left[f\left(p_{\perp}; p_{\|}, k\right) - f\left(q + p_{\perp}; p_{\|}, k\right)\right] \end{aligned} \tag{16.104}$$

Simply, this heating energy generation is derived from the explicit forms of heating energy that maximize the heating energy generation corresponding to its given energy function of $f(p\perp; p\|, k)$ to transform it into heating state energy to heat the **time and space** naturally (Fig. 16.7).

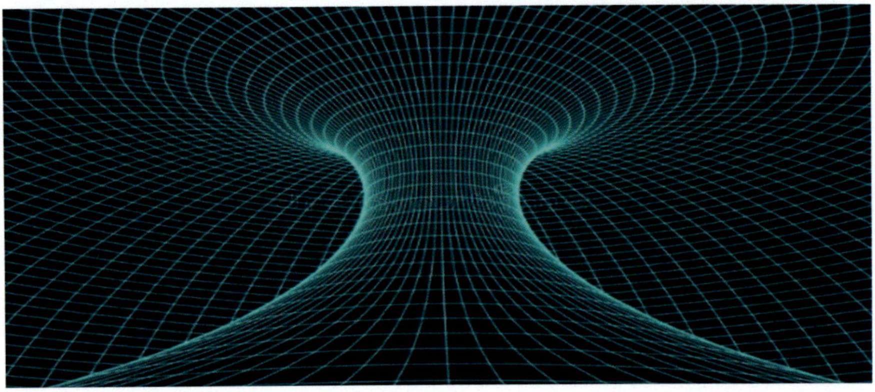

(a)

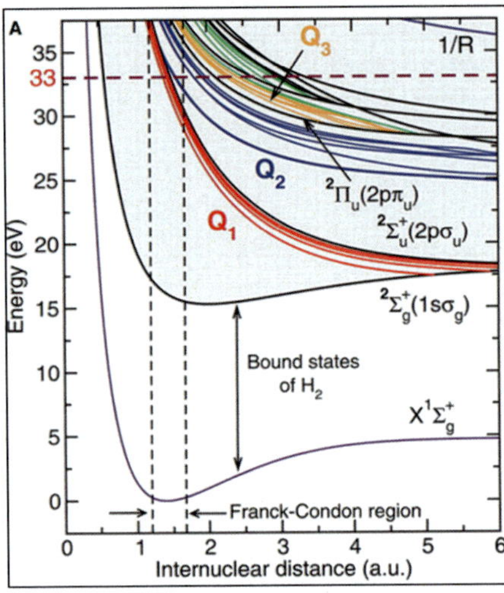

(b)

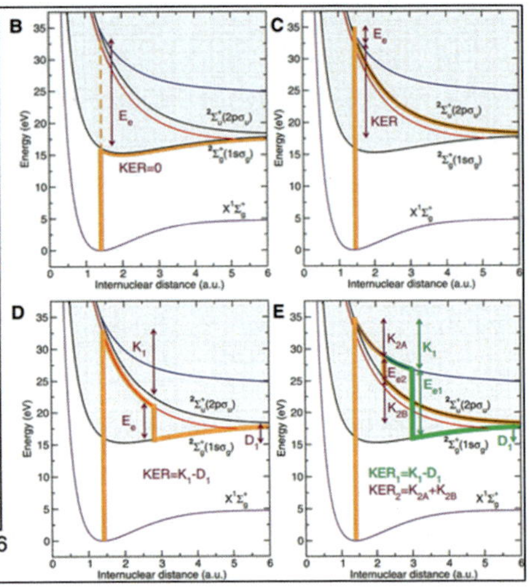

(c)

Fig. 16.7 (**a**) The heating energy formation concept in the time and space, (**b**) the formation of heating energy (*eV*) at its internuclear distance (*a.u.*) of its bound states ($X^1\sum_g$; $^2\sum_g$;$^2\prod_u$) of energy formation region, and (**c**) the formation of heating energy (*eV*) generation at its internuclear distance (*a.u.*) at various spectra ($X^1\sum_g$; $^2\sum_g$;$^2\prod_u$) of its energy amplitude array of time and space. (Martin et al. 2007; Sci)

Conclusions

The presence of light energy (L_e), dark energy (D_e), and antimatter (A_m) is being modeled to confirm that any time and space have enough hidden energy (HhE^-) of light, dark, and antimatter within the time and space to develop a comfortable climate condition anywhere in the multiverse. The results of this computational modeling suggested that the availability HhE^- is very much doable in anywhere in the time and space, which can be transformed into cool-state energy (HcE^-) by implementing *B-E* energy discreet mechanism to cool the time and space naturally. Interestingly, this cool-state energy (HcE^-) can also be reformed into the thermal-state energy (HtE^-) by integrating Higgs-Boson [BR ($H \rightarrow \gamma\gamma^-$)] quantum dynamics of the time and space to heat the time and space naturally. The presence of HhE^- in any time and space and its natural cooling and heating mechanism to cool and heat the time and space at a comfortable level shall indeed be the most innovative scientific discovery to control the climate condition of any time and space to develop a comfortable multiverse to live mankind there in the future.

Acknowledgments The author, Md. Faruque Hossain, declares that any findings, predictions, and conclusions described in this article are solely performed by the author, and it is confirmed that there is no conflict of interest for publishing this research paper in a suitable journal and/or publisher.

References

1. Armani, D. K., Kippenberg, T. J., Spillane, S. M., & Vahala, K. J. (2003). Ultra-high-Q toroid microcavity on a chip. *Nature, 421*(6926), 925–928.
2. Artemyev, N., Jentschura, U. D., Serbo, V. G., & Surzhykov, A. (2012). Strong electromagnetic field effects in ultra-relativistic heavy-ion collisions. *The European Physical Journal C, 72*, 1935.
3. Bakr, W. S., Preiss, P. M., Tai, M. E., Ma, R., Simon, J., & Greiner, M. (2011). Orbital excitation blockade and algorithmic cooling in quantum gases. *Nature, 480*(7378), 500–503.
4. Baur, G., Hencken, K., & Trautmann, D. (2007). Revisiting unitarity corrections for electromagnetic processes in collisions of relativistic nuclei. *Physical Report, 453*(1).
5. Birnbaum, K. M., Boca, A., Miller, R., Boozer, A. D., Northup, T. E., & Kimble, H. J. (2005). Photon blockade in an optical cavity with one trapped atom. *Nature, 436*(7047), 87–90.
6. Boettcher, I., Pawlowski, J. M., & Diehl, S. (2012). Ultracold atoms and the functional renormalization group. *Nuclear Physics B-Proceedings Supplements, 228*, 63–135.
7. Broz, M., Contreras, J. G., & Takaki, J. T. (2020). A generator of forward neutrons for ultra-peripheral collisions: nOOn. *Computer Physics Communications, 253*, 107181.
8. Busch, K., Von Freymann, G., Linden, S., Mingaleev, S. F., Tkeshelashvili, L., & Wegener, M. (2007). Periodic nanostructures for photonics. *Physics Reports, 444*(3–6), 101–202.
9. Cardoso, V., Lemos, J. P., & Yoshida, S. (2004). Quasinormal modes of schwarzschild black holes in four and higher dimensions. *Physical Review D, 69*(4), 044004.
10. Cartwright, J. (2010). Photons meet with three-way split. *Nature*. https://doi.org/10.1038/news.2010.381
11. Chang, D. E., Sørensen, A. S., Demler, E. A., & Lukin, M. D. (2007). A single-photon transistor using nanoscale surface plasmons. *Nature Physics, 3*(11), 807–812.

12. Chen, G., Chen, S., Li, C., & Chen, Y. (2013). Examining non-locality and quantum coherent dynamics induced by a common reservoir. *Scientific Reports, 3*(1), 1–6.
13. Chen, J., Wang, C., Zhang, R., & Xiao, J. (2012). Multiple plasmon-induced transparencies in coupled-resonator systems. *Optics Letters, 37*(24), 5133–5135.
14. Cheng, M., & Song, Y. (2012). Fano resonance analysis in a pair of semiconductor quantum dots coupling to a metal nanowire. *Optics Letters, 37*(5), 978–980.
15. Dayan, B., Parkins, A. S., Aoki, T., Ostby, E. P., Vahala, K. J., & Kimble, H. J. (2008). A photon turnstile dynamically regulated by one atom. *Science, 319*(5866), 1062–1065.
16. Del'Haye, P., Arcizet, O., Schliesser, A., Wilken, T., Holzwarth, R., & Kippenberg, T. J. (2008). Full stabilization of a frequency comb generated in a monolithic microcavity. Paper presented at the 2008 Conference on Lasers and Electro-Optics and 2008 Conference on Quantum Electronics and Laser Science, pp. 1–2.
17. Dobrynina, A., Kartavtsev, A., & Raffelt, G. (2015). Photon-photon dispersion of TeV gamma rays and its role for photon-ALP conversion. *Physical Review D, 91*(8), 083003.
18. Douglas, J. S., Habibian, H., Hung, C., Gorshkov, A. V., Kimble, H. J., & Chang, D. E. (2015). Quantum many-body models with cold atoms coupled to photonic crystals. *Nature Photonics, 9*(5), 326–331.
19. Dupuis, N., Canet, L., Eichhorn, A., Metzner, W., Pawlowski, J. M., Tissier, M., & Wschebor, N. (2021). The nonperturbative functional renormalization group and its applications. *Physics Reports, 910*, 1–114.
20. Eichler, J., & Stöhlker, T. (2007). Radiative electron capture in relativistic ion–atom collisions and the photoelectric effect in hydrogen-like high-Z systems. *Physics Reports, 439*(1–2), 1–99.
21. Englund, D., Majumdar, A., Faraon, A., Toishi, M., Stoltz, N., Petroff, P., & Vučković, J. (2010). Resonant excitation of a quantum dot strongly coupled to a photonic crystal nanocavity. *Physical Review Letters, 104*(7), 073904.
22. Fischer, K. A., Hanschke, L., Wierzbowski, J., Simmet, T., Dory, C., Finley, J. J., et al. (2017). Signatures of two-photon pulses from a quantum two-level system. *Nature Physics, 13*(7), 649–654.
23. Gabor, N. M., Zhong, Z., Bosnick, K., Park, J., & McEuen, P. L. (2009). Extremely efficient multiple electron-hole pair generation in carbon nanotube photodiodes. *Science, 325*(5946), 1367–1371.
24. Gleyzes, S., Kuhr, S., Guerlin, C., Bernu, J., Deleglise, S., Busk Hoff, U., et al. (2007). Quantum jumps of light recording the birth and death of a photon in a cavity. *Nature, 446*(7133), 297–300.
25. Gould, R. J., & Schréder, G. P. (1967). Pair production in photon-photon collisions. *Physical Review, 155*(5), 1404.
26. Grinin, A., Matveev, A., Yost, D. C., Maisenbacher, L., Wirthl, V., Pohl, R., et al. (2020). Two-photon frequency comb spectroscopy of atomic hydrogen. *Science, 370*(6520), 1061–1066.
27. Guerlin, C., Bernu, J., Deleglise, S., Sayrin, C., Gleyzes, S., Kuhr, S., et al. (2007). Progressive field-state collapse and quantum non-demolition photon counting. *Nature, 448*(7156), 889–893.
28. Guo, Y., Al-Jubainawi, A., & Ma, Z. (2019). Performance investigation and optimisation of electrodialysis regeneration for LiCl liquid desiccant cooling systems. *Applied Thermal Engineering, 149*, 1023–1034.
29. Hencken, K., Baur, G., & Trautmann, D. (2006). Transverse momentum distribution of vector mesons produced in ultraperipheral relativistic heavy ion collisions. *Physical Review Letters, 96*(1), 012303.
30. Hübel, H., Hamel, D. R., Fedrizzi, A., Ramelow, S., Resch, K. J., & Jennewein, T. (2010). Direct generation of photon triplets using cascaded photon-pair sources. *Nature, 466*(7306), 601–603.
31. Jahnke, V., Luna, A., Patino, L., & Trancanelli, D. (2014). More on thermal probes of a strongly coupled anisotropic plasma. *Journal of High Energy Physics, 2014*(1), 1–40.

32. Javadi, A., Söllner, I., Arcari, M., Hansen, S. L., Midolo, L., Mahmoodian, S., et al. (2015). Single-photon non-linear optics with a quantum dot in a waveguide. *Nature Communications, 6*(1), 1–5.
33. Johnson, B. R., Reed, M. D., Houck, A. A., Schuster, D. I., Bishop, L. S., Ginossar, E., et al. (2010). Quantum non-demolition detection of single microwave photons in a circuit. *Nature Physics, 6*(9), 663–667.
34. Kaneda, F., & Kwiat, P. G. (2019). High-efficiency single-photon generation via large-scale active time multiplexing. *Science Advances, 5*(10), eaaw8586.
35. Langer, L., Poltavtsev, S. V., Yugova, I. A., Salewski, M., Yakovlev, D. R., Karczewski, G., et al. (2014). Access to long-term optical memories using photon echoes retrieved from semiconductor spins. *Nature Photonics, 8*(11), 851–857.
36. Langford, N. K., Ramelow, S., Prevedel, R., Munro, W. J., Milburn, G. J., & Zeilinger, A. (2011). Efficient quantum computing using coherent photon conversion. *Nature, 478*(7369), 360–363.
37. Naghiloo, M., Foroozani, N., Tan, D., Jadbabaie, A., & Murch, K. W. (2016). Mapping quantum state dynamics in spontaneous emission. *Nature Communications, 7*(1), 1–7.
38. Najjari, B., Voitkiv, A. B., Artemyev, A., & Surzhykov, A. (2009). Simultaneous electron capture and bound-free pair production in relativistic collisions of heavy nuclei with atoms. *Physical Review A, 80*(1), 012701.
39. Reinhard, A., Volz, T., Winger, M., Badolato, A., Hennessy, K. J., Hu, E. L., & Imamoğlu, A. (2012). Strongly correlated photons on a chip. *Nature Photonics, 6*(2), 93–96.
40. Reinhard, P., & Nazarewicz, W. (2021). Nuclear charge densities in spherical and deformed nuclei: Toward precise calculations of charge radii. *Physical Review C, 103*(5), 054310.
41. Shalm, L. K., Hamel, D. R., Yan, Z., Simon, C., Resch, K. J., & Jennewein, T. (2013). Three-photon energy–time entanglement. *Nature Physics, 9*(1), 19–22.
42. Szafron, R., & Czarnecki, A. (2016). High-energy electrons from the muon decay in orbit: Radiative corrections. *Physics Letters B, 753*, 61–64.
43. Tame, M. S., McEnery, K. R., Özdemir, Ş. K., Lee, J., Maier, S. A., & Kim, M. S. (2013). Quantum plasmonics. *Nature Physics, 9*(6), 329–340.
44. Ting, T. (2004). The polarization vector and secular equation for surface waves in an anisotropic elastic half-space. *International Journal of Solids and Structures, 41*(8), 2065–2083.
45. Wang, C., Wang, W., & Chen, Z. (2017). Single-satellite positioning algorithm based on direction-finding. Paper presented at the 2017 Progress in Electromagnetics Research Symposium-Spring (PIERS), 2533–2538.
46. Xiao, Y., Li, M., Liu, Y., Li, Y., Sun, X., & Gong, Q. (2010). Asymmetric Fano resonance analysis in indirectly coupled microresonators. *Physical Review A, 82*(6), 065804.
47. Yan, W., & Fan, H. (2014). Single-photon quantum router with multiple output ports. *Scientific Reports, 4*(1), 1–6.

Part VI
Conclusion

Chapter 17
Modeling of Global Climate Control

Introduction

The massive development of industrialization and the misuse of the natural resources throughout the world quicken the accumulation of atmospheric CO_2 concentration heavily and causing climate change [3, 7]. Numerous studies revealed that current accumulation of CO_2 into the atmosphere increasing at a rate of 2.11% yearly and thus, consequently, the global climate change is increasing accordingly, which will make difficult for the future generation to sustain on Earth [2, 12, 33]. So, it is the time without a doubt to make the global climate in a steady comfortable condition by breaking down the atmospheric CO_2 to confirm the versatility, adaptability, and manageability of our mother Earth, which won't result in maladjustment, but will be presentable for our future generation as a sustainable world. Therefore, an attempt has been conducted to break the CO_2 naturally to mitigate the global climate change by using the activation of the hidden particle of dark energy (D_e), which is abundant everywhere in the Earth [20, 40, 52]. Subsequently, the control mechanism of the global temperature has also been conducted by the process of the activation antimatter (A_m). Simply, in this research, a novel mathematical analysis has been proposed to activate the dark energy (D_e) of the Earth to break the CO_2 and the antimatter (A_m) has been decoded to cool and heat the Earth naturally at a comfort level. Here, the activated dark energy (D_e) will initiate its quantum electrodynamics momentum to breakdown atmospheric CO_2 and transform it into C and O_2, and the energy momentum of the Earth will absorb the nano-point break waveguides of the antimatter (A_m) by using its QED to form cool-state energy to cool the Earth and then convert it into heat-state energy by the induction of its electromagnetic quantum of Higgs-Bosons ($H \rightarrow \gamma\gamma^-$) to heat the Earth, respectively. This simple CO_2 breakdown mechanism and the natural cooling and heating mechanism of the Earth shall indeed be a noble scientific discovery ever to control the global climate naturally.

M. F. Hossain, *Global Sustainability*, https://doi.org/10.1007/978-3-031-34575-3_17

Methods and Simulation

Modeling of Dark Energy ($\mathbf{D}_e$)

The computation of dark energy considering the sphericity and the orbital parameters of the Earth has been conducted to clarify this hidden energy on Earth by expressing the following law of cosines:

$$\rho\cos(c) = \rho\cos(a)\cos(b) + \rho\sin(a)\sin(b)\cos(C) \tag{17.1}$$

Here a, b, and c represent as the *arc* lengths in radians corresponding to the sides of a spherical triangle of the Earth where C represents the angle of the vertex corresponding to the other side of *arc* with the it length c [8, 23, 35]. Naturally, considering the implementation of zenith angle Θ, the following terms can be written as the laws of cosines: $C = h = r$; $c = \Theta$; $a = \frac{1}{2}\pi - \phi$; $b = \frac{1}{2}\pi - \delta$ and thus, the equation can be expressed as:

$$\rho\cos(\Theta) = \sin(\phi)\sin(\delta) + \cos(\phi)\cos(\delta)\cos(r) \tag{17.2}$$

To modify this calculation, a further clarification has been conducted considering a general derivative as below:

$$\begin{aligned}\rho\cos(\theta) &= \rho\sin(\phi)\sin(\delta)\cos(\beta) + \sin(\delta)\cos(\phi)\sin(\beta)\cos(\gamma) \\ &\quad + \cos(\phi)\cos(\delta)\cos(\beta)\cos(h) - \cos(\delta)\sin(\phi)\sin(\beta)\cos(\gamma)\cos(h) \\ &\quad - \cos(\delta)\sin(\beta)\sin(\gamma)\sin(h)\end{aligned} \tag{17.3}$$

Here, β represents the angle corresponding to the horizon and γ corresponding to the azimuth angle, and thus, the sphere of Earth, denoted here R_E and the mean distance is denoted as R_0 represents here as the average distant corresponding of astronomical unit (*a.u.*) and the dark energy constant is S_0. Thus, the dark energy flux onto the plane tangent of the sphere of the Earth is calculated as:

$$Q = \begin{cases} S_o \dfrac{R_0^2}{R_E^2}\cos(\theta) & \cos(\theta) > 0 \\ 0 & \cos(\theta) \leq 0 \end{cases} \tag{17.4}$$

Here, the mean Q over a time is the mean value of Q in relation to its rotation within the time angle of $h = \pi$ to $h = -\pi$, and thus, calculation can be expressed as:

$$Q^{-\text{time}} = -\frac{1}{2\pi}\int_{\pi}^{-\pi} Q\mathrm{d}h \tag{17.5}$$

Since h_0 is the time-related angle once Q is the positive, it will be feasible during the dark energy generation on the Earth when $\Theta = 1/2\ \pi$ or for h_0 and thus the equation can be solved by:

$$\rho\sin(\phi)\sin(\delta) + \cos(\phi)\cos(\delta)\cos\left(h_{0)}\right. = 0$$

or

$$\rho\cos(h_0) = -\tan(\phi)\tan(\delta) \tag{17.6}$$

Since tan(φ)tan(δ) > 1, the dark energy will confirm its existence at $h = \pi$, so $h_0 = \pi$. Subsequently, in a Q^{-}time $= 0$; $\frac{R_0^2}{R_E^2}$ would remain mostly constant during the period of dark energy generation in the Earth, which can be expressed by considering the following integral:

$$\begin{aligned}\int_{\pi}^{-\pi} Q\mathrm{d}h = \int_{h_0}^{-h_0} Q\mathrm{d}h &= S_0\frac{R_0^2}{R_E^2}\int_{h_0}^{-h_0}\cos(\theta)\mathrm{d}h \\ &= S_0\frac{R_0^2}{R_E^2}\left[h\sin(\phi)\sin(\delta) + \cos(\phi)\cos(\delta)\sin(h)\right]_{h=h_0}^{h=-h_0} \\ &= -2S_0\frac{R_0^2}{R_E^2}\left[h_0\sin(\phi)\sin(\delta) + \cos(\phi)\cos(\delta)\sin\left(h_{0)}\right.\right]\end{aligned} \tag{17.7}$$

Therefore:

$$Q^{-\text{time}} = \frac{S_0}{\pi}\frac{R_0^2}{R_E^2}\left[h_0\sin(\phi)\sin(\delta) + \cos(\phi)\cos(\delta)\sin\left(h_{0)}\right.\right] \tag{17.8}$$

Since the θ is representing here as mainstream angle that describes the orbit of the Earth, $\theta = 0$ corresponding to its vernal equinox of δ at its orbital position can be written as:

$$\delta = \varepsilon \sin(\theta) \tag{(17.9)}$$

Here ε represents the mainstream longitude of sphere ϖ that is related to the vernal equinox, and thus, the equation can be written as:

$$R_{\mathrm{E}} = \frac{R_0}{1 + e\cos(\theta - \omega)} \tag{17.10}$$

or

$$\frac{R_0}{R_{\mathrm{E}}} = 1 + e\cos(\theta - \omega) \tag{17.11}$$

Given the above clarification of ϖ, ε, s_0, and e from astrodynamical computation, dark energy L_{e} can be computed for any latitude φ and θ of the Earth (Fig. 17.1). Here, $\theta = 0°$ is representing the precise time of the vernal equinox, $\theta = \infty$ is representing the precise space of the equinox, and thus, the equation can be simplified for dark energy generation on a given time in the Earth that can be expressed as:

$$D_{\mathrm{e}} = S_0\left(1 + \cos\left(2\pi\frac{n}{x}\right)\right) \tag{17.12}$$

Here S_0 is the dark energy constant, n is the unit of time, x is the duration of time, and π is the sphere of the Earth that optimized the generation of dark energy, which is the hidden particles (L_{e}) in the Earth, respectively.

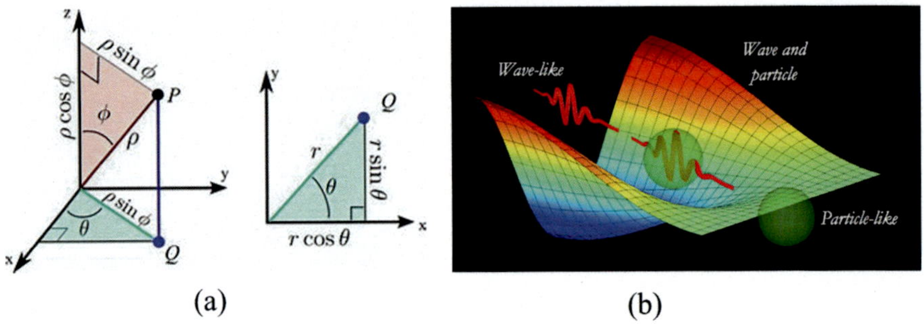

Fig. 17.1 The distribution of dark energy on Earth, (**a**) clarify the presence of dark energy with respect to law of cosines and (**b**) schematic diagram of the dark energy dynamic on the Earth's sphericity and orbital parameter

Activation of Dark Energy ($\mathbf{D}_e$) to Break the CO_2

Since the dark energy is a hidden particle that resembles to the light energy, this particle is being activated by analyzing its minimal way of new U(1) gauge field considering its vectorial (V) dynamics of its tangential (T) field at Z dimension on Earth [1, 9, 10]. Thus, its kinetic energy proliferation corresponding to the dark energy field will release the *hypercharge electron*, which is the striking electron force of the dark energy that can be written as:

$$\overrightarrow{De} = \left|\frac{Edp}{c}, \overrightarrow{Pd}\right| + \overrightarrow{Xef}\left[\frac{Eef}{c}, \overrightarrow{Pef}\right] \quad (17.13)$$

Since dark energy momentum is conserved to its hypercharge electron and is therefore $\Delta P = 0$, the sum of the initial momentums must be equal to the sum of the final momentum, and thus, it can be rewritten as:

$$\overrightarrow{De} + \overrightarrow{Xei} = \overrightarrow{Xp} + \overrightarrow{Xef} \quad (17.14)$$

Here, the isolated momentums of the outgoing hypercharge electron are being calculated and thus the square of both sides of the equation can be rewritten as:

$$\left(\overrightarrow{\overrightarrow{De}} + \overrightarrow{Xei} - \overrightarrow{Xp}\right)^2 = \overrightarrow{Xef}^2 \quad (17.15)$$

Multiplying this dark energy hypercharge electron momentum of its vectors together, the cross term here can be rewritten as:

$$\overrightarrow{De} = 2\,\overrightarrow{Xd}.\overrightarrow{Xei} = 2 * \left[\frac{Ed}{c}, \overrightarrow{Pd}\right].\left[m_e c, \vec{0}\right] = 2Edm_e \quad (17.16)$$

Subsequently, multiplying together all momentums of the initially stationary electron considering the generation of dark energy, the equation can be rewritten to denote hypercharge electron generation as follows:

$$\overrightarrow{De} = m_d^2c^2 - m_e^2c^2 + -2m_eE_p - 2\left(\frac{E_dE_p}{c^2} - \frac{P_dE_p\cos(\theta)}{c}\right) + 2E_dm_e \quad (17.17)$$

which can be simplified as:

$$\overrightarrow{De} = m_d^2c^2 - 2\left(m_eE_p + \left(\frac{E_dE_p}{c^2} - \frac{P_dE_p}{c}\cos(\theta)\right) - E_dm_e\right) \quad (17.18)$$

Here, to confirm this hypercharge electron generation from dark energy, the vector energy momentum has been calculated by conducting the following mathematical probe as:

$$\overrightarrow{De} = 2\left(m_e E_p + \left(\frac{E_d E_p}{c^2}\right) - \frac{P_d E_p}{c}\cos(\theta)\,)\right) = m_d^2 c^2 - 2E_d m_e \tag{17.19}$$

Here, the expression for the hypercharge electron generation is related to the vectorial momentum of dark energy, thus, the limit that $m_d \rightarrow 0$ that can be expressed the case of scattering. Since the $m_d \rightarrow 0$, the $P_d \rightarrow E_c$ and thus, the hypercharge electron generation can be expressed by adding the differential scattering cross section of the configuration of the dark energy which is a striking electron force.

Consequently, this striking electron force of the dark energy ($\overrightarrow{De}$) will have the tremendous active force, which will scatter the electromagnetic wave of its energy momentum at a speed of energy c [11, 24, 60]. Since the hypercharge striking electron force is sensitive to any covalent bond, this electron force will have the extreme momentum to completely break the covalent bond (C=O) of the CO_2 and transform it into C and O_2 sharply. Hence, this derivative process of the hypercharge striking electron force of the dark energy is extremely kinematic in relation to electromagnetic wave and speed of energy, and thus, the equation can be written as:

$$\overrightarrow{De} = 2\left[\frac{Edp}{c}, \overrightarrow{Pd}\right] + \overrightarrow{CO_2} + \left[\overrightarrow{Pemf}\right] \tag{17.20}$$

$$= 2\left[\frac{Edp}{c}, \overrightarrow{Pd}\right] + \left[\overrightarrow{C = O_2}\right] - [\text{Covalent bond}] \tag{17.21}$$

Since dark energy momentum is constant, momentums of the outgoing electron can be rewritten as:

$$\overrightarrow{De} = 2\left[\frac{Ed}{c}, \overrightarrow{Pd}\right] + \left[\overrightarrow{C + O_2}\right] - \left[m_e \overrightarrow{c^2}\right] = 2Edm_e \tag{17.22}$$

Since $\overrightarrow{De} = S_0\left(1 + \cos\left(2\pi\frac{n}{x}\right)\right)$ and CO_2 can be broken down sharply by the activation of this hypercharge electron of dark energy momentums at a speed of energy, the equation can be simplified as:

$$\overrightarrow{De} = 2E_d m_e = S_0\left(1 + \cos\left(2\pi\frac{n}{x}\right)\right) - [C + O_2] \tag{17.23}$$

Here, C will retain with Earth and O_2 will be released to the atmosphere; thus, the dark energy remains constant on Earth as (Fig. 17.2):

$$\overrightarrow{De} = 2E_d m_e = S_0\left(1 + \cos\left(2\pi\frac{n}{x}\right)\right) \tag{17.24}$$

$$V = \sqrt{V_R^2 + (V_L - V_C)^2} \qquad Z = \sqrt{R^2 + (X_L - X_C)^2}$$

$$\phi = \tan^{-1}\frac{V_L - V_C}{V_R} \qquad \phi = \tan^{-1}\frac{X_L - X_C}{R}$$

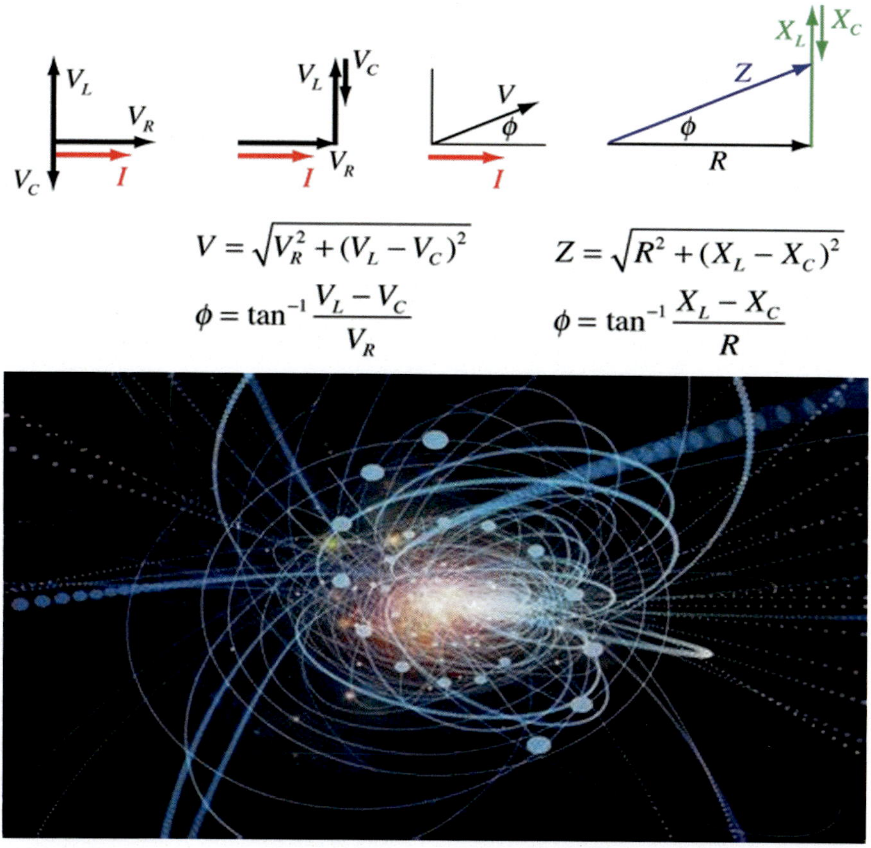

Fig. 17.2 The basic model of dark energy U(1) gauge symmetry in the time and space corresponding to its vectorial dynamics of tangential striking electron within the kinetic energy field parameters

Modeling of Antimatter (A_m)

Since the antimatter is the matter consisting of the anti-particles that represents particles in standard regular matter, nanoscopic numbers of anti-particles will have the standard particle; thus, its electric charge will have opposite quantum numbers in order to confirm that proton has a positive charge and anti-proton has a negative charge [15, 31, 36]. Subsequently, the energy within the antimatter will be released as a mass-energy equation of $E = mc^2$. Simply, it is a confirmation that Earth is consisting entirely of standard matter, which is equal to antimatter (H^*) which is having free quantum fields (n) in both activated and excited states (Fig. 17.3).

Here, the quantum field of the antimatter is the components of all free quantum momentum of Earth; thus, its components (t, x) are being represented into a vector $x = (ct, x)$ considering superposition of the energy momentum in a common metric

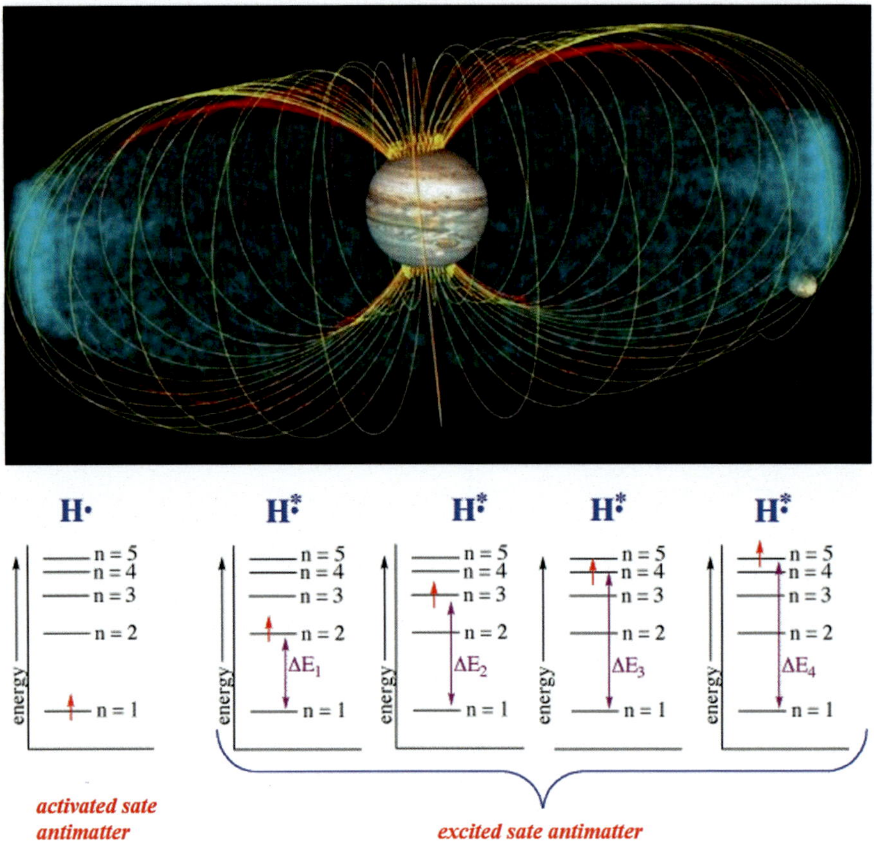

Fig. 17.3 The concept of antimatter distribution energy (H*) in the time and space due to the quantum mechanical functions of its free quantum fields (n) corresponding to its vectorial superposition of activated state and excited state

signature convention of $\eta_{\mu\nu} =$ diag $(\pm 1, \mp 1, \mp 1, \mp 1)$ of Klein–Gordon equations that can be expressed by the following Table 17.1:

Here, $\pm\eta^{\mu\nu}\partial_\mu\partial_\nu$ is representing the d'Alembert function and ∇^2 is representing the Laplace variable, while c is representing the energy speed, and ħ is the Planck constant, while $c = \hbar = 1$.

Here, the Klein–Gordon equation permits two variables where one variable is positive and another is negative, which can be obtained by describing a relativistic wavefunction of the antimatter as:

$$\left[\nabla^2 - \frac{m^2c^2}{\hbar^2}\right]\psi(r) = 0 \qquad (17.25)$$

Table 17.1 Klein–Gordon equation in normal units with metric of $\eta_{\mu\nu} = \text{diag}(\pm 1, \mp 1, \mp 1, \mp 1)$

	Position space $x = (ct, x)$	Fourier transformation $\omega = \frac{E}{\hbar},\ k = \frac{p}{\hbar}$	Momentum space $p = \left(, \frac{E}{c}, p\right)$
Time and space	$\left(\frac{1}{c^2}\frac{\partial^2}{\partial t^2} - \nabla^2 + \frac{m^2c^2}{\hbar^2}\right)\psi(t,x) = 0$	$\psi(t,x) = \int \frac{d\omega}{2\pi\hbar} \int \frac{d^3k}{(2\pi\hbar)^3} e^{\mp i(\omega t - k.x)} \psi(\omega, k)$	$E^2 = p^2c^2 + m^2c^4$
Four-vector form	$(c + \mu^2)\psi(x) = 0,\ \mu = mc/\hbar$	$\psi(x) = \int \frac{d^4p}{(2\pi\hbar)^4} e^{-ip.\frac{x}{\hbar}} \psi(p)$	$p^2 = \pm m^2c^2$

Since the Klein–Gordon function is a standard unit, $(c + m^2)\psi(x) = 0$,with the metric signature of $\eta_{\mu\nu} = \text{diag}(\pm 1, -1, -1, -1)$, it is solved by Fourier transformation as:

$$\psi(x) = \int \frac{d^4p}{(2\pi)^4} \mathrm{e}^{-ip.x}\psi(p) \tag{17.26}$$

and using spherical dimension of Earth, the complex exponentials are described here as:

$$p^2 = (p^0)^2 - p^2 = m^2 \tag{17.27}$$

This result will confirm the momentums of the antimatter's positive and negative energy dynamics as:

$$p^0 = \pm E(p) \text{ where } E(p) = \sqrt{p^2 + m^2}. \tag{17.28}$$

Using a new set of constant $C_{(\mathrm{p})}$,the solution will become

$$\psi(x) = \int \frac{\mathrm{d}^4p}{(2\pi)^4} \mathrm{e}^{ip.x} C(P)\delta(P^0)^2 - E(P)^2) \tag{17.29}$$

It is a standard practice to segregate the positive and negative energy dynamics while working with the positive energy only as:

$$\psi(x) = \int \frac{\mathrm{d}^4p}{(2\pi)^4} \delta\left((p^0)^2 - E(p)^2\right)\left(A(p)\mathrm{e}^{-ip^0x^0 + ip^ix^i} + B(p)\mathrm{e}^{+ip^0x^0 + ip^ix^i}\right)\theta(p^0) \tag{17.30}$$

$$= \int \frac{\mathrm{d}^4p}{(2\pi)^4} \delta\left((p^0)^2 - E(p)^2\right)\left(A(P)\mathrm{e}^{-ip^0x^0 + ip^ix^i} + B(-P)\mathrm{e}^{+ip^0x^0 - ip^ix^i}\right)\theta(p^0)$$

$$\rightarrow \int \frac{\mathrm{d}^4p}{(2\pi)^4} \delta\left((p^0)^2 - E(p)^2\right)\left(A(p)\mathrm{e}^{-ip.x} + B(P)\mathrm{e}^{+ip.x}\right)\theta(p^0) \tag{17.31}$$

Here, $B(p) \rightarrow B(-p)$ can be considered p^0- to pick up the positive frequency from its delta as:

$$\psi(x) = \int \frac{\mathrm{d}^4p}{(2\pi)^4} \frac{\delta(p^0 - E(p))}{2E(p)} \left(A(p)\mathrm{e}^{-ip.x} + B(p)\mathrm{e}^{+ip.x}\right)\theta(p^0) \tag{17.32}$$

$$= \int \frac{\mathrm{d}^3p}{(2\pi)^3} \frac{1}{2E(p)} \left(A(p)\mathrm{e}^{-ip.x} + B(p)\mathrm{e}^{+ip.x}\right)\Big|_{p^0 = +E(p)} \tag{17.33}$$

Since it is a general solution to the Klein–Gordon equation, Lorentz invariant quantities of $p.\ x = p_\mu x^\mu$ is the final variable to solve the Klein–Gordon equation [14, 16, 38]. Subsequently, the Lorentz invariance will absorb the ½$E(p)$ factor into the coefficient $A(p)$ and $B(p)$, and thus, the antimatter energy of a free particle can be written as:

$$\frac{p^2}{2m} = E \tag{17.34}$$

By quantizing this, it can confirm the non-relativistic Schrödinger equation for a free particle energy as:

$$\frac{\widehat{p}^2}{2m}\psi = \widehat{E}\psi \tag{17.35}$$

where $\widehat{p} = -i\hbar\nabla$ is representing the momentum function (∇ is the delta variable) and $\widehat{E} = i\hbar\frac{\partial}{\partial t}$ is the energy dynamic.

Since the Schrödinger equation is a non-relativistic invariant, it is standard to use its free particle as the momentum energy as:

$$\sqrt{p^2c^2 + m^2c^4} = E \tag{17.36}$$

Here, implementing the quantum momentum of the free particle, the energy yields can be written as:

$$\sqrt{(-i\hbar\nabla)^2c^2 + m^2c^4\psi} = i\hbar\frac{\partial}{\partial t}\psi \tag{17.37}$$

Subsequently, the electromagnetic fields of the energy yield can be further clarified by the Klein and Gordon equation of e identity, i.e., $p^2c^2 + m^2c^4 = E^2$, which will give:

$$\left((-i\hbar\nabla)^2c^2 + m^2c^4\right)\psi = \left(i\hbar\frac{\partial}{\partial t}\right)^2\psi \tag{17.38}$$

which simplifies to

$$-\hbar^2c^2\nabla^2\psi + m^2c^4\psi = \hbar^2\frac{\partial^2}{\partial t^2}\psi \tag{17.39}$$

Rearranging terms yields

$$\frac{1}{c^2}\frac{\partial^2}{\partial t^2}\psi - \nabla^2\psi + \frac{m^2c^2}{\hbar^2}\psi = 0 \tag{17.40}$$

Rewriting these functions considering the inverse metric ($-c^2$, 1, 1, 1) in relation to Einstein's summation variable, it can be written as:

$$-\eta^{\mu\nu}\partial_\mu\partial_\nu\psi = \sum_{\mu=0}^{\mu=3}\sum_{\nu=0}^{\nu=3} -\eta^{\mu\nu}\partial_\mu\partial_\nu\psi = \frac{1}{c^2}\partial_0^2\psi - \sum_{\nu=1}^{\nu=3}\partial_\nu\partial_\nu\psi = \frac{1}{c^2}\frac{\partial^2}{\partial t^2}\psi - \nabla^2\psi \tag{17.41}$$

Therefore, the Klein–Gordon equation can be expressed as a covariant function, which is the abbreviation of the form of $(c + \mu^2)\psi = 0$, where $\mu = \frac{mc}{\hbar}$, and $c = \frac{1}{c^2}\frac{\partial^2}{\partial t^2} - \nabla^2$, and thus, the Klein–Gordon equation can be described as a field of energy dynamics $V(\psi)$ as:

$$c\psi + \frac{\partial V}{\partial \psi} = 0 \tag{17.42}$$

Here, the gauge U(1) symmetry is a complex field $\varphi(x) \epsilon \mathbb{C}$ that can satisfy the Klein–Gordon as:

$$\partial_\mu j^\mu(x) = 0, j^\mu(x) = \frac{e}{2m}(\varphi^*(x)\partial^\mu\varphi(x) - \varphi(x)\partial^\mu\varphi^*(x)) \tag{17.43}$$

Since free Klein–Gordon equation is a *classical* Schrödinger field, it is further analyzed by quantum field of Klein–Gordon of the time and space [3, 42, 45]. Subsequently, the quantum field of Klein–Gordon equation is being analyzed to confirm the symmetry of the antimatter energy wavefunction φ is in the local U(1) gauge of the sphere of $\varphi \rightarrow \varphi' = \exp(i\theta)\varphi$ of the Earth. Where $\theta(t, x)$ is a local vectorial angle, which transforms the wavefunction into a complex phase as $\exp(i\theta) = \cos\theta + i\sin\theta$, where derivative ∂_μ can be replaced by gauge-covariant variable $D_\mu = \partial_\mu - ieA_\mu$, while the gauge fields transform as $eA_\mu \rightarrow eA'_\mu + \partial_\mu\theta$, which confirm that generation of (−,+,+,+) metric signature of the antimatter in the Earth as:

$$D_\mu D^\mu\varphi = -(\partial_t - ieA_0)^2\varphi + (\partial_i - ieA_i)^2\varphi = m^2\varphi \tag{17.44}$$

where A is the scalar potential and which can be simplified as functional term of $D_\mu D^\mu\varphi + AF^{\mu\nu}D_\mu\varphi D_\nu(D_\alpha D^\alpha\varphi) = 0$, and thus, the scalar field can be multiplied by i to confirm the for a field of energy dynamics of antimatter (A_m) is being generated sufficiently enough in the Earth by expressing as:

$$A_m = \int_x \eta^{\mu\nu}(\partial_\mu\varphi^* + ieA_\mu\varphi^*)(\partial_\nu\varphi - ieA_\nu\varphi) = \int_x |D\varphi|^2 \tag{17.45}$$

Activation of Antimatter to Cool the Earth

Here, the antimatter is being modeled to deform its cooling-state energy by analyzing its nano-point break waveguide energy momentum of the Earth [5, 17, 39]. Then, the cluster of its energy momentum is being calculated considering its nano-point deflect waveguide to decode the quantum dynamic into cool-state energy (HcE^{-}) by expressing the following *Hossain* equation as:

$$H = C + \sum \omega_{ci} a_i^{\dagger} a_i + \sum_K \omega_k b_k^{\dagger} b_k + \sum_{ik} \left(V_{ik} a^{\dagger} b_k + V_{ik}^{*} b_k^{\dagger} a_i \right), \tag{17.46}$$

where $a_i \left(a_i^{\dagger} \right)$ and $b_k \left(b_k^{\dagger} \right)$ presents the function of the nano-point break mode and its energy-dynamic modules of its cofactor, V_{ik}, denotes the magnitude of the energy mode within its nanostructures.

Simply, this cool-state energy modules will form HcE^{-} from its energy dynamics to cool the Earth considering its internuclear distances of 1D-H($n = 1$) + H ($n = 2,...,\infty$), 2D-H($n = 2$) + H($n = 2,...,\infty$), and 3D-H($n = \infty$) + H ($n =$ unlimited,...,∞), where n and l are the primary and vectorial quanta of the cool-state energy. Thus, the energy state quanta considering time-dependent energy intensity (k) and space-dependent energy wavelength (λ) confirm the mode of energy dynamics within the Earth to activate this energy into cool-state energy (Fig. 17.4).

This cool-state energy can be defined by hidden energy transformation characteristics of the cool energy quanta can be expressed by:

$$E = E_L - E_O \left\{ \exp^{\left[\frac{q\left(V + I_{RS}\right)}{AkTc} \right]} - 1 \right\} - \frac{(V + I_{RS})}{R_S}, \tag{17.47}$$

where E_L presents the cool energy generation, E_O is the energy dynamics, and R_s is the variable function for the adverse condition. Here A, representation of the activated module energy in related to Boltzmann's constant, k (= 1.38×10^{-23} W/m^2K) where q (= 1.6×10^{-19}C) is the electrons of activated energy, and T_C is the dynamic proliferation of the energy and thus, the energy dynamics can be expressed by:

$$\mathrm{E_O} = \mathrm{E_{RS}} \left(\frac{T_C}{T_{ref}} \right)^3 \exp \left[\frac{qEG\left(\frac{1}{T_{ref}} - \frac{1}{T_C} \right)}{KA} \right]. \tag{17.48}$$

Here, E_{RS} denotes the functional energy that relies on the energy speed and qEG denotes the energy flow per unit area of the cool-state energy [4, 6, 37].

Simply, in this cooling energy mode, the energy dynamics is related to *T–R* module of the time and space that can be expressed by:

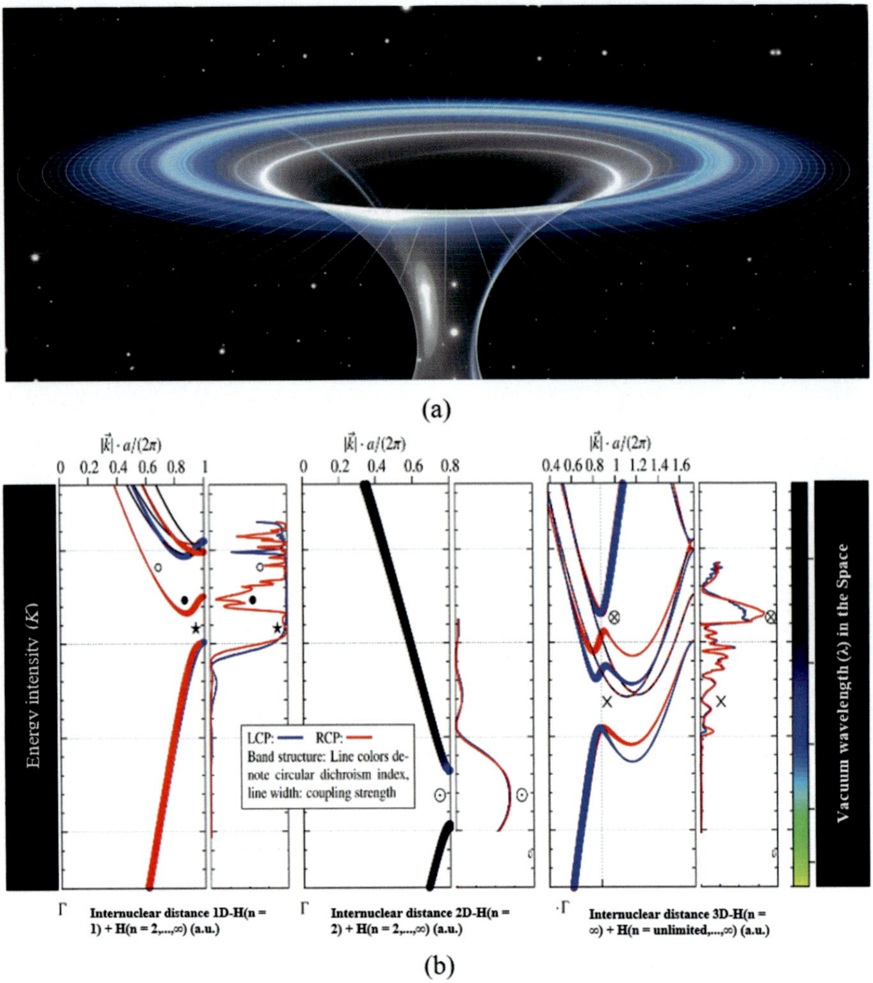

Fig. 17.4 (**a**) The conceptual array of the cooling energy distributions in the time and space and (**b**) semiclassical pathways of energy band structure (LCP and RCP) that generate cool-state energy within the internuclear distance (a.u.) where red and blue represent the excited and activated states of cool-energy corresponding to its wavelength in the space and energy intensity of (k) to the time

$$T = -IR_s + K \log\left[\frac{I_L - I + I_O}{I_O}\right], \tag{17.49}$$

where K represents the constant $\left(= \frac{AkT}{q}\right)$ and I_{mo} and V_{mo} are the energy dynamics. Therefore, the correlation between I_{mo} and V_{mo} is similar as the T–R module of the Earth plane, which can be expressed precisely as:

$$T_{\mathrm{mo}} = -I_{\mathrm{mo}}R_{\mathrm{Smo}} + K_{\mathrm{mo}} \log\left(\frac{I_{\mathrm{Lmo}} - I_{\mathrm{mo}} + I_{\mathrm{Omo}}}{I_{\mathrm{Omo}}}\right) \tag{17.50}$$

where I_{Lmo} denotes the net energy generation, I_{Omo} denotes the activated energy dynamics, R_{smo} is the inactivated energy dynamics, and K_{mo} is the constant cofactor related to $K_{\mathrm{mo}} = N_{\mathrm{S}} \times K$, and thus, net cool energy dynamic in the Earth can be written as:

$$T_{\mathrm{mo}} = -I_{\mathrm{mo}}N_{\mathrm{S}}R_{\mathrm{S}} + N_{\mathrm{S}}K \log\left(\frac{I_{\mathrm{L}} - I_{\mathrm{mo}} + I_{\mathrm{o}}}{I_{\mathrm{o}}}\right). \tag{17.51}$$

Here, energy-generated cooling thermal conductivity relies primarily on energy flux and relativistic thermal state of the Earth; thus, the thermal conductivity could be expressed as follows:

$$I_{\mathrm{L}} = G\left[I_{\mathrm{SC}} + K_{\mathrm{I}}\left(T_{\mathrm{cool}}\right)\right]V_{\mathrm{mo}} \tag{17.52}$$

$$T_{\mathrm{cool}} = T\left(\frac{I_{\mathrm{L}}}{(G * V_{\mathrm{mo}}) \times (I_{\mathrm{sc}} + K_{\mathrm{I}})}\right) \tag{17.53}$$

where I_{sc} is the energy dynamics per unit area, K_{I} is the relativistic energy coefficient, T_{cool} is the cooling thermal of the energy, and G is the cool energy thermal conductivity of the Earth that will confirm to cool the earth naturally.

.

Activation of Antimatter to Heat the Earth

To transform cool-state energy into heat-state energy, the local symmetry of Higgs-Boson quantum field of the Earth has been used. Simply, the Earth has been simulated considering Abelian local symmetries by applying Higgs-Boson electromagnetic field of the time and space to form congenial heat on it [26, 27, 34]. Simply the momentum of energy dynamic will interrupt the gauge-field symmetries of the Earth, and thus, the Goldstone scalar field will act as the longitude modes of the Higgs-Boson field [46, 56, 59]. Therefore, the local symmetries of quantum field of the Earth will be broken down as particle of T^{α} related to the gauge field of $A_{\mu}^{\alpha}(x)$, and thus, the Higgs-Boson quantum fields will then start to generate local U(1) phase symmetries to create heat [13, 21, 30]. So, this transformation process can be consisting of the precise scalar fields $\Phi(x)$ of energically charged q paired with the electromagnetic field $A^{\mu}(x)$ of the Earth that could be written as:

$$\mathcal{L} = -\frac{1}{4} F_{\mu\nu}F^{\mu\nu} + D_{\mu}\Phi^{*} D^{\mu}\Phi - V(\Phi^{*}\Phi), \tag{17.54}$$

where

$$D_\mu \Phi(x) = \partial_\mu \Phi(x) + iqA_\mu(x)\Phi(x)$$
$$D_\mu \Phi^*(x) = \partial_\mu \Phi^*(\mathrm{x}) - iqA_\mu(x)\Phi^*(x), \tag{17.55}$$

and

$$V(\Phi^*\Phi) = \frac{\lambda}{2}(\Phi^*\Phi)^2 + m^2(\Phi^*\Phi). \tag{17.56}$$

Here $\lambda > 0$ but $m^2 < 0$; therefore, $\Phi = 0$ is a local peak factor of the scalar field, and the precise form of quantum dynamics is $\Phi = \frac{v}{\sqrt{2}} * \mathrm{e}^{i\theta}$ that can be expressed as:

$$v = \sqrt{\frac{-2m^2}{\lambda}} \text{ for any real } \theta. \tag{17.57}$$

Consequently, here, the scalar quantum dynamics $\Phi(x)$ will form a nonzero value of $\langle \Phi \rangle \neq 0$, which will form the local U(1) symmetries into electromagnetic field of Earth. Therefore, in this local U(1) symmetries, the quantum dynamics of $\Phi(x)$ will be an activated scalar field of the Earth that can be expressed as:

$$\Phi(\mathrm{x}) = \frac{1}{\sqrt{2}}\Phi_\mathrm{r}(x) * \mathrm{e}^{i\Theta(x)}, \text{real } \Phi_\mathrm{r}(x) > 0, \text{real } \Phi(x). \tag{17.58}$$

Since the scalar field of time and space is a dynamic momentum of the quantum Φ $(x) = 0$, it will meet the principle of $\langle \Phi \rangle \neq 0$ to be considered as the instinctual function of the Earth. Thus, considering the momentum field of $\phi_\mathrm{r}(x)$ and Θ (x), the scalar field related to the radial field of ϕ_r of the Earth can be simplified:

$$V(\phi) = \frac{\lambda}{8}\left(\phi_\mathrm{r}^2 - v^2\right)^2 + \text{const}. \tag{17.59}$$

Subsequently, the momentum field can be shifted by applying the variable scalars, $\Phi_\mathrm{r}(x) = v + \sigma(x)$, and expressed as:

$$\phi_\mathrm{r}^2 - v^2 = (v+\sigma)^2 - v^2 = 2v\sigma + \sigma^2 \tag{17.60}$$

$$V = \frac{\lambda}{8}\left(2v\sigma - \sigma^2\right)^2 = \frac{\lambda v^2}{2} * \sigma^2 + \frac{\lambda v}{2} * \sigma^3 + \frac{\lambda}{8} * \sigma^4. \tag{17.61}$$

Thus, the functional derivative $D_\mu \phi$ will become

$$D_\mu \phi = \frac{1}{\sqrt{2}}\left(\partial_\mu\left(\phi_\mathrm{r}\mathrm{e}^{i\Theta}\right) + iqA_\mu * \phi_r \mathrm{e}^{i\Theta}\right)$$
$$= \frac{\mathrm{e}^{i\Theta}}{\sqrt{2}}\left(\partial_\mu \phi_\mathrm{r} + \phi_\mathrm{r} * i\partial_\mu \Theta + \phi_\mathrm{r} * iqA_\mu\right) \tag{17.62}$$

$$= \frac{1}{2}\left(\partial_\mu \sigma\right)^2 + \frac{(v+\sigma)^2}{2} * \left(\partial_\mu \Theta + qA_\mu\right)^2 \tag{17.63}$$

The Lagrangian is then given by

$$\mathcal{L} = \frac{1}{2}\left(\partial_\mu \sigma\right)^2 - v(\sigma) - \frac{1}{4}F_{\mu\nu}F^{\mu\nu} + \frac{(v+\sigma)^2}{2} * \left(\partial_\mu \Theta + qA_\mu\right)^2. \tag{17.64}$$

Simply, to conduct the heat ($\mathcal{L}_{\text{heat}}$) generation through electromagnetic field of the Earth considering this *Lagrangian*, it has been further expanded $\mathcal{L}_{\text{heat}}$ as the energy dynamics of the electromagnetic field of the Earth that can be expressed as free particles of heat:

$$\mathcal{L}_{\text{heat}} = \frac{1}{2}\left(\partial_\mu \sigma\right)^2 - \frac{\lambda v^2}{2} * \sigma^2 - \frac{1}{4}F_{\mu\nu}F^{\mu\nu} + \frac{v^2}{2} * \left(qA_\mu + \partial_\mu \Theta\right)^2 \tag{17.65}$$

Subsequently, it will initiate to generate heating energy into the quantum field of the Earth considering the scalar particles of λv^2 corresponding to its energy spectrum of singlet and triplet system of the Earth will finally heat the Earth, respectively (Fig. 17.5).

Results and Discussion

Modeling of Dark Energy (D_e)

Hence, the following equation is being computed as the dark energy clarification (P_{pv}) which in fact, the dark energy (D_{e}) irradiance from Earth's gauge field U(1) at vectorial (V) dynamics of its tangential (T) field of the Earth and thus, the equation can be written as:

$$D_{\text{e}} = P_{\text{pv}} = \eta_{\text{pvg}} A_{\text{pvg}} G_{\text{t}} \tag{17.66}$$

Here, η_{pvg} denotes the Earth plane performance rate, A_{pvg} denotes the gauge field (m^2), and G_{t} denotes the dark irradiance intaking rate on the Earth sphere (W/m^2), and thus, the η_{pvg} could be rewritten as follows:

$$\eta_{\text{pvg}} = \eta_{\text{r}}\eta_{\text{pc}}\left[1 - \beta\left(T_{\text{c}} - T_{\text{c ref}}\right)\right] \tag{17.67}$$

η_{pc} denotes the dark energy formation efficiency, when maximum power point tracking (MPPT) is being implemented. Here β denotes as the time cofactor, η_{r} denotes the mode of dark energy efficiency, and $T_{\text{c ref}}$ denotes the condition of tangential, and thus, the equation can be rewritten as:

(a)

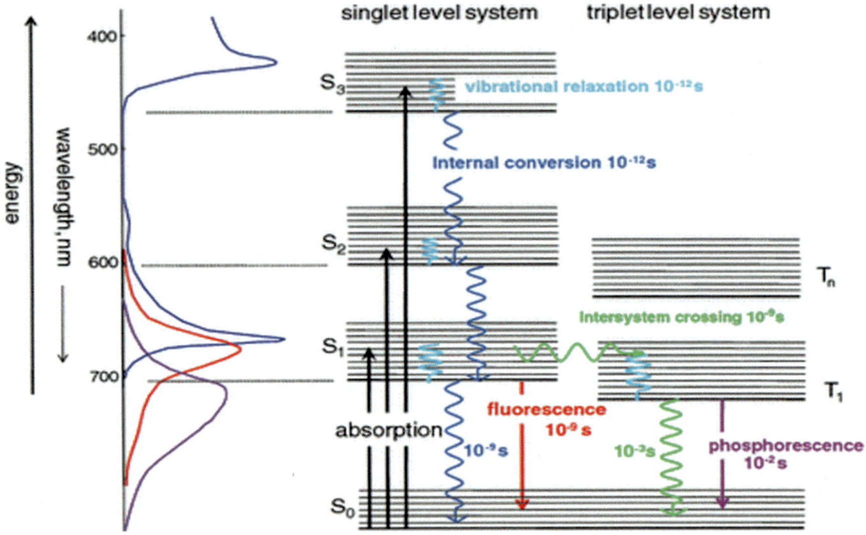

(b)

Fig. 17.5 (**a**) The concept of heating mechanism of time and space by the theoretical function of energy dynamics of optimum energy formation and (**b**) the mode of energy formation corresponding to it wavelength of energy spectrum in relation to its absorbance of singlet and triplet level system

$$T_c = T_a + \left(\frac{\text{NOCT} - 20}{800}\right) G_t \tag{17.68}$$

T_a denotes the encircling tangential, G_t denotes the energy radiation in Earth sphere (W/m^2), and thus, considering this condition, the net dark radiation on Earth surface can be calculated by the equation below:

$$I_t = I_b R_b + I_d R_d + (I_b + I_d) R_r \tag{17.69}$$

The dark energy here is necessarily working as a conceptual P–N junction superconductor to form electricity through the Earth sphere, which is interlinked in a parallel series connection [18, 22, 32]. Thus, a unique conceptual circuit model with respect to the N_s series of Earth surface and N_p parallel arrays has been computed by the following Earth surface dark energy equation based on electron charge and energy relationship

$$I = N_p \left[I_{ph} - I_{rs} \left[\exp\left(\frac{q(V + IR_s)}{\mathrm{AKTN_s}} - 1 \right) \right] \right] \tag{17.70}$$

where

$$I_{rs} = I_{rr} \left(\frac{T}{T_r} \right)^3 \exp\left[\frac{E_G}{AK} \left(\frac{1}{T_r} - \frac{1}{T} \right) \right] \tag{17.71}$$

Hence, in Eqs. (17.5) and (17.6), q denotes the electron charge (1.6×10^{-19} C), K denotes Boltzmann's constant, A denotes the diode standardized efficiency, and T denotes the tangential [47, 49, 58]. Accordingly, I_{rs} denotes the Earth surface reversed dark energy motion at T, where T_r denotes the Earth condition tangential, I_{rr} denotes the reversed dark energy at T_r, and E_G denotes the dark energy band-gap energy of the Earth surface, and thus, the dark energy I_{ph} generation on Earth surface can be expressed by:

$$I_{ph} = \left[I_{SCR} + k_i (T - T_r) \frac{S}{100} \right] \tag{17.72}$$

Here, I_{SCR} denotes the energy motion considering the optimum energy radiation dynamic on the Earth, k_i denotes the energy-motion cofactor, and S denotes the dark radiation calculation in a unit area (mW/cm^2), and thus, the conceptual model of the net dark energy generation can be expressed as:

$$I = I_{ph} - I_D \tag{17.73}$$

$$I = I_{ph} - I_0 \left[\exp\left(\frac{q(V + R_s I)}{\mathrm{AKT}} - 1 \right) \right] \tag{17.74}$$

I_{ph} denotes the dark energy dynamic (A), I_D denotes the originated energy dynamic (A), I_0 denotes the inversed energy dynamic (A), q denotes the charge of the electron (1.6×10^{-19} C), K denotes Boltzmann's constant, T denotes the Earth tangential, R_s denotes the energy cofactor, R_{sh} denotes the energy coefficient, I denotes the dark energy motion (A), and V denotes the energy velocity (V). Therefore, the net dark energy flow into the Earth surface can be determined by conducting the following equation:

$$I = I_{PV} - I_{D1} - \left(\frac{V + IR_S}{R_{SH}} \right) \tag{17.75}$$

where

$$I_{\mathrm{D1}} = I_{01}\left[\exp\left(\frac{V + IR_{\mathrm{s}}}{a_1 V_{\mathrm{T1}}}\right) - 1\right] \tag{17.76}$$

Here, I and I_{01} are being denoted as the reversed dark energy flow into the conceptual dark energy vector, and V_{T1} and V_{T2} are being denoted as the optimum energy condition [48, 50, 57]. Thus, the vector of Earth is being presented by a_1 and a_2, and then it has normalized the mode of Earth surface by expressing the following equation to confirm the net dark energy generation on any array of Earth as:

$$v_{\mathrm{oc}} = \frac{V_{\mathrm{oc}}}{cK\,T/q} \tag{17.77}$$

$$P_{\max} = \frac{\frac{V_{\mathrm{oc}}}{cK\,T/q} - \ln\left(\frac{V_{\mathrm{oc}}}{cK\,T/q} + 0.72\right)}{\left(1 + \frac{V_{\mathrm{oc}}}{K\,T/q}\right)}\left(1 - \frac{V_{\mathrm{oc}}}{\frac{V_{\mathrm{oc}}}{I_{\mathrm{SC}}}}\right) \times \left(\frac{V_{\mathrm{oc0}}}{1 + \beta \ln \frac{G_0}{G}}\right)\left(\frac{T_{\mathrm{o}}}{T}\right)^{y} I_{\mathrm{sc0}}\left(\frac{G}{G_o}\right)^{a} \tag{17.78}$$

where ν_{oc} denotes the standard point of the open-energy flow, V_{oc} denotes the energy condition $V_{\mathrm{t}} = nkT/q$, c denotes the constant energy motion, K denotes Boltzmann's constant, T denotes the tangential of Earth surface, α denotes the function that represents the nonlinear motion of energy, q denotes the electron charge, γ denotes the function acting for all nonlinear energy, while β denotes the Earth surface mode for specific dimensionless function for enhancing energy flowing rate.

Activation of Dark Energy ($\mathbf{D}_e$) to Break the CO_2

To activate dark energy to break the CO_2 on Earth, the global Albanian symmetries of gauge field have been analyzed to confirm the functional test of the dark energy on CO_2. Thus, the dark energy $\overrightarrow{De} = S_0\left(1 + \cos\left(2\pi \frac{n}{x}\right)\right)$ has been analyzed considering the net dark energy particle T^{α} at the global symmetrical array of Earth sphere by initiating the gauge field of $A_{\mu}^{\alpha}(x)$. The results suggested that the global U(1) phase symmetry to deliver dark energy in relation to the complex vector field of $\Phi(x)$ of Earth sphere where dark energy charge q will couple with the EM field of $A^{\mu}(x)$ and thus, the equation can be expressed by ɧ:

$$P_{\max} = \frac{\frac{V_{\mathrm{oc}}}{cK\,T/q} - \ln\left(\frac{V_{\mathrm{oc}}}{cK\,T/q} + 0.72\right)}{\left(1 + \frac{V_{\mathrm{oc}}}{K\,T/q}\right)}\left(1 - \frac{V_{\mathrm{oc}}}{\frac{V_{\mathrm{oc}}}{I_{\mathrm{SC}}}}\right) \times \left(\frac{V_{\mathrm{oc0}}}{1 + \beta \ln \frac{G_0}{G}}\right)\left(\frac{T_o}{T}\right)^{y} I_{\mathrm{sc0}}\left(\frac{G}{G_o}\right)^{a} \tag{17.79}$$

$$= -\frac{1}{4}F_{\mu\nu}F^{\mu\nu} + D_\mu\Phi^*\, D^\mu\Phi - V(\Phi^*\Phi) \tag{17.80}$$

where

$$D_\mu\Phi(x) = \partial_\mu\Phi(x) + iqA_\mu(x)\Phi(x) \tag{17.81}$$

$$D_\mu\Phi^*(x) = \partial_\mu\Phi^*(x) - iqA_\mu(x)\Phi^*(x) \tag{17.82}$$

and

$$V(\Phi^*\Phi) = \frac{\lambda}{2}(\Phi^*\Phi)^2 + m^2(\Phi^*\Phi) \tag{17.83}$$

Here $\lambda > 0m^2 < 0$; therefore $\Phi = 0$ is a local optimum vector quantity, while the minimum form of degenerated scalar circle is clarified as $\Phi = \frac{v}{\sqrt{2}} * e^{i\theta}$,

$$v = \sqrt{\frac{-2m^2}{\lambda}}, \text{ any real } \theta \tag{17.84}$$

Subsequently, the vector field Φ of the global Earth sphere will form a nonzero functional value $\langle\Phi\rangle \neq 0$, which will simultaneously determine the U(1) symmetrical net dark energy generation. Therefore, the global U(1) net symmetrical dark energy of $\Phi(x)$ will be delivered as expected value of $\langle\Phi\rangle$ by confirming the x-dependent state of the symmetrical $\Phi(x)$ array of Earth sphere and can be expressed by the following equation:

$$\Phi(\mathrm{x}) = \frac{1}{\sqrt{2}}\,\Phi_r(x) * e^{i\Theta(x)}, \text{real } \Phi_\mathrm{r}(\mathrm{x}) > 0, \text{real } \Phi(x) \tag{17.85}$$

Then, the net calculation of the hypercharge energy generation from dark energy is being determined considering the vector $\Phi(x) = 0$, and it is first-order function of $\langle\Phi\rangle \neq 0$, considering the peak level of dark energy emission on the Earth sphere of $\Phi\langle x\rangle \neq 0$ [19, 25, 28]. Thus, the net hypercharge energy generation from the dark energy calculation $\phi_\mathrm{r}(x)$ and $\Theta(x)$, its vector on the Earth field ϕ_r has been confirmed by conducting the following equation:

$$V(\phi) = \frac{\lambda}{8}\left(\phi_\mathrm{r}^2 - v^2\right)^2 + \text{const}, \tag{17.86}$$

or the resultant hypercharge energy generation is shifted by its VEV, $\Phi_\mathrm{r}(x) = v + \sigma(x)$,

$$\phi_\mathrm{r}^2 - v^2 = (v+\sigma)^2 - v^2 = 2v\sigma + \sigma^2 \tag{17.87}$$

$$V = \frac{\lambda}{8}\left(2v\sigma - \sigma^2\right)^2 = \frac{\lambda v^2}{2} * \sigma^2 + \frac{\lambda v}{2} * \sigma^3 + \frac{\lambda}{8} * \sigma^4 \tag{17.88}$$

Simultaneously, the functional derivative $D_\mu \phi$ will become

$$\begin{aligned} D_\mu \phi &= \frac{1}{\sqrt{2}}\left(\partial_\mu\left(\phi_{\mathrm{r}} \mathrm{e}^{i\Theta}\right) + iqA_\mu * \phi_r \mathrm{e}^{i\Theta}\right) = \frac{\mathrm{e}^{i\Theta}}{\sqrt{2}} \\ &\times \left(\partial_\mu \phi_{\mathrm{r}} + \phi_{\mathrm{r}} * i\partial_\mu \Theta + \phi_{\mathrm{r}} * iqA_\mu\right) \end{aligned} \tag{17.89}$$

$$\begin{aligned} \left|D_\mu \phi\right|^2 &= \frac{1}{2}\left|\partial_\mu \phi_{\mathrm{r}} + \phi_{\mathrm{r}} * i\partial_\mu \Theta + \phi_{\mathrm{r}} * iqA_\mu\right|^2 \\ &= \frac{1}{2}\left(\partial_\mu \phi_{\mathrm{r}}\right) + \frac{\phi_{\mathrm{r}}^2}{2} * \left(\partial_\mu \Theta q A_\mu\right)^2 \\ &= \frac{1}{2}\left(\partial_\mu \sigma\right)^2 + \frac{(v+\sigma)^2}{2} * \left(\partial_\mu \Theta + qA_\mu\right)^2 \end{aligned} \tag{17.90}$$

Altogether,

$$= \frac{1}{2}\left(\partial_\mu \sigma\right)^2 - v(\sigma) - \frac{1}{4} F_{\mu\upsilon} F^{\mu\upsilon} + \frac{(v+\sigma)^2}{2} * \left(\partial_\mu \Theta + qA_\mu\right)^2 \tag{17.91}$$

To determine the formation of this net hypercharge energy from dark energy referred as ($ɧ_{\mathrm{sef}}$) into the Earth, the function of the hyper-static fields has been quantified by conducting the quadratic calculation and described by the following equation:

$$_{\mathrm{sef}} = \frac{1}{2}\left(\partial_\mu \sigma\right)^2 - \frac{\lambda v^2}{2} * \sigma^2 - \frac{1}{4} F_{\mu\upsilon} F^{\mu\upsilon} + \frac{v^2}{2} * \left(qA_\mu + \partial_\mu \Theta\right)^2 \tag{17.92}$$

Here this net hypercharge energy generation ($ɧ_{\mathrm{free}}$) function certainly will admit a realistic vector particle of positive mass$^2 = \lambda v^2$ integrating the areal $A_\mu(x)$ function, which is extremely sensitive to CO_2, and thus, the equation can be written as:

$$\begin{aligned} _{\mathrm{sef}} &= \frac{1}{2}\left(\partial_\mu \sigma\right)^2 - \frac{\lambda v^2}{2} * \sigma^2 - \frac{1}{4} F_{\mu\upsilon} F^{\mu\upsilon} + \frac{v^2}{2} * \left(qA_\mu + \partial_\mu \Theta\right)^2 \\ &= 2\left[\frac{Ed}{c}, \overrightarrow{Pd}\right] + \left[\overrightarrow{\mathrm{CO}_2}\right] - \left[m_{\mathrm{e}} \overrightarrow{c^2}\right] \end{aligned} \tag{17.93}$$

So,

$$_{\mathrm{sef}} = 2\left[\frac{Ed}{c}, \overrightarrow{Pd}\right] + \left[\overrightarrow{\mathrm{CO}_2}\right] - \left[m_{\mathrm{e}} \overrightarrow{c^2}\right] \tag{17.94}$$

Here, the hypercharge energy of dark energy releases the electron, and thus, the equation can be rewritten as:

$$_{\text{sef}} = 2\left[\frac{Ed}{c}, \overrightarrow{Pd}\right] + \left[\overrightarrow{\text{C} + \text{O}_2}\right] - \left[\overrightarrow{m_e c^2}\right] \tag{17.95}$$

Since $\text{ɧ}_{\text{sef}} = D_e$, dark energy finally can be written as:

$$D_e = 2\left[\frac{Ed}{c}, \overrightarrow{Pd}\right] + \left[\overrightarrow{\text{C} + \text{O}_2}\right] - \left[\overrightarrow{m_e c^2}\right] \tag{17.96}$$

Simply, the result suggested that dark energy can play vital role to breakdown CO_2 by the activation of hypercharge energy momentum within its energy generation fields $\Theta(x)$ that indeed will confirm to mitigate the global CO_2 dramatically from Earth.

Modeling of the Antimatter

To computationally confirm the formation of cool-state energy (HcE^-) in the time and space, the vectorial momentum of its steady weak coupling energy field has been clarified in order to determine its absorbed energy rate per unit area of $J(\omega)$. Here, $J(\omega)$ is thus considered as the quantum field of the density of states (DOS) energy proliferation, which, in fact, is the deformed cooling energy magnitudes $V(\omega)$, which is formed from its nano-point break waveguide energy momentum [19, 29, 44]. Thus, cooling energy formation follows the Weisskopf–Wigner theory is therefore, all the generated HcE^- shall pass through all dimensional modes (1D, 2D, and 3D) within the Earth [12, 16, 18]. (Table 17.2).

Subsequently, proof Klein–Gordon equation using algebraic manipulations form a complex field $\varphi(x)$ of mass m can be expressed as a covariant notation as follows:

$$\left(c + \mu^2\right)\varphi(x) = 0 \tag{17.97}$$

and its complex conjugate

$$\left(c + \mu^2\right)\varphi^*(x) = 0 \tag{17.98}$$

Then, multiplying by the vectorial variables of $\varphi^*(x)$ and $\varphi(x)$, it can be expressed as:

$$\varphi^*\left(c + \mu^2\right)\varphi = 0 \tag{17.99}$$

Table 17.2 Cooling energy dynamics considering its densities of states (DOS) in the time and space

Energy particle	Unit area $J(\omega)$ for different DOS*	Reservoir-induced self-energy induction $\Sigma(\omega)$*
1D	$\frac{C-1}{\pi}\frac{1}{\sqrt{\omega-\omega_e}}\Theta(\omega-\omega_e)$	$-\frac{C-1}{\sqrt{\omega_e-\omega}}$
2D	$C-\eta\left[\ln\left\|\frac{\omega-\omega_0}{\omega_0}\right\|-1\right]\Theta(\omega-\omega_e)\Theta(\Omega_d-\omega)$	$C-\eta\left[\mathrm{Li}_2\left(\frac{\Omega_d-\omega_0}{\omega-\omega_0}\right)-\mathrm{Li}_2\left(\frac{\omega_0-\omega_e}{\omega_0-\omega}\right)-\ln\frac{\omega_0-\omega_e}{\Omega_d-\omega_0}\ln\frac{\omega_e-\omega}{\omega_0-\omega}\right]$
3D	$\chi\sqrt{\frac{\omega-\omega_e}{\Omega_C}}\exp\left(-\frac{\omega-\omega_e}{\Omega_C}\right)\Theta(\omega-\omega_e)$	$\chi\left[\pi\sqrt{\frac{\omega_e-\omega}{\Omega_C}}\exp\left(-\frac{\omega-\omega_e}{\Omega_C}\right)\mathrm{erfc}\sqrt{\frac{\omega_e-\omega}{\Omega_C}}-\sqrt{\pi}\right]$

The unit areas $J(\omega)$ and the generation of self-energy inductions in the time and space reservoir $\Sigma(\omega)$ demonstrated by the cooling energy corresponding to its variables of C, η, and χ as coupled forces among three dimensional (1D, 2D, and 3D) point break of energy band

$$\varphi(c+\mu^2)\varphi^*=0 \tag{17.100}$$

Subtracting the former from the latter, it can be obtained

$$\varphi^* c\varphi - \varphi c\varphi^* = 0 \tag{17.101}$$

$$\varphi^* \partial_\mu \partial^\mu \varphi - \varphi \partial_\mu \partial^\mu \varphi^* = 0 \tag{17.102}$$

Then it will confirm

$$\partial_\mu(\varphi^* \partial^\mu \varphi) = \partial_\mu \varphi^* \partial^\mu \varphi + \varphi^* \partial_\mu \partial^\mu \varphi \tag{17.103}$$

From which it can be obtained the conservation law for the Klein–Gordon field:

$$\partial_\mu j^\mu(x) = 0, j^\mu(x) \equiv \varphi^*(x)\partial^\mu \varphi(x) - \varphi(x)\partial^\mu \varphi^*(x) \tag{17.104}$$

Here, implementing the vectorial variable to the Klein–Gordon field, the scalar field-derived stress–energy tensor can be calculated as:

$$T^{\mu\nu} = \frac{\hbar^2}{m}\left(\eta^{\mu\alpha}\eta^{\nu\beta} + \eta^{\mu\beta}\eta^{\nu\alpha} - \eta^{\mu\nu}\eta^{\alpha\beta}\right)\partial_\alpha \overline{\psi}\partial_\beta \psi - \eta^{\mu\nu} mc^2 \overline{\psi}\psi \tag{17.105}$$

Here, the time–time component T^{00} will confirm the positive frequency related to the particles with free energy of antimatter as $\psi(x,t) = \varnothing(x,t)e^{-\frac{i}{\hbar}mc^2 t}$ where $\varnothing(x,t) = u_E(x)e^{-\frac{i}{\hbar}Et}$. Thus, defining the kinetic energy $E' = E - mc^2 = \sqrt{m^2c^4 + c^2p^2} - mc^2 \approx \frac{p^2}{2m}, E' \ll mc^2$ in the non-relativistic limit $v \sim p \ll c$, and hence $\left|i\hbar \frac{\partial \varnothing}{\partial t}\right| = E'\varnothing \ll mc^2 \varnothing$; thus, this yield of the derivative ψ can be written as:

$$\begin{aligned} \frac{\partial \psi}{\partial t} &= \left(-i\frac{mc^2}{\hbar}\varnothing + \frac{\partial \varnothing}{\partial t}\right)e^{-\frac{i}{\hbar}mc^2 t} \approx -i\frac{mc^2}{\hbar}\varnothing e^{-\frac{i}{\hbar}mc^2 t} \\ \frac{\partial^2 \psi}{\partial t^2} &\approx -\left(i\frac{2mc^2}{\hbar}\frac{\partial \varnothing}{\partial t} + \left(\frac{mc^2}{\hbar}\right)^2 \varnothing\right)e^{-\frac{i}{\hbar}mc^2 t} \end{aligned} \tag{17.106}$$

Substituting it with the free Klein–Gordon equation, $c^{-2}\partial_t^2\psi = \nabla^2\psi - m^2\psi$, yield $-\frac{1}{c^2}\left(i\frac{2mc^2}{\hbar}\frac{\partial \varnothing}{\partial t} + \left(\frac{mc^2}{\hbar}\right)^2 \varnothing\right)e^{-\frac{i}{\hbar}mc^2 t} \approx \left(\nabla^2 - \left(\frac{mc^2}{\hbar}\right)^2 \varnothing e^{-\frac{i}{\hbar}mc^2 t}\right.$ can be simplified as (Table 17.3):

$$-i\hbar\frac{\partial \varnothing}{\partial t} = -\frac{\hbar^2}{2m}\nabla^2 \varnothing \tag{17.107}$$

Table 17.3 Klein–Gordon equation in natural unit metric of $\eta_{\mu\nu} = \text{diag}\,(\pm 1, \mp 1, \mp 1, \mp 1)$

	Position space	Fourier transformation $\omega = E,\ k = p$	Momentum space $p = (E, p)$
Time and space	$\left(\partial_t^2 - \nabla^2 + m^2\right)\psi(t,x) = 0$	$\psi(t,x) = \int \frac{dw}{2\pi} \int \frac{d^3k}{(2\pi)^3} e^{\mp i(wt - k.x)} \psi(w,k)$	$E^2 = p^2 + m^2$
Four-vector form	$(c + m^2)\psi(x) = 0$	$\psi(x) = \int \frac{d^4p}{(2\pi)^4} e^{-ip.x} \psi(p)$	$p^2 = \pm m^2$

Activation of Antimatter to Cool the Earth

Subsequently, all dimensions of 1D, 2D, and 3D of the time and space are being further clarified by implanting frequency cutoff to avoid the bifurcation of the DOS to pave the path for cooling energy formation [41, 51]. Thus, the DOS analyzed at various dimensions of the time and space has been calculated, here named as $\varrho_{PC}(\omega)$, which is being confirmed by clarifying cooling energy frequency formation of Maxwell's theory of nanostructure corresponding to the time and space (Fig. 17.6). The DOS is thus here expressed by $\varrho_{PC}(\omega) \propto \frac{1}{\sqrt{\omega - \omega_e}} \Theta(\omega - \omega_e)$, where $\Theta(\omega - \omega_e)$ where ω_e is denoted as the frequency of the PBE at the used DOS.

Then, the density of state (DOS) has been implemented to confirm the qualitative form of the non-Weisskopf–Wigner module that represents the precise cooling-state energy, calculated by the dimensional clarification of the time and space [19, 42, 43]. This DOS is thereafter determined considering the electromagnetic field of time and space, which is determined by $\varrho_{PC}(\omega) \propto \frac{1}{\sqrt{\omega - \omega_e}} \Theta(\omega - \omega_e)$. Simply, here, the cooling energy of DOS reveals a perfect algorithmic function close to the PBE, which is $\varrho_{PC}(\omega) \propto -[\ln|(\omega - \omega_0)/\omega_0| - 1]\Theta(\omega - \omega_e)$, where ω_e defines the prime tip point of the DOS distribution in the time and space.

Subsequently, this distributed DOS energy will transform it into the cool state, which will confirm the total deliberation of energy dynamics of the time and space corresponding to its quantum fields, and thus, defined above, $J(\omega)$ will confirm the quantum field of cooling energy magnitudes $V(\omega)$ corresponding to its photonic band (PB), photonic band edge (PBE), and photonic band gap (PBG) in the time and space as:

$$J(\omega) = \varrho(\omega)|V(\omega)|^2. \tag{17.108}$$

Therefore, considering here the energy frequency of ω_c and the generated cooling energy dynamic of $\langle a(t) \rangle = u(t, t_0)\langle a(t_0) \rangle$, the function $u(t, t_0)$ has been described as the photon structure as $u(t, t_0)$ by clarifying the integral–differential equation as:

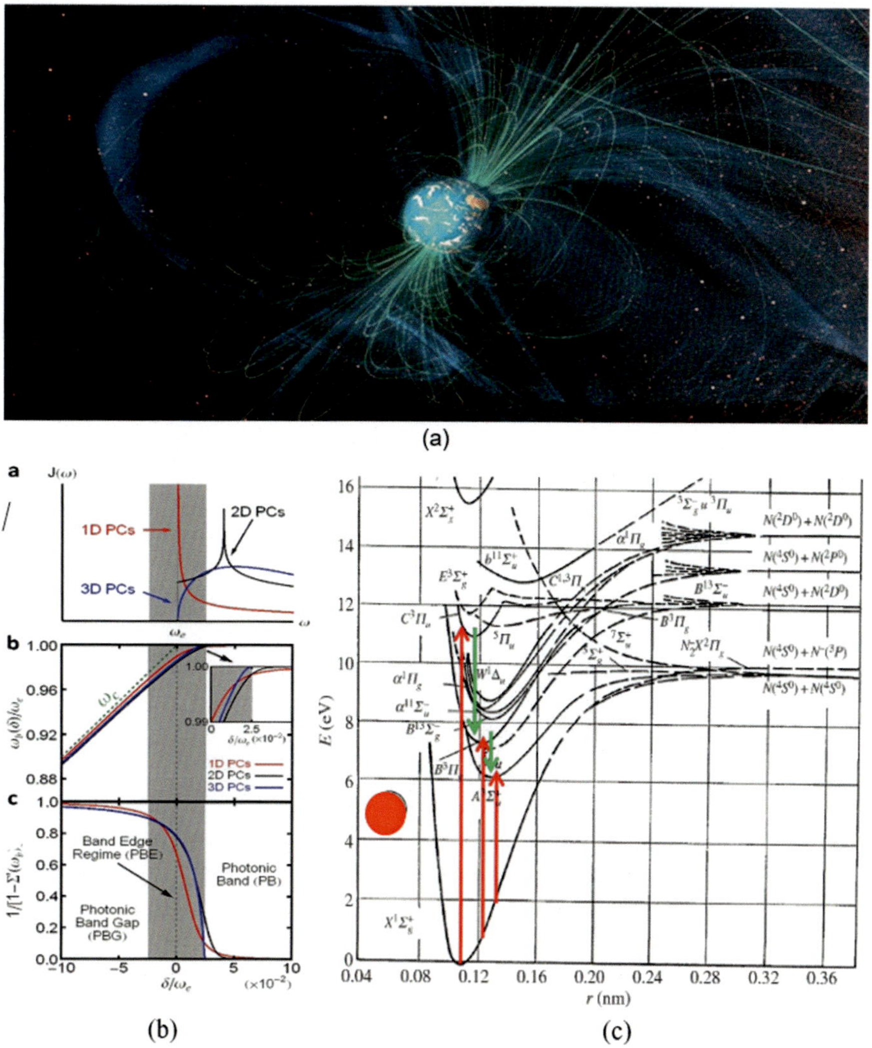

Fig. 17.6 (**a**) The concept of cooling-state energy dynamics in the time and space, (**b**) the cooling energy band structures and energy transformation mode at various DOSs of 1D, 2D, and 3D in the time and space and (**c**) the cooling energy formation magnitude (eV) in relation to the radial (r) dimension of time and space

$$u(t, t_0) = \frac{1}{1 - \Sigma'(\omega_b)} e^{-i\omega(t - t_0)} + \int_{\omega_e}^{\infty} d\omega \frac{J(\omega) e^{-i\omega(t - t_0)}}{[\omega - \omega_c - \Delta(\omega)]^2 + \pi^2 J^2(\omega)}, \quad (17.109)$$

where $\Sigma'(\omega_\mathrm{b}) = [\partial\Sigma(\omega)/\partial\omega]_{\omega=\omega_\mathrm{b}}$ and $\Sigma(\omega)$ confirm the self-energy induction into the reservoir,

$$\Sigma(\omega) = \int_{\omega_\mathrm{e}}^{\infty} \mathrm{d}\omega' \frac{J(\omega')}{\omega - \omega'}. \tag{17.110}$$

Here, the frequency ω_b represents the cooling energy frequency mode ($0 < \omega_\mathrm{b} < \omega_\mathrm{e}$),calculated as $\omega_\mathrm{b} - \omega_\mathrm{c} - \Delta(\omega_\mathrm{b}) = 0$, where $\lesssim\Delta(\omega) = \mathcal{P}\left[\int \mathrm{d}\omega' \frac{\mathrm{J}(\omega')}{\omega - \omega'}\right]$ is a primary-valued integral of the time and space. In detail, it can be explained that energy dynamics (eV) is in the time and space is considered as the Fock cooling state n_0, i. e. , $\rho(t_0) = |n_0\rangle\langle n_0|$, which is obtained computationally from the real-time quantum field of its radial (r) dimension of the space at a time t and expressed as:

$$\rho(t) = \sum_{n=0}^{\infty} \mathcal{P}_n^{(n_0)}(t)|n_0\rangle\langle n_0| \tag{17.111}$$

$$\mathcal{P}_n^{(n_0)}(t) = \frac{[v(t,t)]^n}{[1 + v(t,t)]^{n+1}} [1 - \Omega(\mathrm{t})]^{n_0} \times \sum_{k=0}^{\min\{n_0,n\}} \binom{n_0}{k} \times \binom{n}{k} \left[\frac{1}{v(t,t)} \frac{\Omega(\mathrm{t})}{1 - \Omega(\mathrm{t})}\right]^k \tag{17.112}$$

where $\Omega(\mathrm{t}) = \frac{|u(t,t_0)|^2}{1+v(t,t)}$. Thus, this result suggested that the Fock state cooling energy is indeed generated into dynamic states of the time and space $\mathcal{P}_n^{(n_0)}(t)$ of $|n_0\rangle$ where the proliferation of energy deliberation $\mathcal{P}_n^{(n_0)}(t)$ is in the prime state $|n_0 = 5\rangle$ and in the steady-state limits $\mathcal{P}_n^{(n_0)}(t \to \infty)$, which will eventually reach at a non-equilibrium cool-state energy in order to cool the Earth finally.

Activation of Antimatter to Heat the Earth

Since the electromagnetic field of the Earth is being modeled by Higgs-Boson quantum dynamics, the local symmetry of U(1) here will allow to form gauge-variable QED, in terms of gauge particles $\varnothing' \to \mathrm{e}^{i\alpha(x)}\varnothing$, which is the cooling-state energy that can be transformed into thermal state energy in order to heat the Earth at a comfort level [51, 53, 55]. This mechanism is being confirmed by the following variable derivatives considering the specific transformational laws of the scalar field, written as:

$$\begin{aligned} &\partial_\mu \to D_\mu = \partial_\mu = ieA_\mu && \text{[covariant derivatives]} \\ &A'_\mu = A_\mu + \frac{1}{e}\,\partial_\mu \alpha && [A_\mu \text{ derivatives}] \end{aligned} \tag{17.113}$$

Here, the local U(1) gauge-invariant Lagrangian is being considered as the perplex scalar field that is expressed by

$$\mathcal{L} = (D^\mu)^\dagger\,(D_\mu \varnothing) - \frac{1}{4}\,F_{\mu\nu}F^{\mu\nu} - V(\varnothing). \tag{17.114}$$

The term $\frac{1}{4}\,F_{\mu\nu}F^{\mu\nu}$ is the kinetic motion of the gauge field for considering thermal energy and $V(\varnothing)$ is denoted as the kinetic term written as $V(\varnothing^*\varnothing) = \mu^2(\varnothing^*\varnothing) + \lambda\,(\varnothing^*\varnothing)^2$.

Thus, the equation of the Lagrangian $\mathcal{L}$ is being considered as the quantum field of scalar particle ϕ_1 and ϕ_2 and a mass μ thus, here, $\mu^2 < 0$ will confirm an infinite number of quanta to satisfy the equation $\phi_1^2 + \phi_2^2 = -\mu^2/\lambda = v^2$. So, the quantum field can be clarified as $\phi_0 = \frac{1}{\sqrt{2}}[(v+\eta) + i\xi]$, and then the derivative of the Lagrangian can be expressed as kinetic mode of:

$$\begin{aligned} \mathcal{L}_{\text{kin}}(\eta,\xi) &= (D^\mu\phi)^\dagger (D^\mu\phi) \\ &= (\partial^\mu + ieA^\mu)\phi^*\,(\partial_\mu - ieA_\mu)\,\phi \end{aligned} \tag{17.115}$$

Here, $V(\eta,\xi) = \lambda\, v^2\eta^2$ is the final term to the second order and thus the full Lagrangian can be expressed as:

$$\begin{aligned} \mathcal{L}_{\text{kin}}(\eta,\xi) = \frac{1}{2}(\partial_\mu\eta)^2 - \lambda v^2\eta^2 + \frac{1}{2}(\partial_\mu\xi)^2 - \frac{1}{4}F_{\mu\nu}F^{\mu\nu} + \frac{1}{2}\,e^2v^2A_\mu^2 \\ - evA_\mu(\partial^\mu\xi) \end{aligned} \tag{17.116}$$

Here, η represents the mass, ξ represents the massless, μ represents the mass for the quanta, and thus A_μ is defined as the term $\partial_\mu\alpha$, as is the function of the quantum field. Naturally, A_μ and ϕ can be changed spontaneously, so the equation can be rewritten as a Lagrangian scalar to confirm the formation of the thermal energy particles within the quantum field of the time and space:

$$\begin{aligned} \mathcal{L}_{\text{scalar}} &= (D^\mu\phi)^\dagger (D^\mu\phi) - V(\phi^\dagger\phi) \\ &= (\partial^\mu + ieA^\mu)\frac{1}{\sqrt{2}}\,(v+h)\,(\partial_\mu - ieA_\mu)\frac{1}{\sqrt{2}}\,(v+h) - V\,(\phi^\dagger\phi) \end{aligned} \tag{17.117}$$

$$= \frac{1}{2}(\partial_\mu h)^2 + \frac{1}{2}e^2A_\mu^2\,(v+h)^2 - \lambda v^2h^2 - \lambda v h^3 - \frac{1}{4}\lambda h^4 + \frac{1}{4}\,\lambda h^4 \tag{17.118}$$

Here, the Lagrangian scalar thus revealed that the Higgs-Boson quantum field can certainly be initiated to form thermal energy in the Earth [22, 54]. To confirm to determine this heat-energy generation in the time and space, a further calculation considering its isotropic distributed kinetic energy has been conducted with respect to the angle θ from the dimensional axis of the time and space and the differential density of energy $\in$ and written as:

$$\mathrm{d}n = \frac{1}{2} n(\in) \sin\theta \mathrm{d} \in \mathrm{d}\theta. \tag{17.119}$$

Here, the speed of high-energy frequency is being implemented as $c\ (1 - \cos\theta)$ considering the absorption of energy per unit area by expressing the following equation:

$$\frac{\mathrm{d}\tau_{\mathrm{abs}}}{\mathrm{d}x} = \int\int \frac{1}{2} \sigma n\ (\in)(1 - \cos\theta)\ \sin\theta \mathrm{d} \in \mathrm{d}\theta. \tag{17.120}$$

Rewriting these variables as integral over s instead of θ, by (17.119) and (17.120), it has been determined as:

$$\frac{\mathrm{d}\tau_{\mathrm{abs}}}{\mathrm{d}x} = \pi r_0^2 \left(\frac{m^2 c^4}{E}\right)^2 \int_{\frac{m^2 c^4}{E}}^{\infty} \in^{-2} n(\in)\ \overline{\phi}[s_0(\in)]\mathrm{d}e, \tag{17.121}$$

where

$$\overline{\phi}[s_0(\in)] = \int_1^{s_0(\in)} s\overline{\sigma}\ (s)\mathrm{d}s, \overline{\sigma}(s) = \frac{2\sigma(s)}{\pi r_0^2}. \tag{17.122}$$

Simply, this result suggests the representation of the heating energy generation in that is readily allow the confirmation of $\overline{\phi}[s_0]$ to the determine the net heating energy formation of s_0 [20, 24, 28] in the time and space. Therefore, the net functional asymptotic formula is expressed as follows to confirm the heating energy formation:

$$\overline{\phi}[s_0] = 2s_0(\ln 4s_0 - 2) + \ln 4s_0(\ln 4s_0 - 2) - \frac{(\pi^2 - 9)}{3} + s_0^{-1}\left(\ln 4s_0 + \frac{9}{8}\right) \tag{17.123}$$
$$+ \ldots (s_0 \gg 1);$$

$$\overline{\phi}[s0] = \left(\frac{2}{3}\right)(S_0 - 1)^{\frac{3}{2}} + \left(\frac{5}{3}\right)(S_0 - 1)^{\frac{5}{2}} - \left(\frac{1507}{420}\right)(S_0 - 1)^{\frac{7}{2}}$$
$$+ \ldots (s_0 - 1 \ll 1). \tag{17.124}$$

The function $\frac{\overline{\phi}[s_0]}{(s_0-1)}$ is revealed as $1 < s_0 < 10$; at larger s_0, it confirms a standard algorithm function of s_0. Thus, the energy spectra of the thermal energy can be written as $n(\in) \propto \in^m$, and thus, the calculation of the thermal energy in terms energy spectra in the time and space can be expressed as:

$$\begin{aligned} n(\in) &= 0,\ \in < \in_0 \\ &= C_{\in}^{-\alpha} \text{ or } D_{\in}^{\beta}, \in_0 < \in < \in_m \\ &= 0,\ \in > \in_m \end{aligned} \tag{17.125}$$

Then, it can be transformed as:

$$\left(\frac{\mathrm{d}\tau_{\text{abs}}}{\mathrm{d}x}\right)_{\alpha} = \pi r_0^2 C \left(\frac{m^2c^4}{E}\right)^{1-\alpha} \times \begin{cases} 0, & E < E_m \\ [F_\alpha(1) - F_\alpha(\sigma_m)], & E_m < E < E_0 \\ [F_\alpha(\sigma_0) - F_\alpha(\sigma_m)], & E > E_0; \end{cases} \tag{17.126}$$

$$\left(\frac{\mathrm{d}\tau_{\text{abs}}}{\mathrm{d}x}\right)_{\beta} = \pi r_0^2 D \left(\frac{m^2c^4}{E}\right)^{1+\beta} \times \begin{cases} 0, & E < E_m \\ [F_\beta(\sigma_m), & E_m < E < E_0 \\ [F_\beta(\sigma_m) - F_\beta(\sigma_0)], & E > E_0. \end{cases} \tag{17.127}$$

In these functional variables, the heating energy spectra on the time and space can be properly defined by asymptotic formula. Thus, the term $\Gamma_\gamma^{\text{LPM}}$ defines the energy generation in relation to the irradiated energy dynamics of per unit area as:

$$\Gamma_\gamma \equiv \frac{\mathrm{d}n_\gamma}{\mathrm{d}V\mathrm{d}t}. \tag{17.128}$$

Here, the contributions $\Gamma_\gamma^{\text{LPM}}$ and the rate of thermal energy generation are being confirmed as $O(\alpha_{\text{EM}}\ \alpha_s)$. Thus, it has been confirmed by implementing the contributed thermal energy rate $\Gamma_\gamma^{\text{LPM}}$ to thermodynamically control the temperature T and energy-physical reaction μ of the time and space by expressing the following equation:

$$\frac{\mathrm{d}\Gamma_\gamma^{\text{LPM}}}{\mathrm{d}^3k} = \frac{d_F q_s^2 \alpha_{EM}}{4\pi^2 k} \int_{-\infty}^{\infty} \frac{\mathrm{d}p_\parallel}{2\pi} \int \frac{\mathrm{d}^2 p_\perp}{(2\pi)^2} A\left(p_\parallel, k\right) \text{Re}\left\{2P_\perp \cdot f\left(p_\perp; p_\parallel, k\right)\right\} \tag{17.129}$$

where d_F is the variable state of the energy particles [N_c in SU(N_c)] and q_s is the Abelian charge of the energy quark, $k \equiv |k|$; and thus, the kinetic functional mode $A(p\|, k)$ is being expressed by:

$$A\left(p_{\|},k\right) \equiv \begin{cases} \dfrac{n_b\left(k+p_{\|}\right)\left[1+n_b\left(p_{\|}\right)\right]}{2p_{\|}\left(p_{\|}+k\right)} & \text{scalars} \\ \dfrac{n_f\left(k+p_{\|}\right)\left[1-n_f\left(p_{\|}\right)\right]}{2\left[p_{\|}\left(p_{\|}+k\right)\right]^2}\left[{p_{\|}}^2+\left(p_{\|}+k\right)^2\right], & \text{fermions} \end{cases} \tag{17.130}$$

with

$$n_b(p) \equiv \frac{1}{\exp[\beta(p-\mu)]-1}, n_f(p) \equiv \frac{1}{\exp[\beta(p-\mu)]+1} \tag{17.131}$$

The function $f(p\perp;\ p\|,\ k)$ is then integrated to resolve the below equation, which suggested that the heating energy proliferation in the time and space are very much possible, which can be expressed as the following derivative of energy generation:

$$\begin{aligned} 2\mathrm{p}_{\perp} &= i\delta E f\left(\mathrm{p}_{\perp};p_{\|},k\right) + \frac{\pi}{2}C_{\mathrm{F}}g_{\mathrm{s}}^2m_{\mathrm{D}}^2\int\frac{d^2q_{\perp}}{(2\pi)^2}\frac{dq_{\|}}{2\pi}\frac{dq^0}{2\pi}2\pi\delta\left(q^0-q_{\|}\right) \\ &\times\frac{T}{|q|}\left[\frac{2}{|q^2-\Pi_L(Q)|^2}+\frac{\left[1-\left(q^0/\left|\mathrm{q}_{\|}\right|\right)^2\right]^2}{\left|(q^0)^2-\mathrm{q}^2-\Pi_T(\mathrm{Q})\right|^2}\right]\left[f\left(p_{\perp};p_{\|},k\right)-f\left(q+p_{\perp};p_{\|},k\right)\right] \end{aligned} \tag{17.132}$$

Simply, this heating energy generation is derived from the explicit forms of heating energy that maximize the heating energy generation corresponding to its given energy function of $f(p\perp;\ p\|,\ k)$ to transform it into heating state energy to heat the Earth naturally (Fig. 17.7).

(a)

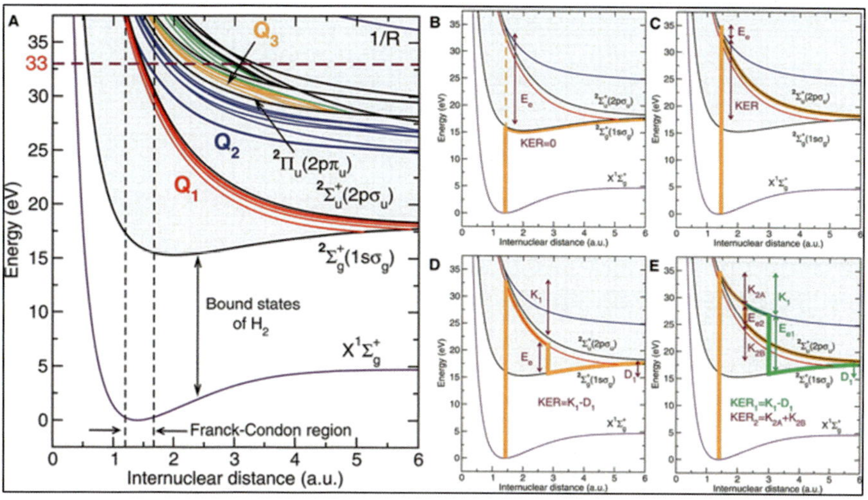

(b) (c)

Fig. 17.7 (**a**) The heating energy formation concept in the time and space, (**b**) the formation of heating energy (eV) at its internuclear distance (a.u.) of its bound states ($X^1\sum_g$; $^2\sum_g$;$^2\prod_u$) of energy formation region, and (**c**) the formation of heating energy (eV) generation at its internuclear distance (a.u.) at various spectra ($X^1\sum_g$; $^2\sum_g$;$^2\prod_u$) of its energy amplitude array of time and space

Conclusions

The activation of dark energy (D_e) and antimatter (A_m) is being modeled to confirm that Earth has enough hidden energy of dark and antimatter (D_e, A_m) within the Earth to secure a comfortable climate condition on it. The results suggested that the availability D_e energy is very much doable anywhere in the Earth, which can play vital role to break global CO_2 naturally and mitigate global climate change dramatically. Subsequently, the computational results of the antimatter suggested that it can be transformed into cool-state energy (HcE^-) by implementing nano-point defect of the electromagnetic field to cool the Earth naturally and this cool-state energy (HcE^-) can also be reformed into the thermal state energy (HtE^-) by integrating Higgs-Boson [BR ($H \rightarrow \gamma\gamma^-$)] quantum dynamics to heat the Earth naturally. The activation of hidden energy of dark and antimatter (D_e, A_m) of the Earth and its vital function to mitigate global climate change crisis naturally shall indeed be the most innovative scientific discovery ever.

Acknowledgments All this research work is solely performed by the author. This research has no conflict of interest to publish any suitable journal and/or publisher.

References

1. Artemyev, N., Jentschura, U., Serbo, V., & Surzhykov, A. (2012). Strong electromagnetic field effects in ultra-relativistic heavy-ion collisions. *European Physical Journal C: Particles and Fields, 72*, 1935.
2. Baur, G., Hencken, K., & Trautmann, D. (2007). Revisiting unitarity corrections for electromagnetic processes in collisions of relativistic nuclei. *Physics Reports, 453*, 1–27.
3. Baur, G., Hencken, K., Trautmann, D., Sadovsky, S., & Kharlov, Y. (2002). Dense laser-driven electron sheets as relativistic mirrors for coherent production of brilliant X-ray and γ-ray beams. *Physics Reports, 364*, 359–450.
4. Birnbaum, K. M., et al. (2005). Photon blockade in an optical cavity with one trapped atom. *Nature, 436*, 87–90.
5. Boettcher, I., Pawlowski, J. M., & Diehl, S. (2012). Ultracold atoms and the functional renormalization group. *Nuclear Physics B - Proceedings Supplements, 228*, 63–135.
6. Broz, M., et al. (2020). A generator of forward neutrons for ultra-peripheral collisions. *Computer Physics Communications, 253*, 107181.
7. Busch, K., et al. (2007). Periodic nanostructures for photonics. *Physics Reports, 444*, 101–202.
8. Cardoso, V., Lemos, J. P., & Yoshida, S. (2004). Quasinormal modes of Schwarzschild black holes in four and higher dimensions. *Physical Review D, 69*, 044004.
9. Cartwright, J. (2010). Photons meet with three-way split. *Nature*. https://doi.org/10.1038/news.2010.381
10. Chang, D. E., Sørensen, A. S., Demler, E. A., & Lukin, M. D. (2007). A single-photon transistor using nanoscale surface plasmons. *Nature Physics, 3*, 807–812.
11. Chen, G., et al. (2013). Examining non-locality and quantum coherent dynamics induced by a common reservoir. *Scientific Reports, 3*, 2514.
12. Chen, J., Wang, C., Zhang, R., & Xiao, J. (2012). Multiple plasmon-induced transparencies in coupled-resonator systems. *Optics Letters, 37*, 5133–5135.

13. Cheng, M. T., & Song, Y. Y. (2012). Fano resonance analysis in a pair of semiconductor quantum dots coupling to a metal nanowire. *Optics Letters, 37*, 978–980.
14. Dayan, B., et al. (2008). A photon turnstile dynamically regulated by one atom. *Science, 319*, 1062–1065.
15. Dobrynina, A., Kartavtsev, A., & Raffelt, G. (2015). Photon-photon dispersion of TeV gamma rays and its role for photon-ALP conversion. *Physical Review D, 91*, 083003.
16. Douglas, J. S., et al. (2015). Quantum many-body models with cold atoms coupled to photonic crystals. *Nature Photonics, 9*, 326–331.
17. Dupuis, N. L., et al. (2021). The nonperturbative functional renormalization group and its applications. *Physics Reports, 910*, 1.
18. Eichler, J., & Stöhlker, T. (2007). Radiative electron capture in relativistic ion–atom collisions and the photoelectric effect in hydrogen-like high-Z systems. *Physics Reports, 439*, 1–99.
19. Englund, D., et al. (2010). Resonant excitation of a quantum dot strongly coupled to a photonic crystal nanocavity. *Physical Review Letters, 104*, 073904.
20. Fernandez, J., & Martín, F. (2009). Electron and ion angular distributions in resonant dissociative photoionization of H2 and D2 using linearly polarized light. *New Journal of Physics, 11*, 043020.
21. Gabor, M., et al. (2009). Extremely efficient multiple electron-hole pair generation in carbon nanotube photodiodes. *Science, 325*, 1367–1371.
22. Gleyzes, S., et al. (2007). Quantum jumps of light recording the birth and death of a photon in a cavity. *Nature, 446*, 297–300.
23. Gould, R. J. (1967). Pair production in photon-photon collisions. *Physics Review, 155*, 1404–1406.
24. Guerlin, C., et al. (2007). Progressive field-state collapse and quantum non-demolition photon counting. *Nature, 448*, 889–893.
25. Guo, Y., Al-Jubainawi, A., & Ma, Z. (2019). Performance investigation and optimisation of electrodialysis regeneration for LiCl liquid desiccant cooling systems. *Applied Thermal Engineering, 149*, 1023–1034.
26. Hencken, K., Baur, G., & Trautmann, D. (2006). Transverse momentum distribution of vector mesons produced in ultraperipheral relativistic heavy ion collisions. *Physical Review Letters, 96*, 012303.
27. Hübel, H., et al. (2010). Direct generation of photon triplets using cascaded photon-pair sources. *Nature, 466*, 601–603.
28. Igor, B., et al. (2012). Ultracold atoms and the functional renormalization group. *Nuclear Physics B - Proceedings Supplements, 228*, 63–135.
29. Javadi, A., et al. (2015). Single-photon non-linear optics with a quantum dot in a waveguide. *Nature Communications, 6*, 8655.
30. Johnson, B. R. (2010, June 20). Quantum non-demolition detection of single microwave photons in a circuit. *Nature Physics, 6*, 663–667.
31. Kaneda, F., et al. (2019). High-efficiency single-photon generation via large-scale active time multiplexing. *Science Advances, 5*(10), eaaw8586.
32. Kevin, A., et al. (2017). Signatures of two-photon pulses from a quantum two-level system. *Nature Physics, 3*, 649–654.
33. Langford, K., et al. (2021). Efficient quantum computing using coherent photon conversion. *Nature, 478*, 360–363.
34. Langer, L., et al. (2014). Access to long-term optical memories using photon echoes retrieved from semiconductor spins. *Nature Photonics, 8*, 851–857.
35. Naghiloo, M., et al. (2016). Mapping quantum state dynamics in spontaneous emission. *Nature Communications, 7*, 11527.
36. Najjari, B., Voitkiv, A. B., Artemyev, A., & Surzhykov, A. (2009). Simultaneous electron capture and bound-free pair production in relativistic collisions of heavy nuclei with atoms. *Physical Review A, 80*, 012701.
37. Reinhard, A., et al. (2012). Strongly correlated photons on a chip. *Nature Photonics, 6*, 93–96.

38. Reinhard, P., et al. (2021). Nuclear charge densities in spherical and deformed nuclei: Toward precise calculations of charge radii. *Physical Review C, 103*, 054310.
39. Shalm, L. K., et al. (2012). Three-photon energy–time entanglement. *Nature Physics, 9*, 19.
40. Szafron, R., & Czarnecki, A. (2016). High-energy electrons from the muon decay in orbit: Radiative corrections. *Physics Letters B, 753*, 61–64.
41. Tame, M. S., et al. (2013). Quantum plasmonics. *Nature Physics, 9*, 329–340.
42. Ting, T. C. (2004). The polarization vector and secular equation for surface waves in an anisotropic elastic half-space. *International Journal of Solids and Structures, 41*, 2065–2083.
43. Stehle, C., Zimmermann, C., & Slama, S. (2014). Cooperative coupling of ultracold atoms and surface plasmons. *Nature Physics, 10*, 937.
44. Prasad, A. S., Hinney, J., Mahmoodian, S., Hammerer, K., et al. (2020). Correlating photons using the collective nonlinear response of atoms weakly coupled to an optical mode. *Nature Photonics, 14*, 719–722.
45. Lukas, H., Fischer, K. A., Appel, S., Lukin, D., et al. (2018). Quantum dot single-photon sources with ultra-low multi-photon probability. *npj Quantum Information, 4*, 43.
46. Liu, K., Hong, X., Zhou, Q., Jin, C., Li, J., Zhou, W., Liu, J., Wang, E., Zettl, A., & Wang, F. (2013). High-throughput optical imaging and spectroscopy of individual carbon nanotubes in devices. *Nature Nanotechnology, 8*, 917.
47. Broz, M., Contreras, J. G., & Tapia Takaki, J. D. (2020). A generator of forward neutrons for ultraperipheral collisions. *Computer Physics Communications, 253*, 107181.
48. Jiaqi, L., Yi, Z., Chengkai, T., & Xingxing, Z. (2019). *INS aided high dynamic single-satellite position algorithm*. 2019 IEEE international conference on signal processing, communications and computing (ICSPCC).
49. Wang, C., Wang, W., & Chen, Z. (2017). *Single-satellite positioning algorithm based on direction-finding*. 2017 progress in electromagnetics research symposium – Spring (PIERS).
50. Gazi, V., & Passino, K. M. (2005). Stability of a one dimensional discrete-time asynchronous swarm. *IEEE Transactions on Systems, Man and Cybernetics, Part B (Cybernetics), 35*, 834.
51. Berges, J., & Mesterházy, D. (2012). Introduction to the nonequilibrium functional renormalization group. *Nuclear Physics B. Proceedings Supplements, 228*, 37–60.
52. Yan, J. Z., & Chang, S.-J. (2018). The contingent effects of political strategies on firm performance: A political network perspective. *Strategic Management Journal*. https://doi.org/10.1002/smj.2908
53. Dupuis, N., Canet, L., Eichhorn, A., Metzner, W., Pawlowski, J. M., Tissier, M., & Wschebor, N. (2021). The nonperturbative functional renormalization group and its applications. *Physics Reports, 910*, 1.
54. Boettcher, I., Pawlowski, J. M., & Diehl, S. (2012). Ultracold atoms and the Functional Renormalization Group. *Nuclear Physics B - Proceedings Supplements, 228*, 63.
55. Bertozzi, E. (2010, November 1). Hunting the ghosts of a 'strictly quantum field': The Klein–Gordon equation. *European Journal of Physics, 31*(6), 1499.
56. Arnold, P. (2001, November 27). Photon emission from ultrarelativistic plasmas. *Journal of High Energy Physics*. https://doi.org/10.1088/1126-6708/2001/11/057
57. Tu, M. W., & Zhang, W. M. (2008). Non-Markovian decoherence theory for a double-dot charge qubit. *Physical Review B, 78*, 235311.
58. Viktor, J., et al. (2014). More on thermal probes of a strongly coupled anisotropic plasma. *Journal of High Energy Physics, 2014*, 149.
59. Wang, C. et al. (2017). *Single-satellite positioning algorithm based on direction-finding*. 2017 Progress in electromagnetics research symposium – Spring (PIERS).
60. Waseem, S. B. (2011, December 12). Orbital excitation blockade and algorithmic cooling in quantum gases. *Nature, 480*(7378), 500–503.

Index

M. F. Hossain, *Global Sustainability*, https://doi.org/10.1007/978-3-031-34575-3